Nonparametric Statistics and Mixture Models

A Festschrift in Honor of Thomas P Hettmansperger

Nonparametric Statistics and Mixture Models

A Festschrift in Honor of Thomas P Hettmansperger

The Pennsylvania State University, USA, 23 – 24 May 2008

editors

David R Hunter

Donald St P Richards

James L Rosenberger

The Pennsylvania State University, USA

World Scientific

NEW JERSEY • LONDON • SINGAPORE • BEIJING • SHANGHAI • HONG KONG • TAIPEI • CHENNAI

Published by

World Scientific Publishing Co. Pte. Ltd.
5 Toh Tuck Link, Singapore 596224
USA office: 27 Warren Street, Suite 401-402, Hackensack, NJ 07601
UK office: 57 Shelton Street, Covent Garden, London WC2H 9HE

British Library Cataloguing-in-Publication Data
A catalogue record for this book is available from the British Library.

NONPARAMETRIC STATISTICS AND MIXTURE MODELS
A Festschrift in Honor of Thomas P Hettmansperger

ISBN-13 978-981-4340-55-7
ISBN-10 981-4340-55-3

Printed in Singapore by World Scientific Printers.

PREFACE

This volume is a tribute to Thomas P. Hettmansperger on the occasion of his retirement in 2008 from the faculty of the Department of Statistics at The Pennsylvania State University. It contains a collection of papers by some of Tom's closest friends, students, and colleagues, covering a wide range of topics in nonparametric statistics and mixture models.

Most of these papers were presented at an international conference to celebrate Tom's scientific contributions entitled "Nonparametric Statistics and Mixture Models: Past, Present, and Future." The conference was hosted by Penn State's Department of Statistics on May 23–24, 2008 and drew over 100 participants from six continents. The papers in this volume, ordered alphabetically by first author, have undergone reviews by external referees and by the editors. We are confident that a broad swath of readers will find much of interest in them.

The depth of Tom's knowledge about nonparametrics has served many of his colleagues and students well; the breadth of his statistical knowledge is an indication of his ceaseless curiosity and his collegiality, qualities that extend far beyond the statistical realm. Over the years he has compiled a lengthy record of service to the international statistics community through his work as teacher, author, editor, reviewer, thesis advisor, and department head. Referring to this record, Bill Harkness, Tom's colleague of over 40 years and his department head for nearly half that time, observes "I consider Tom to be the model university professor as reflected by this assessment." Tom is a fellow of the American Statistical Association and the Institute of Mathematical Statistics. He has received awards for both his research and his teaching, including the 2004 Gottfried E. Noether Senior Scholar Award from the American Statistical Association for his work in nonparametric statistics and the 1986 C. I. Noll Award for Teaching in the Eberly College of Science at Penn State University. As one of his former students, Ryan Elmore, put it, Tom is the rare academic who excels in all aspects of academia: He is an outstanding scholar, teacher, and departmental citizen.

However, Tom's list of honors and accomplishments does not capture his generosity, friendliness, and good humor. For us, he is more than a colleague; he is a loyal friend, a wise mentor. Not having him around the Statistics Department leaves a huge gap in the place, though this loss is made slightly tolerable by the fact that we remain good friends with him and his wife Ann, so we still see them frequently when they are not off on tandem bicycle adventures or visiting grandchildren in Colorado.

In addition to thanking Tom, we wish to express our gratitude to the authors of the papers in this volume for their contributions and also their patience during the preparation of this festschrift. We also thank the helpful team at World Scientific Press, particularly Professor K. K. Phua, the editor-in-chief; Ms. Rok Ting Tan, the editor who has guided us through the publishing process; and Rajesh Babu, the LaTeX wizard who answered many emails from us and always knew just how to fix whatever technical problem we were facing. Finally, we thank Bruce Lindsay, the department head, and Barb Freed, Bonnie Cain, Laurie Roan, and Jenn Parkes, the staff of the Statistics Department at Penn State, all of whom helped in many ways to bring about a very well-run conference in 2008.

In closing, we wish Tom and his family all the best in his retirement years, and (selfishly) many more years of contributing to his still-growing department and university.

David R. Hunter
Donald St. P. Richards
James L. Rosenberger
(Department of Statistics,
The Pennsylvania State University)

State College, PA
August 2010

A BIT OF HISTORY

Editors' Note: Tom was an early statistics faculty recruit to Penn State University in 1967 and became a member of the newly formed Department of Statistics the following year, as shown in the group photograph in Figure 1. What follows is Tom's brief first-person synopsis of his work over the years.

Fig. 1. Penn State's statistics department 1968–69, left to right: Department secretary Ingrid James, G. P. Patil, Bob Hultquist, Dean of the College of Science Clarence Noll, Jim Bartoo, Tom Hettmansperger, Charles Antle, Tom Ryan, Marlyn Boswell, Bill Harkness.

After graduation with an A.B. degree in mathematics from Indiana University, Bloomington, I worked for two years as an actuarial assistant in Indianapolis. While studying for the probability and statistics actuarial exam, I developed an interest in statistics and successfully applied for an assistantship at the University of Iowa in the mathematics department. In August of 1963 I, with wife Ann and daughter Kathy, arrived in Iowa City and began graduate work. In 1965 the department of statistics was formed and I became the first Ph.D. in statistics when I graduated in June, 1967. Bob Hogg, the first department head, and Tim Robertson were my joint thesis advisors while providing moral support and much needed advice. Nonparametric statistics was the topic of my thesis, and I have maintained a loyal interest in the area throughout my academic life.

Jim Bartoo, department head in mathematics at Penn State University in 1967, offered me a job, and I accepted without ever having been in Pennsylvania—not to mention State College. We arrived in June, 1967 with a new department of statistics on the horizon. Later that summer Bill Harkness returned to State College after a sabbatical stay at Stanford. Bill, who replaced Bartoo as the department head soon after the department was formed in 1968, has been a long-time friend and mentor.

In 1970 Hoben Thomas from the psychology department came over for some consulting on a statistical problem. We began a long-term friendship and collaboration. We recently completed ten years of NSF support for the development of nonparametric and semi-parametric mixture models motivated by modeling cognitive development in children. This was a very rewarding part of my career, with five Ph.D. students writing dissertations on the topic. My colleague Dave Hunter joined us as a co-principal investigator and contributed his expertise on statistical computing.

The years 1974-1977 were busy with travel and living in interesting places, a wonderful benefit of sabbaticals and leaves of absences. We (Ann, Kathy, Bob, Rebecca and I) spent a year in Berkeley, where I strengthened my background in traditional statistics in seminars with Peter Bickel, Fred Mosteller, Lucien LeCam, and Erich Lehmann, and then I spent another year at Princeton, where I learned about robustness challenges to those traditional approaches from John Tukey. During those years I also had the privilege of supervising Joe McKean's dissertation on the topic of rank tests in linear models. We have continued to be close friends and colleagues. We wrote a monograph, Robust Nonparametric Statistical Methods (1998), and will publish a second edition in 2010.

We spent 1984-85 in Melbourne, Australia at LaTrobe and Melbourne Universities. I met Simon Sheather at LaTrobe University as he was finishing his thesis under the direction of Stephan Maritz, and Simon and I began a long-term friendship and collaboration on various projects in nonparametric statistics, concentrating on rank methods in linear models. I also met Bruce Brown at the University of Tasmania, and we began working on projects related to extending the ideas of sign and rank to multivariate settings. The academic life has made these long-time friendships and collaborations possible.

Early papers with Bruce Brown on rank methods in multivariate models were based on work by Hannu Oja in Finland. In 1989 Hannu and Jukka Nyblom visited Penn State and Hannu and I began a series of collaborations on affine invariant and equivariant sign and rank methods in multivariate models. Over the period 1992–2002 we published several papers, and I had the privilege of collaborating with the fine young Finnish Ph.D. students Jyrki Möttönen and Esa Ollila, who were working with Hannu during that time.

Our first trip to Europe was in 1987, where I met Jürg Hüsler and Hans Riedwyl at the University of Bern after a European Meeting of Statisticians in Greece. Jürg and I subsequently worked on minimum distance methods and at the same time developed a long-term friendship and collaboration.

Throughout my academic life I have been interested in the development of statistical methods that are nonparametric and robust, beginning with location models, extending to linear models, then to multivariate models, and finally multivariate mixture models. Along the way I have had the great good fortune to work with colleagues around the world who have become close friends as well and to supervise the work of twenty-six amazing and dedicated Ph.D. students.

PH.D. STUDENTS SUPERVISED BY THOMAS P. HETTMANSPERGER

Name (date), thesis title. (First job)

- Peter V. Tryon (1970), Non-Parametric Tests of Homogeneity Against Restricted Alternatives in a One-Way Classification. (National Bureau of Standards, Boulder, Colorado)
- Timothy J. Killeen (1971), Bivariate Tests for Location and Their Bahaur Efficiencies. (University of Connecticut)
- George E. Policello (1974), Adaptive Robust Procedures for the One-Sample Location Problem. (Bell Laboratories, Chicago)
- Joseph W. McKean (1975), Tests of Hypothesis in the General Linear Model Based on Ranks. (Western Michigan University)
- Ronald M. Schrader (1976), Robust Inference Based on M-Estimates in the Linear Model. (University of New Mexico)
- Jessica M. Utts (1978), A Robust Class of Tests and Estimates for Multivariate Location Based on Ranks. (University of California, Davis)
- Kuo-Sheuan Cheng (1979), On Weighted Least Squares Rank Estimates. (General Software Corporation)
- Lewis H. Shoemaker (1980), Robust Estimates and Tests for the One-Sample Scale Model with Applications to Variance Component Models. (Millersville State University)
- Edward P. Markowski (1980), Simple Estimation Procedures Based on Rank Tests for the One- and Two-Sample Location Models. (Old Dominion University)
- James C. Aubuchon (1982), Rank Tests in the Linear Model: Asymmetric Errors. (Minitab, Inc.)
- Mara Tableman (1984), Two-Sample Procedures Based on One-Sample Linear Signed Rank Statistics. (Portland State University)
- Marianthi Markatou (1988), Robust Bounded Influence Tests in Linear Models. (Columbia University)
- Joshua Naranjo (1989), Bounded-Influence Regression: A Modified

Wilcoxon Procedure. (Western Michigan University).

- Clint W. Coakley (1991), Advances in the Study of Breakdown and Resistance. (Virginia Tech and State University).
- George F. Anderson (1993), Generalized Wilcoxon Methods (Astra/Merck Pharmaceuticals).
- Ömer Öztürk (1994), Minimum Distance Estimation (Ohio State University, Marion Campus)
- Suzanne R. Dubnicka (1998), Rank-Based Procedures for Combined Paired and Unpaired Data. (Kansas State University)
- Yong-Xin Yang (1998), Statistical Inference for Symmetries based on Ranks. (Deluxe Corporation, Minneapolis, Minn.)
- Mustafa Nadar (2000), Multivariate Sign and Rank Methods Based on the Oja Criterion Function. (Gebze Technical University, Turkey)
- I. Roberto Cruz-Medina (2001), Nonparametric Inference in Mixture Models. (Instituto Tecnolgico de Sonora)
- Ryan P. Elmore (2003), Semiparametric Analysis of Mixture Models with Repeated Measures. (Post-doc: Institute for Advanced Studies, Canberra, Australia, then Colorado State University)
- Xiaojiang Zhan (2005), Bayesian Semiparametric Inference Based on Ranks (Merck)
- Fengjuan Xuan (2005), Semiparametric Finite Mixture Models When Training Samples are Available. (Schering-Plough Pharmaceutical)
- Makhtar Sarr (2006), Robust Nonparametric Inference Based on the Multivariate Trimmed Mean (Abu Dhabi University, United Arab Emirates)
- Tatiana Benaglia (2008), Nonparametric Estimation in Multivariate Finite Mixture Models. (Post Doc, Cambridge University)
- Tracey W. Hammel (2010), Semiparameteric Estimation for Finite Mixture Models using an Exponential Tilt. (Penn State, World Campus)

CONTENTS

Estimation of Location and Scale Parameters Based on Kernel Functional Estimators

Ibrahim A. Ahmad

Department of Statistics, Oklahoma State University
Stillwater, OK 74078, USA
E-mail: iahmad@okstate.edu

Mohamod Amezziane

Department of Mathematical Sciences, DePaul University
Chicago, IL 60604, USA
E-mail: mamezzia@depaul.edu

In this paper we present new estimators of location and scale parameters that are constructed based on kernel estimators of the distribution functionals. We investigate the asymptotic behavior of the estimators under different sets of weak moment conditions and we devise a simple method to obtain the optimal bandwidth for the kernel function. The theoretical results are supported by a simulation study of the estimators' small sample performance.

Keywords: Asymptotic distribution; Kernel estimation; Location and scale parameters; Smoothing parameter

1. Introduction

Since Parzen wrote his 1962 paper (Parzen, 1962) on nonparametric density function estimation, kernels have had a multitude of uses in statistical inference, especially in the area of functional estimation. In fact, during the last few decades, a large number of kernel-based techniques have been developed to solve recurrent problems in inferential statistics such as regression function estimation (Watson, 1964), goodness-of-fit tests (Bickel and Rosenblatt, 1973), and tests of independence (Rosenblatt, 1975). The texts by Silverman (1986), Härdle (1990), and Scott (1992) provide details on the theory and practice of these estimators plus numerous references. Also, textbooks by Wand and Jones (1995), Bowman and Azzalini (1997), Simonoff (1996), and Hart (1997) are good sources of information and references. Of particular interest to us is the subject of kernel estimation of density functionals of the form $\int \gamma(x)f^2(x)dx$, where $\gamma(\cdot)$ is some continuous function and $f(\cdot)$ is a density function. Such estimators have been extensively studied over the last four decades and used for a variety of reasons ranging from evaluating the power of nonparametric tests, as is the case in Aubuchon and Hettmansperger (1984), to obtaining estimates of smoothing parameter selectors in Park and Marron (1992) and Sheather and Jones (1991). However, the usefulness of kernel functional estimators should be extended beyond the nonparametric domain and can be employed, as we will show, to obtain consistent and often unbiased estimators of population parameters. It will also be shown that the proposed estimators approach normality under conditions that are weaker than usual.

Let us define a location family and scale family of distributions as:

$$f_\mu(x) = f_o(x-\mu),$$
$$f_\sigma(x) = \frac{1}{\sigma} f_1\left(\frac{x}{\sigma}\right),$$

where $f_o(.)$ and $f_1(.)$ are known, bounded, almost everywhere continuous, probability density functions (*pdf*'s).

Consider the following functionals of the above densities:

$$\int x f_\mu^2(x)dx = \int x f_o^2(x-\mu)dx$$
$$= \int u f_o^2(u)du + \mu \int f_o^2(u)du,$$
$$\int f_\sigma^2(x)dx = \frac{1}{\sigma}\int f_1^2(u)du.$$

The location and scale parameters can be written as functionals of the densities:

$$\mu = \frac{\int_{-\infty}^{\infty} x f_\mu^2(x)dx - \int_{-\infty}^{\infty} x f_o^2(x)dx}{\int_{-\infty}^{\infty} f_o^2(x)dx},$$
$$\sigma = \frac{\int_{-\infty}^{\infty} f_1^2(x)dx}{\int_{-\infty}^{\infty} f_\sigma^2(x)dx}.$$

Functionals of $f_o(.)$ and $f_1(.)$ being fully known, the only quantities that require estimation are $I_1 = \int_{-\infty}^{\infty} x f_\mu^2(x)dx$ and $I_2 = \int_{-\infty}^{\infty} f_\sigma^2(x)dx$. This can be done by replacing the densities with their nonparametric kernel estimators in the following way:

$$\hat{I}_1 = \int_{-\infty}^{\infty} x\hat{f}_\mu(x)dF_{\mu,n}(x),$$

$$\hat{I}_2 = \int_{-\infty}^{\infty} \hat{f}_\sigma(x)dF_{\sigma,n}(x),$$

where $\hat{f}(x) = \frac{1}{n}\sum_{i=1}^{n} K_h(X_i - x)$ is a kernel estimator of $f(.)$ and $F_n(x) = \frac{1}{n}\sum_{i=1}^{n} 1(X_i \leq x)$ is the empirical distribution of $F(.)$. The kernel $K_h(.) = \frac{1}{h}K(\frac{\cdot}{h})$ is a sufficiently smooth symmetric density with decaying tails that integrates to 1 and the bandwidth h is a smoothing parameter that will be used later to control the properties of the parameter estimators.

2. The Location Estimator and its Properties

By replacing the functional of $f_\mu(.)$ by its estimator, one obtains the following estimator of μ:

$$\tilde{\mu} = \frac{1}{A_o(0)}\binom{n}{2}^{-1} \sum\sum_{1\leq i<j\leq n} X_i K_h(X_i - X_j) - \frac{A_1(0)}{A_o(0)}$$

where $A_i(t) = \int_{-\infty}^{\infty} u^i f_o(u) K_t * f_o(u)du$, for some $i, t \geq 0$ and $A_i(0)$ denotes the limit of $A_i(t)$ as t converges to zero.

Note that the diagonal terms of the double sum have been discarded since they do not make use of the data and add bias to the estimator.

The estimator $\tilde{\mu}$ can be shown to be consistent and asymptotically unbiased under simple conditions on the bandwidth h. However, one can improve on the performance of this estimator through bias correction. In fact:

$$\begin{aligned} E[X_1K_h(X_1 - X_2)] &= \iint xK_h(x-y)f_\mu(x)f_\mu(y)dxdy \\ &= \iint (u+\mu)K_h(u-v)f_o(u)f_o(v)dudv \\ &= \mu A_o(h) + A_1(h) \end{aligned}$$

and hence $E[\tilde{\mu}] = \mu\frac{A_o(h)}{A_o(0)} + \frac{A_1(h)}{A_o(0)} - \frac{A_1(0)}{A_o(0)}$.

This leads us to define the following unbiased estimator:

$$\hat{\mu} = \frac{1}{A_o(h)}\binom{n}{2}^{-1} \sum\sum_{1\leq i<j\leq n} X_i K_h(X_i - X_j) - \frac{A_1(h)}{A_o(h)}.$$

Rewriting the estimator as a U-statistic makes it easy to study its asymptotic properties:

$$\hat{\mu} = \frac{1}{A_o(h)} \binom{n}{2}^{-1} \sum\sum_{1 \le i < j \le n} \frac{X_i + X_j}{2} K_h(X_i - X_j) - \frac{A_1(h)}{A_o(h)}. \tag{1}$$

The contribution of the average of each pair of points (X_i, X_j) to the estimator is weighted by $K_h(X_i - X_j)$, a function of the distance between the two points, that increases as the points get closer and decreases when they are farther apart. Hence, outliers will contribute less to the estimator than observations belonging to the main bulk. The magnitude of these weights can be controlled by choosing an appropriate bandwidth h. Such a choice depends on the tail thickness of the *pdf* $f_o(.)$. For example, the bandwidths for short-tailed distributions such as the Gaussian and sub-Gaussians should be quite large so that $\hat{\mu}$ becomes the sample mean, which is MSE-optimal for this class of distributions. At the other extreme, the bandwidths for distributions with very long tails should be arbitrarily small for $\hat{\mu}$ to get closer to the median. This estimator behaves similarily to L-, R-, and M-estimators, except that it does not utilize ranks or order statistics nor does it require minimization of a loss function. The behavior of $\hat{\mu}$ and the effect of the bandwidth can be seen in the next theorem. But first, we need to define the following functionals of f_o:

$$B_{i,j,k}(t) = \int_{-\infty}^{\infty} u^i f_o(u) \left[K_t * f_o\right]^j (u) \left[L_t * f_o\right]^k (u) du, \tag{2}$$

for some $i, j, k, t \ge 0$, where $L(u) = -uK(u)$. The symbol $B_{i,j,k}(0)$ denotes the limit of $B_{i,j,k}(t)$ as t converges to zero.

Theorem 2.1. *If $\int u f_o^2(u)du < \infty$ and $\int u^2 f_o(u)du < \infty$, then for every $h > 0$ such that $nh \longrightarrow \infty$ as $n \longrightarrow \infty$, we have the following result:*

$$\sqrt{\frac{n}{4V_\mu(h)}}(\hat{\mu} - \mu) \xrightarrow{d} N(0, 1) \tag{3}$$

as $n \to \infty$, where

$$V_\mu(h) = \mu^2 \frac{B_{0,2,0}(h)}{B_{0,1,0}(h)^2} - \left(\mu + \frac{B_{1,1,0}(h)}{B_{0,1,0}(h)}\right)^2 + 2\mu \frac{B_{1,2,0}(h)}{B_{0,1,0}(h)^2} + \mu h \frac{B_{0,1,1}(h)}{B_{0,1,0}(h)^2}$$
$$+ \frac{B_{2,2,0}(h)}{B_{0,1,0}(h)^2} + h \frac{B_{1,1,1}(h)}{B_{0,1,0}(h)^2} + \frac{h^2}{4} \frac{B_{0,0,2}(h)}{B_{0,1,0}(h)}.$$

Proof. Let

$$\varphi(X_i, X_j) = \frac{1}{A_o(h)} \frac{X_i + X_j}{2} K_h(X_i - X_j).$$

Then $\hat{\mu}$ is a U-statistic whose variance can be written as

$$\mathrm{var}(\hat{\mu}) = \frac{4}{n}\mathrm{cov}(\varphi(X_1, X_2), \varphi(X_1, X_3)) + \frac{2}{n(n-1)}\mathrm{var}(\varphi(X_1, X_2)).$$

It can easily be shown that $\mathrm{var}(\varphi(X_1, X_2)) = O(h^{-1})$ and, since $\frac{1}{nh} = o(1)$, the second term of $\mathrm{var}(\hat{\mu})$ can be neglected. If we denote the predominant term of $\mathrm{var}(\hat{\mu})$ by $4V_\mu(h)/n$, we obtain:

$$\begin{aligned}
V_\mu(h) &+ \frac{1}{A_o^2(h)}\mu \iiint (u+v)K_h(u-v)K_h(u-w)f_o(u)f_o(v)f_o(w)dudvdw \\
&+ \frac{1}{4A_o^2(h)} \iiint (u+v)(u+w)K_h(u-v)K_h(u-w)f_o(u)f_o(v)f_o(w)dudvdw \\
&- \left(\mu + \frac{A_1(h)}{A_0(h)}\right)^2 \\
= &\frac{1}{A_o^2(h)}\mu^2 \int f_o(u)\left[K_h * f_o(u))\right]^2 du \\
&+ \frac{1}{A_o^2(h)}\mu \iiint \left[2u - (u-v)\right] K_h(u-v)K_h(u-w)f_o(u)f_o(v)f_o(w)dudvdw \\
&+ \frac{1}{4A_o^2(h)} \iiint \Big\{\left[2u - (u-v)\right]\left[2u - (u-w)\right] K_h(u-v)K_h(u-w) \\
&\qquad f_o(u)f_o(v)f_o(w)\Big\} dudvdw \\
&- \left(\mu + \frac{A_1(h)}{A_0(h)}\right)^2 \\
= &\mu^2 \frac{B_{0,2,0}(h)}{B_{0,1,0}(h)^2} + 2\mu \frac{B_{1,2,0}(h)}{B_{0,1,0}(h)^2} + \mu h \frac{B_{0,1,1}(h)}{B_{0,1,0}(h)^2} + \frac{B_{2,2,0}(h)}{B_{0,1,0}(h)^2} \\
&+ h\frac{B_{1,1,1}(h)}{B_{0,1,0}(h)^2} + \frac{h^2}{4}\frac{B_{0,0,2}(h)}{B_{0,1,0}(h)^2} - \left(\mu + \frac{B_{1,1,0}(h)}{B_{0,1,0}(h)}\right)^2.
\end{aligned}$$

The finiteness of V_μ results from that of the $B_{i,j,k}(h)$'s. In fact, since the density f_o is bounded and the kernel K is smooth enough, there exists a $C > 0$ such that for $i, j, k = 0, 1, 2$,

$$|B_{i,j,k}(h)| \leqslant C \int u^2 f_o(u)du.$$

Since $\hat{\mu}$ is a U-statistic, the finiteness of V_μ guarantees $\hat{\mu}$'s asymptotic normality. □

The fact that the variance is a function of μ might seem to be a hindrance to the performance of the estimator $\hat{\mu}$, but by choosing the bandwidth that minimizes $V_\mu(h)$, the magnitude order of μ will have a minimal effect. For example, when the underlying distribution is normal, a sufficiently large bandwidth reduces the factor of μ^2 to almost zero, as can be seen in Figure 1. In addition to its dependence on the bandwidth h, $V_\mu(h)$ also depends on the $B_{i,j,k}(.)$'s. Hence, the variability of $\hat{\mu}$ depends via these functionals on both the tail thickness and smoothness of the *pdf* f_o.

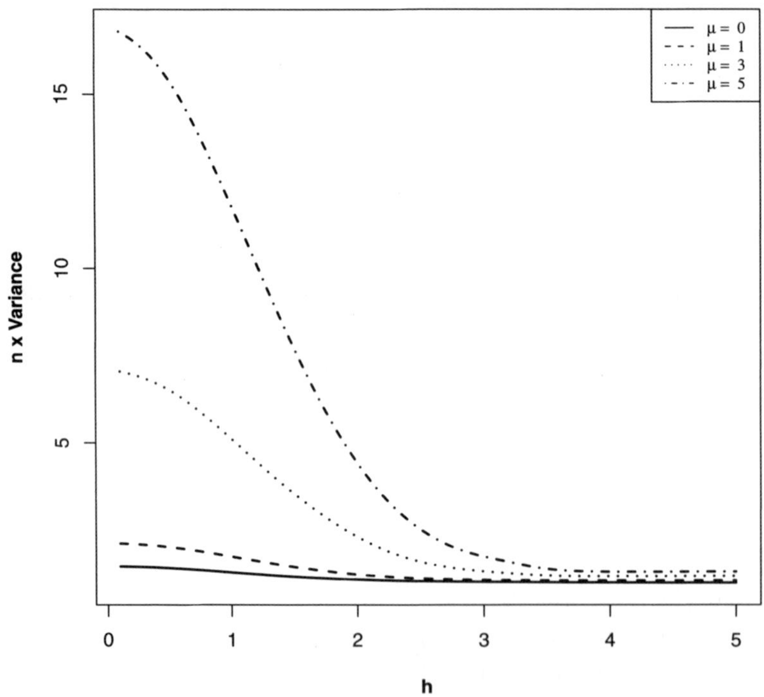

Fig. 1. The variance of $\hat{\mu}$ as a function of the bandwidth.

Another method of reducing the impact of the magnitude of μ on the variability of the estimator is to subtract from each observation some rough estimate of the location, such as the sample mean or preferably the sample median, and then apply the new methodology.

Because of the complexity of $\hat{\mu}$'s variance, it is hard to deduce a meaningful result from a general formula of the asymptotic relative efficiency (ARE) of $\hat{\mu}$ with respect to the maximum likelihood estimator (MLE). However, it is easy to show that the ARE of $\hat{\mu}$ with respect to the MLE of μ is equal to 1 for the normal, Laplace, logistic, and Cauchy distributions.

The asymptotic normality of $\hat{\mu}$, as stated in Theorem 1, requires that the second moment of f_o exists. This excludes distributions with tails equivalent to or longer than a Cauchy distribution's, for example. To overcome this obstacle, we put an additional condition on the bandwidth h in order to obtain asymptotic normality with a weaker moment condition on f_o:

Theorem 2.2. *If $h \to 0$, $nh \to \infty$, $\int u f_o^2(u)du < \infty$, and $\int u^2 f_o^3(u)du < \infty$, then*

$$\sqrt{\frac{n}{4V_\mu(h)}}(\hat{\mu} - \mu) \xrightarrow{d} N(0,1) \tag{4}$$

as $n \to \infty$.

Proof. The argument is similar to that of Theorem 1, except that since h converges to zero as n gets larger, we can obtain different bounds on the $B_{i,j,k}(h)$'s for $i, j, k = 0, 1, 2$:

$$\begin{aligned}|B_{i,j,k}(h)| &\leqslant C \int u^j \left[K_h * f_o\right]^j(u) f_o(u)du \\ &\leqslant C \int u^j \left[f_o(u) + O(h)\right]^j f_o(u)du \\ &\leqslant C_1 \int u^j f_o^{j+1}(u)du,\end{aligned}$$

where C and C_1 are some positive constants. □

Remark 2.1. The moment conditions in Theorem 2 imply that asymptotic normality holds when the underlying distribution tails decay at any rate that is faster than x^{-1}.

3. The Scale Estimator and its Properties

When we replace $I_2 = \int_{-\infty}^{\infty} f_\sigma^2(x)dx$ by its estimator $\hat{I}_2 = \int_{-\infty}^{\infty} \hat{f}_\sigma(x)dF_{\sigma,n}(x)$ in the formula of σ, we obtain an estimator of the scale parameter.

The statistical smoothing and rank-based nonparametrics literature is rich with articles concerning the estimation of integrated squared density (Aubuchon and Hettmansperger, 1984) and its derivatives (Wand and

Jones, 1995). We will use an estimator of I_2 similar to the one used by Aubuchon and Hettmansperger (1984) and Hall and Marron (1987), which enjoys a mean-squared error of order n^{-1}:

$$\hat{I}_2 = \binom{n}{2}^{-1} \sum\sum_{1\leq i<j\leq n} K_h(X_i - X_j).$$

Since $K_h(X_i - X_j)$ is a decreasing function of the distance between two sample points, with the right choice of the bandwidth, sample points closer to the central tendency of the distribution will contribute more to the estimation of I_2, and hence of σ, than outliers. The asymptotic properties of the estimator are presented in the next lemma along with guidelines on how to select an appropriate h.

Lemma 3.1. *If* $nh^4 \longrightarrow 0$, $nh \longrightarrow \infty$, $I_3 = \int_{-\infty}^{\infty} f_\sigma^3(x)dx < \infty$ *and if* $f_\sigma(.)$ *is twice differentiable, then*

$$\sqrt{n}(\hat{I}_2 - I_2) \xrightarrow{d} N(0, 4\left(I_3 - I_2^2\right))$$

as $n \to \infty$.

Proof. $\hat{I}_2$ is a $U-$statistic, so its mean and variance can be written as follows:

$$\begin{aligned} E(\hat{I}_2) &= \int\int K_h(x-y) f_\sigma(x) f_\sigma(y) dx dy \\ &= \int K_h * f_\sigma(x) f_\sigma(x) dx \\ &= \int f_\sigma^2(x)dx + \frac{h^2}{2}\int u^2 K(u)du \int f''_\sigma(x) f_\sigma(x) dx + o(h^2) \end{aligned}$$

and

$$\begin{aligned} \frac{n}{4}\text{var}(\hat{I}_2) &\approx \text{var}[E(K_h(X_2 - X_1)|X_1)] \\ &= \text{var}[K_h * f_\sigma(X_1)] \\ &= \int [K_h * f_\sigma]^2(x) f_\sigma(x)dx - \left[\int K_h * f_\sigma(x) f_\sigma(x) dx\right]^2 \\ &\longrightarrow \int f_\sigma^3(x)dx - \left[\int f_\sigma^2(x)dx\right]^2. \end{aligned}$$

Since $\sqrt{n}(E\hat{I}_2 - I_2) = O(\sqrt{n}h^2)$ converges to zero, asymptotic normality of $\hat{I}_2$ ensues from the finiteness of the I_3. □

By substituting $\hat{I}_2$ in the formula of σ, we can write the estimator of the scale parameter as

$$\hat{\sigma} = \hat{I}_2^{-1} \int_{-\infty}^{\infty} f_1^2(x)dx,$$

whose asymptotic properties are shown in the next result:

Theorem 3.1. *Under the conditions and assumptions of Lemma 1,*

$$\sqrt{n}(\hat{\sigma} - \sigma) \xrightarrow{d} N(0, 4V_\sigma) \tag{5}$$

as $n \to \infty$, *where* $V_\sigma = \sigma^2 \left(\frac{I_3}{I_2^2} - 1 \right)$.

Proof. This result is obtained by using the delta method and the asymptotic distribution of $\hat{I}_2$ from the previous lemma. □

Remark 3.1. The $\sqrt{n}$ rate of convergence is guaranteed by the conditions on the bandwidth h. These conditions are somewhat vague in nature and do not ensure the optimality of the variance. To obtain the optimal bandwidth, we need to minimize the mean-squared error of $\hat{I}_2$ below. Figure 2 shows how the MSE varies as a function of the bandwidth.

$$MSE(\hat{I}_2) = \left[\int K_h * f_\sigma(x) f_\sigma(x) - f_\sigma^2(x) dx \right]^2 + \frac{4}{n} \int [K_h * f_\sigma]^2(x) f_\sigma(x) dx$$
$$- \frac{4}{n} \left[\int K_h * f_\sigma(x) f_\sigma(x) dx \right]^2 .$$

Remark 3.2. The term I_3/I_2^2 is scale-invariant, making the variance of $\hat{\sigma}$ proportional to σ^2. This can also be noticed in Figure 2: When the bandwidth and consequently the bias are small, the MSEs corresponding to three different scale parameters, $0.5, 0.75$, and 1, behave in a similar fashion.

To compare the performance of $\hat{\sigma}$ to that of the MLE of σ, we compute the ARE for the normal, Laplace, logistic, and Cauchy distributions. The results are shown in Table 1.

4. Simulations

To assess the performance of the new estimator for small size data, we compare it to the maximum likelihood estimator (MLE) and Huber's M-estimator using samples of size 25 that are generated from the Gaussian,

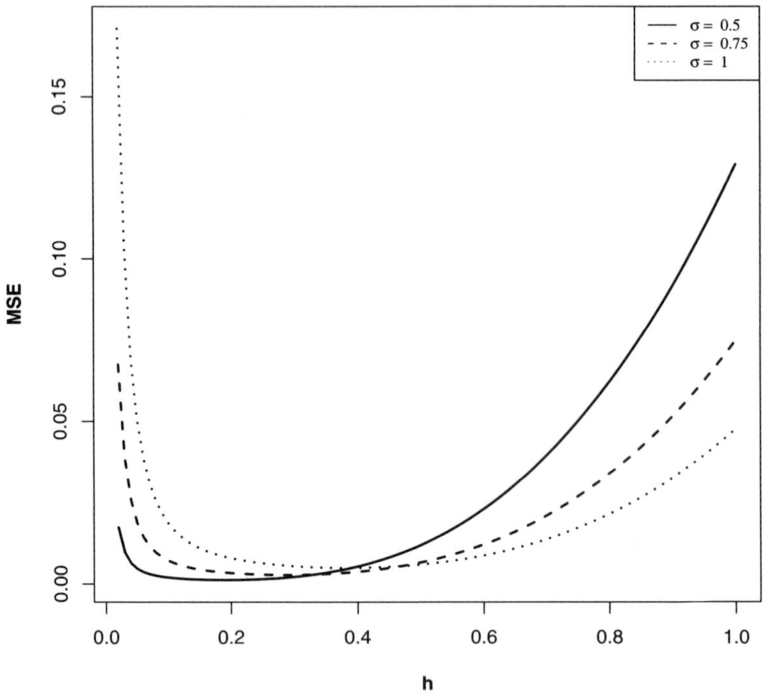

Fig. 2. The variance as a function of the bandwidth.

Table 1. Asymptotic Relative Efficiency for the normal, Laplace, logistic, and Cauchy distributions.

Distribution	Normal	Laplace	Logistic	Cauchy
ARE	1.19	1.33	1.15	0.86

the Laplace, the logistic, and the Cauchy distributions. The choice of these distributions is motivated by their tail thickness and density smoothness.

First, we need to find the optimal bandwidths corresponding to each of the four distributions. Figure 3 shows four plots of the variance of $\hat{\mu}$ against the bandwidth h. Notice that the optimal bandwidth converges to infinity when the underlying distribution is normal or logistic, whereas it is finite for the less smooth Laplace and the long-tailed Cauchy distributions.

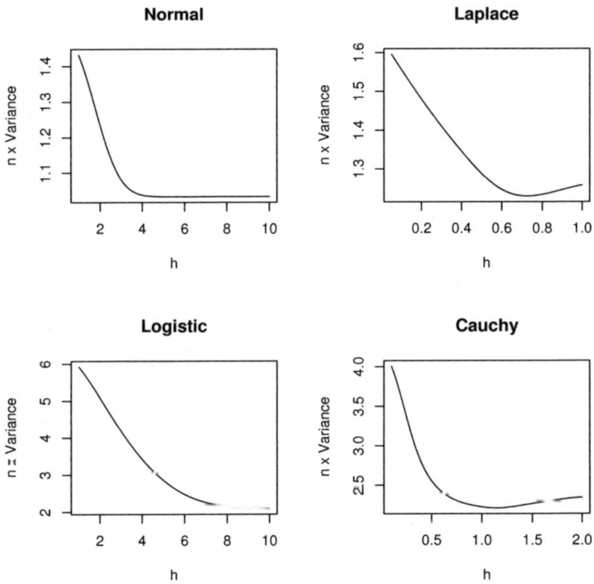

Fig. 3. The variance of $\hat{\mu}$ as a function of the bandwidth.

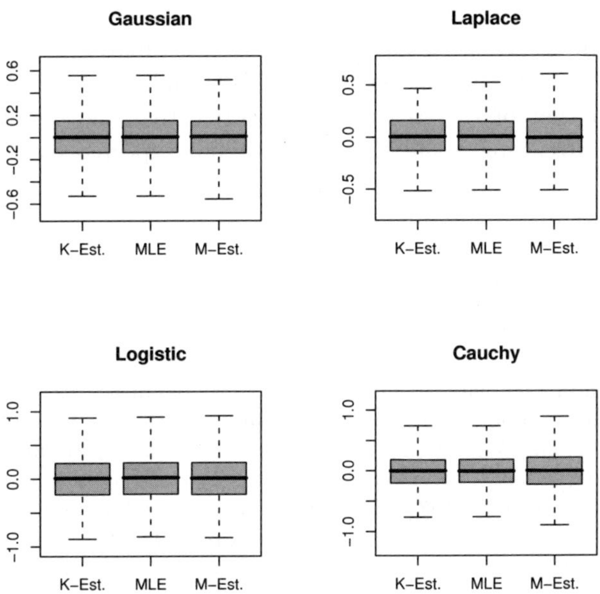

Fig. 4. Boxplots of the location estimators.

Different settings for the distribtuion parameters will not affect the small-sample behavior of the estimator and will only lead to different optimal bandwidths, as discussed in Section 2.

After plugging the optimal bandwidth into Equation (1), we simulate 1000 realizations of $\hat{\mu}$ for each of the distributions and construct corresponding boxplots, displayed in Figure 4. Overall, the kernel location estimation clearly has the same variability as the MLE and outperforms the M-estimator. When the underlying distribution has short tails, it mimics the behavior of a sample mean, which has a slightly smaller variance than the M-estimator. For longer-tailed distributions, where the M-estimator is usually preferred to the sample mean, the kernel estimator enjoys a smaller dispersion as shown in the boxplots of the Laplace and Cauchy distributions. Although the choice of the bandwidth affects the estimator's variability, we noticed throughout the simulation that a moderate increase or decrease in the bandwidth magnitude has a very small effect on the estimator's performance.

We repeat the same simulation process to investigate the small sample behavior of $\hat{\sigma}$ by obtaining the optimal bandwidths for the four distributions (Figure 5) and then generating 1000 realizations of the new scale estimator and comparing it to the MLE and the M-estimator through their centered and scaled boxplots (Figure 6).

For a sample of size 25, the kernel scale estimator has a variability that is very comparable to that of the MLE and, for most distrbutions, smaller than that of the M-estimator. Notice that the boxplots for the kernel estimators show a slightly noticeable but expected bias.

References

Aubuchon, J. C. and Hettmansperger, T. P. (1984). A note on the estimation of the integral of $f^2(x)$, *Journal of Statistical Planning and Inference* **9**, pp. 321–331.

Bickel, P. J. and Rosenblatt, M. (1973). On some global measures of the deviations of density function estimates, *The Annals of Statistics* **1**, pp. 1071–1095.

Bowman, A. W. and Azzalini, A. (1997). *Applied Smoothing Techniques for Data Analysis* (Oxford University Press).

Hall, P. and Marron, J. S. (1987). Estimation of integrated squared density derivatives, *Statistics and Probability Letters* **6**, pp. 109–115.

Härdle, W. (1990). *Applied Nonparametric Regression* (Cambridge University Press, Cambridge), ISBN 0-521-38248-3.

Hart, J. D. (1997). *Nonparametric Smoothing and Lack-of-Fit Tests* (Springer-Verlag, New York), ISBN 0-387-94980-1.

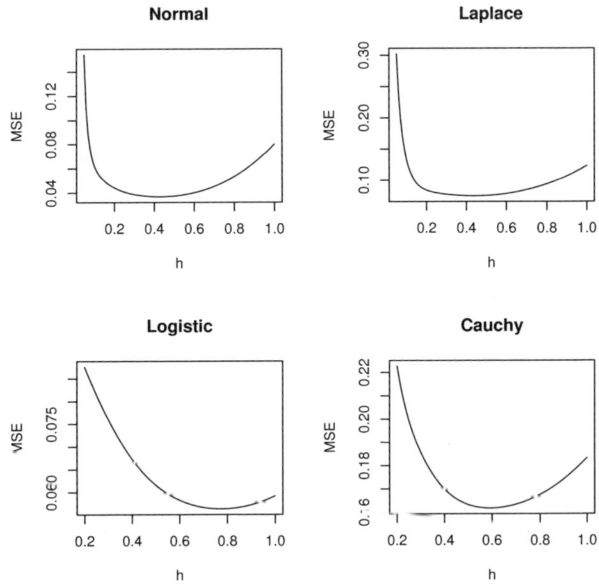

Fig. 5. The MSE of $\hat{\sigma}$ as a function of the bandwidth.

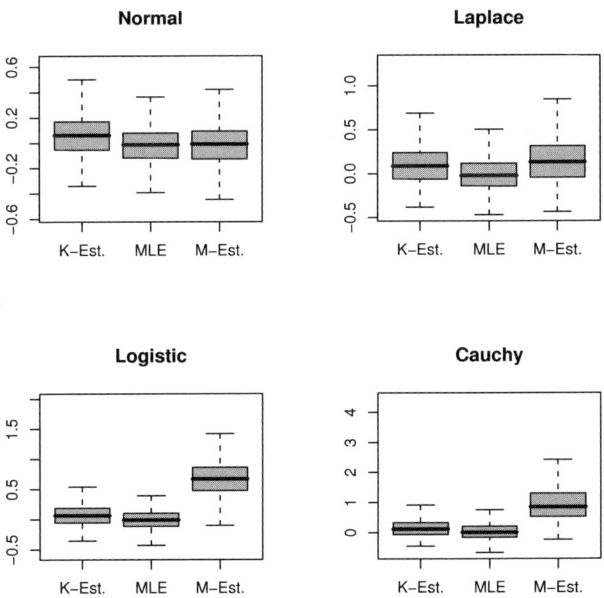

Fig. 6. Boxplots of the scale parameter estimators.

Park, B. U. and Marron, J. S. (1992). On the use of pilot estimators in bandwidth selection, *Journal of Nonparametric Statistics* **1**, pp. 231–240.

Parzen, E. (1962). On estimation of a probability density function and mode, *Annals of Mathematical Statistics* **33**, pp. 1065–1076.

Rosenblatt, M. (1975). A quadratic measure of deviation of two-dimensional density estimates and a test of independence, *The Annals of Statistics* **3**, pp. 1–14.

Scott, D. W. (1992). *Multivariate Density Estimation: Theory, Practice, and Visualization* (John Wiley & Sons Inc., New York), ISBN 0-471-54770-0.

Sheather, S. J. and Jones, M. C. (1991). A reliable data-based bandwidth selection method for kernel density estimation, *Journal of the Royal Statistical Society, Series B* **53**, pp. 683–690.

Silverman, B. W. (1986). *Density Estimation for Statistics and Data Analysis* (Chapman & Hall, London), ISBN 0-412-24620-1.

Simonoff, J. S. (1996). *Smoothing Methods in Statistics* (Springer, New York), ISBN 0-387-94716-7.

Wand, M. P. and Jones, M. C. (1995). *Kernel Smoothing* (Chapman & Hall, New York).

Watson, G. S. (1964). Smooth regression analysis, *Sankhya Ser. A* **26**, pp. 359–372.

Bandwidth Selection in an EM-Like Algorithm for Nonparametric Multivariate Mixtures

Tatiana Benaglia

MRC Biostatistics Unit, Institute of Public Health, Cambridge University
Robinson Way, Cambridge CB2 0SR, United Kingdom
E-mail: tatiana.benaglia@mrc-bsu.cam.ac.uk

Didier Chauveau

Laboratoire MAPMO, UMR 6628, Université d'Orléans
BP 6759, 45067 Orléans cedex 2, France
E-mail: didier.chauveau@univ-orleans.fr

David R. Hunter

Department of Statistics, Pennsylvania State University
University Park, PA 16801, USA
E-mail: dhunter@stat.psu.edu

This article extends the nonparametric EM (npEM) algorithm of Benaglia *et al.* (2009a) and describes a method to select the bandwidths used in this extended algorithm. The extension allows for a distinct bandwidth for each mixture component and each block of conditionally independent and identically distributed repeated measures, and the bandwidth selection method is a generalization of Silverman's rule of thumb.

Keywords: EM algorithm; Kernel density estimation; Multivariate mixture; Nonparametric mixture; Weighted quantile estimate

1. Introduction

Suppose the r-dimensional vectors $\mathbf{X}_1, \ldots, \mathbf{X}_n$ are a simple random sample from a finite mixture of m distributions, where $m > 1$. For nonparametric mixtures, we relax the common assumption that the component distributions come from a family of densities that may be indexed by a finite-dimensional parameter vector. However, it is still necessary to restrict the family $\mathcal{F}$ of multivariate density functions from which the component densities are drawn somewhat in order to avoid the problem of non-identifiability of model parameters, as discussed by Benaglia *et al.* (2009a) and several of the references therein. To this end, we assume throughout this article that $\mathcal{F}$ consists of densities equal to the product of their r univariate marginal densities, which implies that the coordinates of each $\mathbf{X}_i$ vector are independent, conditional on the subpopulation or component (1 through m) from which $\mathbf{X}_i$ is drawn. This *conditional independence assumption* has appeared in a growing body of literature on non- and semi-parametric multivariate mixture models; see Benaglia *et al.* (2009a) for a discussion of the relevant literature. See also Kasahara and Shimotsu (2008) and especially Allman *et al.* (2009) for recent breakthroughs related to the identifiability of these models. We do not discuss the identifiability question here, except to say that Allman *et al.* (2009) give mild sufficient conditions for identifiability whenever $r \geq 3$.

We let $\boldsymbol{\theta}$ denote the vector of parameters, including the mixing proportions $\lambda_1, \ldots, \lambda_m$ and the univariate densities f_{jk} (here and throughout this article, j indexes the component and k indexes the coordinate, so $1 \leq j \leq m$ and $1 \leq k \leq r$). Thus, under the assumption of conditional independence, the mixture density evaluated at $\mathbf{x}_i = (x_{i1}, \ldots, x_{ir})^t$ is

$$g_{\boldsymbol{\theta}}(\mathbf{x}_i) = \sum_{j=1}^{m} \lambda_j \prod_{k=1}^{r} f_{jk}(x_{ik}). \tag{1}$$

In many models, there is reason to assume that *some* or *all* of the r coordinate densities $f_{j1}(\cdot), \ldots, f_{jr}(\cdot)$ are the same for all j. For instance, Elmore *et al.* (2004) and related articles assume that the coordinates are conditionally independent and identically distributed (i.i.d.), i.e., $f_{j1}(\cdot) = \cdots = f_{jr}(\cdot)$ for all j. In order to encompass both the conditionally i.i.d. case and the more general case simultaneously in the model and our algorithm, we allow that the coordinates of $\mathbf{X}_i$ are conditionally independent, and that there exist *blocks* of coordinates that are also identically distributed. These blocks may all be of size one, so that the general case is still treated here; or there may exist only a single block of size r, which is the conditionally i.i.d. case. If we let b_k denote the block to which the kth coordinate belongs, where $1 \leq b_k \leq B$ and B is the total number of such blocks, the general model is

$$g_{\boldsymbol{\theta}}(\mathbf{x}_i) = \sum_{j=1}^{m} \lambda_j \prod_{k=1}^{r} f_{jb_k}(x_{ik}). \tag{2}$$

As an example in which model (2) arises naturally, Benaglia *et al.* (2009a) cite a dataset from Thomas *et al.* (1993) containing 8-dimensional observations on $n = 405$ individuals that logically may be broken into four blocks of two observations each.

Benaglia *et al.* (2009a) construct an algorithm for estimating $\boldsymbol{\theta}$—i.e., the mixing weights λ_j and the densities $f_{jb_k}(\cdot)$—that is based on the well-known family of EM algorithms for parametric mixture models. After introducing the original algorithm, the current article presents several extensions of the algorithm that make it more flexible.

2. The Nonparametric EM Algorithm

The algorithm described in Benaglia *et al.* (2009a), which is implemented as a function called "npEM" in the `mixtools` package (Benaglia *et al.*, 2009b) for R (R Development Core Team, 2010), operates as follows: Given initial values $\boldsymbol{\theta}^0 = (\boldsymbol{\lambda}^0, \mathbf{f}^0)$, iterate the following three steps for $t = 0, 1, \ldots$:

- **E-step:** Letting Z_{ij} denote the (unobserved) indicator of the event that the ith observation is drawn from the jth component, calculate the "posterior" probabilities (conditional on the data and $\boldsymbol{\theta}^t$) of component inclusion,

$$p_{ij}^t \stackrel{\text{def}}{=} P_{\boldsymbol{\theta}^t}(Z_{ij} = 1|\mathbf{x}_i) \tag{3}$$

$$= \frac{\lambda_j^t \prod_{k=1}^r f_{jb_k}^t(x_{ik})}{\sum_{j'=1}^m \lambda_{j'}^t \prod_{k=1}^r f_{j'b_k}^t(x_{ik})}, \tag{4}$$

  for all $i = 1, \ldots, n$ and $j = 1, \ldots, m$.
- **M-step:** Set

$$\lambda_j^{t+1} = \frac{1}{n} \sum_{i=1}^n p_{ij}^t \tag{5}$$

  for $j = 1, \ldots, m$.
- **Nonparametric density estimation step:** For each component $j \in \{1, \ldots, m\}$ and each block $\ell \in \{1, \ldots, B\}$, define the function

$$f_{j\ell}^{t+1}(u) = \frac{\frac{1}{h} \sum_{k=1}^r \sum_{i=1}^n p_{ij}^t I\{b_k = \ell\} K\left(\frac{u - x_{ik}}{h}\right)}{\sum_{k=1}^r \sum_{i=1}^n p_{ij}^t I\{b_k = \ell\}}$$

$$= \frac{1}{nhC_\ell \lambda_j^{t+1}} \sum_{k=1}^r \sum_{i=1}^n p_{ij}^t I\{b_k = \ell\} K\left(\frac{u - x_{ik}}{h}\right), \tag{6}$$

  where $K(\cdot)$ is a kernel density function, h is a bandwidth (see section 3), and

$$C_\ell = \sum_{k=1}^r I\{b_k = \ell\} \tag{7}$$

  is the number of coordinates in the ℓth block.

Benaglia *et al.* (2009a) also construct modified versions of the npEM algorithm tailored for specific situations (i.i.d. coordinates, components differing only by a location or scale parameter, univariate symmetric components, and others). The unmodified npEM algorithm is an extension of a similar idea first presented by Bordes *et al.* (2007) in a slightly different context. As both Bordes *et al.* (2007) and Benaglia *et al.* (2009a) point out, this algorithm is not a true EM algorithm—each of these articles refers to the algorithm it presents as "EM-like"—because there is no likelihood that is being maximized. In particular, there is no analogue of the well-known "ascent property" of a true EM algorithm, which asserts that every iteration of an EM algorithm must increase the current value of the likelihood function. Furthermore, since even a true EM algorithm cannot guarantee convergence to a global maximizer of the likelihood (in general a local maximum may occur instead), there is certainly no sense in which the npEM algorithm can be claimed to achieve anything like a global maximum of some objective function. Nonetheless, empirical evidence shows that the npEM algorithm does produce reasonable answers and, like an EM algorithm, using multiple starting points does appear to protect the npEM algorithm against the problem of locally optimum solutions.

3. Bandwidth Selection

The nonparametric density estimation step (6) of the npEM algorithm described above is a modified version of kernel density estimation, a well-studied topic in statistics (see, for example Silverman, 1986; Scott, 1992; Härdle *et al.*, 2004). The central decision in this step is the selection of an appropriate value for the bandwidth h, or smoothing parameter, since this choice affects density estimates dramatically. The choice of kernel function $K(\cdot)$ is not as influential, so we simply take the kernel to be the standard normal density function.

There exist standard rules in the literature for choosing the bandwidth in the univariate case in which there is no mixture structure. We will use some of these standard ideas and extend them to the mixture case. However, selecting a bandwidth in a mixture setting like this one appears to be a fundamentally more complicated problem than in the corresponding non-mixture case due to the fact that we do not obtain direct information about the individual densities $f_{j\ell}$ from a mixture sample.

Benaglia *et al.* (2009a) use a single fixed bandwidth h for all components and blocks, selected by default according to a rule of thumb found on page 48 of Silverman (1986). The entire $n \times r$ dataset is treated as a vector of length nr and then

$$h = 0.9 \min \left\{ \mathrm{SD}, \frac{\mathrm{IQR}}{1.34} \right\} (nr)^{-1/5}, \tag{8}$$

where SD and IQR are, respectively, the standard deviation and interquartile range of all nr data values.

The nonparametric mixture setting makes bandwidth selection challenging, and Benaglia *et al.* (2009a) suggest several reasons why selecting h using equation (8) might produce an under- or over-estimate for the bandwidth. First, pooling all of the data implicitly treats all components as though they are from the same distribution. This can lead to an overestimate of the bandwidth, particularly if the mixture components' centers are well-separated, for in that case, the variability of the pooled dataset will be larger than that of the individual components. Similarly, if the vector coordinates are not identically distributed within each component, the bandwidth could be biased upward for the same reason. On the other hand, note that the expression nr in equation (8) is an overestimate of the "true" sample size obtained from one of the components or one of the i.i.d. coordinate blocks, especially when each of the r coordinates forms its own block as in equation (1), in which case it may be better to eliminate r from equation (8) entirely. But regardless of whether or not the coordinates are combined in blocks, the fact remains that the "true" sample size from each component is actually some fraction of n, namely, about $n\lambda_j$ for the jth component.

A further limitation of the algorithm of Benaglia *et al.* (2009a) is that equation (6) forces each component and block to employ the same bandwidth, when in fact there is no reason to assume *a priori* that these bandwidths should be the same. In other words, we may simply modify the algorithm from Benaglia *et al.* (2009a) by replacing h by $h_{j\ell}$ in equation (6), thus allowing a different value of the bandwidth for each component and each block density function. The remainder of this section describes how to modify the algorithm of Benaglia *et al.* (2009a) to allow for this greater flexibility.

As noted earlier, the challenge of bandwidth selection for a mixture model is that a sample from $g_{\boldsymbol{\theta}}(\mathbf{x})$ does not give direct information about the individual densities $f_{j\ell}$. However, after each iteration of the npEM algorithm, we do have density estimates $f_{j\ell}^{t+1}$. This suggests an iterative scheme whereby re-estimation of the $h_{j\ell}$ values becomes part of a modified npEM algorithm at each iteration.

To adapt Silverman's rule of thumb (Silverman, 1986, p. 46) to select the value of $h_{j\ell}$ in an iterative procedure, we require at each iteration of the npEM algorithm estimates of the sample size, the sample standard deviation, and the interquartile range for each component and each block. Once these estimates are in place, the estimated bandwidths at the $(t+1)$th iteration, calculated just before the density estimation step of the algorithm, are given by:

$$h_{j\ell}^{t+1} = 0.9 \min\left\{\sigma_{j\ell}^{t+1}, \frac{IQR_{j\ell}^{t+1}}{1.34}\right\}(nC_\ell\lambda_j^{t+1})^{-1/5}, \tag{9}$$

where $nC_\ell\lambda_j^{t+1}$ estimates the sample size for the ℓth block of coordinates in the jth component, and $\sigma_{j\ell}^{t+1}$ and $IQR_{j\ell}^{t+1}$ are the weighted standard deviation and empirical interquartile range for the jth component and ℓth block. Calculation of $\sigma_{j\ell}^{t+1}$ is fairly straightforward if we augment each M-step to include

$$\mu_{j\ell}^{t+1} = \frac{\sum_{i=1}^{n}\sum_{k=1}^{r} p_{ij}^t I\{b_k = \ell\} x_{ik}}{\sum_{i=1}^{n}\sum_{k=1}^{r} p_{ij}^t I\{b_k = \ell\}} = \frac{\sum_{i=1}^{n}\sum_{k=1}^{r} p_{ij}^t I\{b_k = \ell\} x_{ik}}{n\lambda_j^{t+1}C_\ell} \tag{10}$$

and

$$\sigma_{j\ell}^{t+1} = \left[\frac{1}{nC_\ell\lambda_j^{t+1}}\sum_{i=1}^{n}\sum_{k=1}^{r} p_{ij}^t I\{b_k = \ell\}(x_{ik} - \mu_{j\ell}^{t+1})^2\right]^{1/2}. \tag{11}$$

We compute $IQR_{j\ell}^{t+1}$ as the difference between the estimated 0.75 and 0.25 quantiles of the ℓth block of the jth component. To accomplish this, we first introduce the notion of a weighted quantile estimate, as follows:

Let $a_1, \ldots, a_\nu$ be real numbers and $w_1, \ldots, w_\nu$ be associated (nonnegative) weights, with $W = w_1 + \cdots + w_\nu$. The first step in finding the weighted quantile estimate is to sort the a_i in non-decreasing order. To this end, let $\tau(\cdot)$ be a permutation on the integers $\{1, \ldots, \nu\}$ such that

$$a_{\tau(1)} \le a_{\tau(2)} \le \cdots \le a_{\tau(\nu)}.$$

(The τ permutation need not be unique if there are ties among the a_i.) Then for $\alpha \in (0,1)$, we define the *weighted α-quantile estimate* to be $a_{\tau(i_\alpha)}$, where

$$i_\alpha = \min\left\{s : \sum_{i=1}^{s} w_{\tau(i)} \ge \alpha W\right\}$$

is the smallest integer that gives at least a proportion α of the total sum of weights W. Note that in the special case in which all w_i are the same, the weighted quantile estimate is simply a particular way to define the regular sample quantile. The `mixtools` package includes a `wquantile` function, based on the `findInterval` function in `R`, for finding the weighted quantile.

To find $IQR_{j\ell}^{t+1}$, we first calculate the weighted 0.25 and 0.75 quantile estimate of the nC_ℓ data values in block ℓ, with corresponding weights given by the posterior probabilities p_{ij}^t. The weighted interquartile range is then the difference between these two quantiles. In the `mixtools` package, the function `wIQR` calculates the weighted interquartile range. Note that when this calculation is performed as part of the npEM algorithm, the permutation τ need only be calculated once due to the fact that the data and block structure do not change during the running of the algorithm. Furthermore, the sum W of the weights at iteration t is equal to $n\lambda_j^{t+1}C_\ell$ because of equation (5).

Note that there are alternatives to the bandwidth selection method of Silverman (1986) given by equation (9). For instance, simply replacing 0.9 by 1.06 in equation (9) yields the method found in Section 6.5 of Scott (1992). Further bandwidth selection methods are detailed in the documentation for the `bw.nrd0` function in R (R Development Core Team, 2010), and these might be adapted to the case of a weighted sample just as the Silverman (1986) and Scott (1992) methods have been.

4. An Example

A simple example serves to illustrate the effect of allowing different bandwidths in the npEM algorithm. We generated 300 observations from a two-component mixture from the general model (2) with trivariate ($r = 3$) observations, one block (i.e., $B = 1$ and $b_1 = b_2 = b_3 = 1$, which means that we have three conditionally i.i.d. repeated measures), with

$$\lambda_1 = 0.4, \; f_{11} \equiv \mathcal{N}(0,1), \text{ and } f_{21} \equiv \mathcal{N}(15,25). \tag{12}$$

We apply the npEM algorithm using both a single bandwidth given by (8) and different bandwidths computed by (9). The code used to generate the data and then apply the npEM algorithm in R using the `mixtools` package may be found in the Appendix. Note that in the actual npEM output, the component labels were switched—i.e., actually $\hat{h}_{11} = 1.246$ and $\hat{h}_{21} = 0.266$—but we have relabeled the components to agree with the definitions of equation (12).

If we were to consider each component separately and apply Silverman's rule of thumb, the bandwidths would be $h_1 = 0.9(0.4 \times 300)^{-1/5} \approx 0.35$ and $h_2 = 0.9 \times 5 \times (0.6 \times 300)^{-1/5} \approx 1.593$ for the first and second components, respectively. The npEM algorithm found $\hat{h} = 1.932$ for a single bandwidth, and $\hat{h}_{11} = 0.266$ and $\hat{h}_{21} = 1.246$ when allowing different bandwidths. The left-hand plot of Figure 1 shows that the large value $\hat{h}$ results in an over-smoothed density estimate for the first component. Allowing separate bandwidth estimates $\hat{h}_{11}$ and $\hat{h}_{21}$, as in the right-hand plot of Figure 1, gives better results.

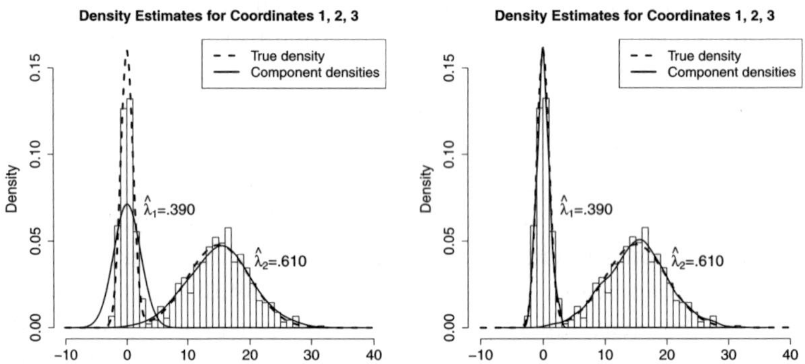

Fig. 1. Effect of bandwidth in the component density estimates. On the left are estimates from the original npEM algorithm using the same bandwidths; on the right are estimates using the modified npEM algorithm in which individual component bandwidths are updated at each iteration.

In Figure 1, we see that the final estimates of λ_1 and λ_2 are the same for the two algorithms, i.e., $\hat{\lambda}_1 = 0.390$ and $\hat{\lambda}_2 = 0.610$. This is not guaranteed to be the case: Since the λ estimates are ultimately functions of the density estimates in the npEM algorithm, the original (fixed, single h) npEM algorithm can lead to different estimates that the modified algorithm we introduce here. However, the particular example we have chosen involves mixture components that are "well-separated" in the sense that it is fairly easy to assign each observation to one component or the other. Thus, the final estimate of λ_1 is simply the proportion of observations that are classified as belonging to the first component, which is 117/300 in this example. We may observe that this is true by noting that all of the posterior probabilities p_{ij} at the final iteration are very close to zero or one. (In fact, the smallest value of $\max_j p_{ij}$ is larger than 0.9999 for both algorithms of Figure 1.)

5. Further Extensions

Among the special cases of the npEM algorithm developed by Benaglia *et al.* (2009a) are those in which component or block density estimates may differ from one another, but they differ only in (say) a location or scale parameter and otherwise they have the same shape. In cases such as these, an extension of the npEM algorithm as we have suggested here—one that recalculates each h_{jk} value at each iteration—should be designed to exploit this structure.

For instance, the most restrictive case discussed in Benaglia *et al.* (2009a) is the case in which the density function for each component and block shares exactly the same shape, and the different components and blocks differ only by a possible location and scale parameter, so that for every component j and block ℓ, the (j, ℓ) density function becomes

$$f_{j\ell}(x) \;=\; \frac{1}{\sigma_{j\ell}} f\left(\frac{x-\mu_{j\ell}}{\sigma_{j\ell}}\right). \tag{13}$$

In this case, because we wish to estimate only a single density $f(x)$, the density estimation step in the algorithm would specify for any $u \in \mathbb{R}$ that

$$f^{t+1}(u) = \frac{1}{nrh}\sum_{i=1}^{n}\sum_{j=1}^{m}\sum_{k=1}^{r} p_{ij}^{t} K\left(\frac{u - x_{ik} + \mu_{jb_k}}{h\sigma_{jb_k}}\right). \tag{14}$$

In particular, note that there is no reason to allow a different bandwidth $h_{j\ell}$ for each component j and block ℓ as in equation (9). In effect, the different bandwidths for each component and block have been entirely accounted for by the location and scale parameters. Indeed, in the example of Section 4, the component and block densities all have exactly the same true shape — normal — so that by accounting for the location and scale changes, we eliminate the need in that example for different bandwidths. Even in this case, however, the updating of h at each iteration as in equation (9) is still useful, since it is still difficult for the simplistic bandwidth-selection rule of thumb employed by Benaglia *et al.* (2009a) to make a good selection before anything is known about the mixture structure.

To update h at each iteration under model (13), we cannot merely use a slightly modified version of equation (9) since the $\mu_{j\ell}$ and $\sigma_{j\ell}$ parameters now play a more integral role in the model. Indeed, the whole point of iteratively updating h is to ignore the location and scale parameters for each component and block. In other words, we should apply the bandwidth-selection rule of thumb only *after* each data point x_{ik} has been appropriately shifted and scaled. Since each x_{ik} will actually be shifted and scaled according to *each* of the m components, naturally each $(x_{ik} - \mu_{jb_k})/\sigma_{jb_k}$ must be weighted according to the corresponding weight p_{ij}. As the sum of all of these weights is exactly n, the effective sample size is nr, leading to the following formula for updating h:

$$h^{t+1} = \frac{0.9}{(nr)^{1/5}} \times \min\left\{1, \frac{IQR^{t+1}}{1.34}\right\}, \tag{15}$$

where IQR^{t+1} is the weighted interquartile range of the $n \times m \times r$ values $(x_{ik} - \mu_{jb_k})/\sigma_{jb_k}$, with corresponding weights p_{ij}^t, for all $1 \le i \le n$, $1 \le j \le m$, and $1 \le k \le r$. Note that the weighted mean and weighted variance of these same values are 0 and 1 due to the fact that each x_{ik} is properly normalized within each of the m components, which explains why the standard deviation of equation (9) has been simply replaced by one in equation (15).

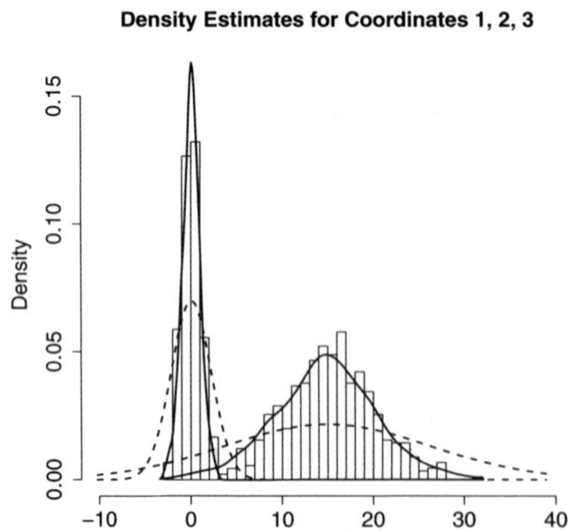

Fig. 2. Two algorithms fitting model (13) are compared, one in which the bandwidth is fixed throughout (dashed lines) and one in which the bandwidth is updated at each iteration (solid lines). For each of these two methods, the two components' density estimates are identical in shape and differ only in location and scale.

In Figure 2, we can see the effect of iteratively updating the bandwidth. Each of the two algorithms compared in that figure starts at the same default bandwidth value of 1.932; yet the iteratively updated bandwidth decreases dramatically starting at the second iteration, ultimately settling on a final value of 0.220. The dashed lines show that the algorithm that fails to update this bandwidth produces a far poorer density estimate than the one that does the updating. This is not surprising, since failing to update means that the bandwidth must be chosen prior to knowing anything about the mixture structure, at which point it is very difficult to properly choose a bandwidth.

For some applications, model (13) may be too restrictive in its assumption that all components and blocks have exactly the same shape of density function. As Benaglia *et al.* (2009a) discuss, it is also possible to allow each component or each block to have its own density shape, which leads to replacing (13) by

$$f_{j\ell}(x) \equiv \frac{1}{\sigma_{j\ell}} f_j \left(\frac{x - \mu_{j\ell}}{\sigma_{j\ell}} \right) \qquad \text{or} \qquad f_{j\ell}(x) \equiv \frac{1}{\sigma_{j\ell}} f_\ell \left(\frac{x - \mu_{j\ell}}{\sigma_{j\ell}} \right),$$

respectively. Implementation of our algorithm for either of these models is relatively straightforward using the same techniques described earlier; and, in either case, the bandwidth (h_j or h_ℓ) may once again be allowed to depend upon either the component or the block, as the case may be.

6. Summary

This article extends the bandwidth-selection scheme of Benaglia *et al.* (2009a) by allowing for (1) iteratively updating bandwidths; and (2) component- and block-specific bandwidths. Each of these two extensions improves the density estimations for a different reason: The first solves the problem that it is difficult to estimate a bandwidth before knowing about the mixture structure, while the second takes care of cases in which different components or blocks have very different properties. In cases in which "very different properties" includes *only* location and/or scale differences, it is possible (as we explain in Section 5) to correct for these differences in the model without component- and block-specific bandwidths; yet in more general cases in which these different bandwidths are desirable, implementation of common bandwidth selection methods such as those of Scott (1992) or Silverman (1986) requires weighted versions of the interquartile range and the standard deviation as we present here.

Acknowledgments

This work was generously supported by NSF Award SES-0518772 and by le Studium, CNRS Orléans, France.

Appendix A. Computer code

The following computer code can be used in R, once `mixtools` version 0.3.3 is installed, to produce the examples seen in this article.

```
library(mixtools) # mixtools package must be installed first
set.seed(123) # Ensure that results are exactly reproducible

#Generate data:
mu <- matrix(c(0, 15), 2, 3)
sigma <- matrix(c(1, 5), 2, 3)
x <- rmvnormmix(300, lambda = c(.4,.6), mu = mu, sigma = sigma)

# npEM algorithm results:
a <- npEM(x, mu0 = 2, blockid = rep(1,3), samebw = TRUE)
b <- npEM(x, mu0 = 2, blockid = rep(1,3), samebw = FALSE)

# Produce plots like in Figure 1:
s <- seq(-10, 40, len = 200)
plot(a, xlim=c(-10, 40), ylim = c(0, .16), xlab = "", breaks = 30)
lines(s, dnorm(s)*.4 + dnorm((s-15)/5)/5*.6, lwd = 2, lty = 2)
plot(b, xlim=c(-10, 40), ylim = c(0, .16), xlab = "", breaks = 30)
lines(s, dnorm(s)*.4 + dnorm((s-15)/5)/5*.6, lwd = 2, lty = 2)

# Display npEM bandwidths, minimum values of max_j p_{ij}
a$bandwidth
b$bandwidth
min(apply(a$posteriors, 1, max))
min(apply(b$posteriors, 1, max))

# spEM algorithm results:
d <- spEM(x, mu0 = 2, blockid = rep(1,3), constbw = FALSE)
d2 <- spEM(x, mu0 = 2, blockid = rep(1,3), constbw = TRUE)

# Produce plot like in Figure 2:
plot(d, xlim=c(-10, 40), ylim = c(0, .16), xlab = "", breaks = 30,
     addlegend=FALSE)
plot(d2, newplot=FALSE, addlegend=FALSE, lty=2, dens.col=1)

# Display spEM bandwidths
d$bandwidth
d2$bandwidth
```

References

Allman, E., Matias, C. and Rhodes, J. (2009). Identifiability of parameters in latent structure models with many observed variables, *The Annals of Statistics* **37**, 6A, pp. 3099–3132.

Benaglia, T., Chauveau, D. and Hunter, D. R. (2009a). An EM-like algorithm for semi-and nonparametric estimation in multivariate mixtures, *Journal of Computational and Graphical Statistics* **18**, 2, pp. 505–526.

Benaglia, T., Chauveau, D., Hunter, D. R. and Young, D. S. (2009b). mixtools: An R package for analyzing mixture models, *Journal of Statistical Software* **32**, 6, pp. 1–29, URL `http://www.jstatsoft.org/v32/i06`.

Bordes, L., Chauveau, D. and Vandekerkhove, P. (2007). A stochastic EM algorithm for a semiparametric mixture model, *Computational Statistics and Data Analysis* **51**, 11, pp. 5429–5443.

Elmore, R. T., Hettmansperger, T. P. and Thomas, H. (2004). Estimating component cumulative distribution functions in finite mixture models, *Communications in Statistics—Theory and Methods* **33**, 9, pp. 2075–2086.

Härdle, W., Müller, M., Sperlich, S. and Werwatz, A. (2004). *Nonparametric and Semiparametric Models*, Springer Series in Statistics (Springer-Verlag, New York), ISBN 3-540-20722-8.

Kasahara, H. and Shimotsu, K. (2008). Nonparametric identification and estimation of multivariate mixtures, Tech. Rep. 1153, Queen's University Department of Economics, URL `http://qed.econ.queensu.ca/faculty/shimotsu/papers/nonparaid_revised.pdf`.

R Development Core Team (2010). *R: A Language and Environment for Statistical Computing*, R Foundation for Statistical Computing, Vienna, Austria, URL `http://www.R-project.org`, ISBN 3-900051-07-0.

Scott, D. W. (1992). *Multivariate Density Estimation*, Wiley Series in Probability and Mathematical Statistics: Applied Probability and Statistics (John Wiley & Sons Inc., New York), ISBN 0-471-54770-0, theory, practice, and visualization, A Wiley-Interscience Publication.

Silverman, B. W. (1986). *Density Estimation for Statistics and Data Analysis*, Monographs on Statistics and Applied Probability (Chapman & Hall, London), ISBN 0-412-24620-1.

Thomas, H., Lohaus, A. and Brainerd, C. J. (1993). Modeling growth and individual differences in spatial tasks, *Monographs of the Society for Research in Child Development* **58**, 9.

Dealing with More Variables than the Sample Size: An Application to Shape Analysis

Chiara Brombin[†] and Fortunato Pesarin[‡]

Department of Statistics, University of Padova
Via Cesare Battisti 241-243, 35121 Padova, Italy
[†]*E-mail: chiara.brombin@stat.unipd.it*
[‡]*E-mail: pesarin@stat.unipd.it*

Luigi Salmaso

Department of Management and Engineering, University of Padova
Stradella S. Nicola 3, 36100 Vicenza, Italy
E-mail: salmaso@gest.unipd.it

In statistical shape analysis, as well as in several other application fields, researchers have to cope with a large number of variables and few available specimens. With reference to parametric inferential methods based on multivariate normality, the most natural way to compare two mean shapes is by using the Hotelling's T^2 test. Despite its widespread use, this test may not be appropriate unless there is a large number of specimens available (Dryden and Mardia, 1998; Blair *et al.*, 1994). For these reasons, we propose a nonparametric permutation counterpart, stressing the case in which the number of variables is larger than the permutation sample space, and find it very powerful. On the basis of these results, we perform a simulation study to evaluate the power of multivariate permutation tests based on Pesarin's combining functions (Pesarin, 2001). In particular, we show that the power of the suggested tests increases when the number of processed variables increases, provided that the induced noncentrality parameter δ increases, and this result holds even when the number of variables is larger than the permutation sample space. Such findings look very relevant for analysing multivariate small-sample problems when shapes are the object of study.

Keywords: Hotelling's T^2; Nonparametric combination (NPC) approach; Shape analysis

1. Introduction

In genetics and many observational studies, it may happen that the number of observed variables is very much larger than the number of subjects. An interesting and relatively new application field, statistical shape analysis, relates to the study of objects' shapes, i.e., all the geometric information that remains when location, scale, and rotational effects are filtered out. A substantial role in shape reasoning has been played by "landmark-based" analysis, where shapes are represented by a discrete sampling of an object's contours (Dryden and Mardia, 1998; Small, 1996).

Bookstein and his colleagues recommended the use of landmarks for the analysis of biological features, constraining the choice of landmarks to prominent features of the organism or biological structure (Dryden and Mardia, 1998; Bookstein, 1986, 1997). Landmarks are defined as *points of correspondence on each object that match between and within populations.* These loci have the same name, i.e., they are homologues, as well as Cartesian coordinates, and they correspond in some sensible way over the forms of a data set (Slice *et al.*, 1996). In Srivastava *et al.* (2005), some limitations of landmark-based representations are emphasized. Despite the effectiveness of this approach in applications where landmarks are easily available (e.g., physician-assisted medical image analysis), automatic detection of landmarks is not straightforward and the resulting analysis is to a great extent determined by the choice of landmarks.

Along with landmarks, it is possible to collect semi-landmark points. The term "semi" is used because the landmarks lie in a lower number of dimensions than other types of landmarks, e.g., along a one dimensional curve in a two dimensional image (Dryden and Mardia, 1998). Semi-landmarks are defined relative to other landmarks, for example, "midway between landmarks 1 and 2". Hence they do not enjoy the homology property; while they lie on homologous curves, their exact position along these usually smooth regions or curves is unclear.

Defining semi-landmarks could be useful to study substantial regions in a object that cannot be defined simply using anatomical or mathematical landmarks, or a region comprised between two or more real landmark points (Adams *et al.*, 2004).

On the basis of these considerations, Katina *et al.* (2007) proposed another landmark classification, including the information carried by semi-landmarks on curves and surfaces. By means of two- or three-dimensional landmarks and semi-landmarks, i.e., the variables of the multivariate problem, it is possible to accurately describe the shape of an object. Rohlf (2000) reviews the main statistical tests proposed in shape analysis, comparing their statistical power. For example, in the case of two independent samples, a practical method for comparing the mean shapes of two groups is to use the Procrustes tangent space coordinates, if data are concentrated (i.e., close in shape or presenting small variation in shape), calculating the Mahalanobis distance between the two data sets and then applying Hotelling's T^2 test statistic. Under the assumption of isotropy, another simple approach is to work with statistics based on squared Procrustes distance and then apply Goodall's F test statistic.

As is well-known, many useful shape models contain a large number of (nuisance) parameters lying in nonlinear spaces. In the presence of concentrated data, it is possible to make a good linear approximation to the shape space (Kent, 1994). The resulting space is the so-called tangent space, which is of the same dimensionality as the shape space. Actually, the tangent space is the linearized version of the shape space in the neighborhood of a particular point of shape space (the pole of the tangent projection). Hence, Euclidean distances in the tangent space can be used to approximate Procrustes distances in the shape space in the vicinity of the pole. Since the tangent space is linear, it is possible to apply standard multivariate statistical methods for studying shape variations (Slice *et al.*, 1996). That is why concentrated data are required in order to correctly apply Hotelling's T^2 and Goodalls F tests.

Coming back to the general case, it is well known that, despite their widespread use, Hotelling's T^2 and Goodall's F tests may be problematic. On one hand, Hotelling's T^2 test may not be appropriate unless there is available a number of specimens larger than that of observed landmarks (Dryden and Mardia, 1998; Blair *et al.*, 1994); on the other hand, Goodall's F test assumes that configurations of landmarks are isotropic normal perturbations from mean configurations and that the distributions of the squared Procrustes distances are approximately chi-squared distributed. Because of the sphericity assumption, the required model is rather demanding (Dryden and Mardia, 1998).

For these reasons, it is important to develop powerful statistical methods that guarantee good accuracy (which means many landmarks and semilandmarks, hence many variables) in describing shapes even in presence of small sample sizes.

After introducing the NonParametric Combination (NPC) methodology (Pesarin, 2001) and highlighting the advantages of distribution-free approaches, we present a Hotelling's T^2 permutation counterpart (Section 2) and the simulation results (Section 3) that motivate a wide simulation study (Section 4) in which we evaluate the power of multivariate permutation tests when the number of variables is larger than the permutation sample space. Finally, we illustrate the flexibility of the NPC method through an application to a biological case-study in a shape analysis framework (Section 5).

2. An NPC solution to multivariate small-sample problems

As pointed out in Good (2000), although in the null hypothesis the assumption of equal covariance matrices can often be justified, it may happen that, in the alternative hypothesis, it becomes unreasonable in applications where it is easy to justify stochastic dominance. Moreover, the multinormal model in the tangent space (Rohlf, 1999) may be doubtful and sometimes there are few individuals and many landmarks, implying over-dimensioned spaces, loss of power or even inadequacy for the Hotelling's T^2 test. Hence, an alternative procedure is to consider a permutation approach. Permutation methods are distribution-free, allow for efficient solutions, are fully robust when the variance-covariance matrix is not of full rank, and do not fail when the number of "cases" (e.g., specimens) is less than the number of variables. Furthermore, permutation tests may be tailored for sensitivity to specific treatment alternatives and to treat tests of hypotheses for one-sided (i.e., order-restricted) alternatives or two-sided (i.e., unrestricted) alternatives (Blair *et al.*, 1994).

In light of these considerations, we propose the Hotelling's T^2 permutation counterpart within the NPC approach (Pesarin, 2001). The NPC methodology is based on: (a) a decomposition of the hypotheses into k, $k > 1$, sub-hypotheses, where for each sub-hypothesis, a suitable partial permutation test is available; (b) a simulation procedure, conditional on the set of observed data, that provides an estimate of the null multivariate permutation distribution of the whole set of test statistics; and (c) a combination of the partial tests into a second-order statistic whose null permutation distribution is estimated by using the same simulation results of the first step (Celant *et al.*, 2000). The NPC methodology is a conditional testing procedure that, under very mild and reasonable conditions, provided that exchangeability of data with respect to groups is satisfied in the null hypothesis, is found to be at least consistent and unbiased (Pesarin, 2001). Moreover, for large sample sizes, its power behavior under multivariate normality is comparable to that of Hotelling's T^2, the two being asymptotically equivalent.

Let us consider the two independent sample case and assume that the response variables are related according to the following multivariate model:

$$X_{hji} = \mu_h + \delta_{hj} + Z_{hji},$$

$i = 1, \ldots, n_j$, $j = 1, 2$, $h = 1, \ldots, k$, where μ_h represents a population constant for the hth variable; δ_{hj} represents the treatment effect (i.e., the noncentrality parameter) in the jth group on the hth variable which, without loss of generality, is assumed to be δ_{h1}=0, $\delta_{h2} \leq$ (or $\geq$) 0, e.g., $\delta_{h2} \geq 0$ for all $h = 1, \ldots, k$ and for at least one h is strictly positive; and Z_{hji} are random errors assumed to be exchangeable with respect to treatment levels and independent with respect to units, with null mean vector ($\mathbb{E}(\mathbf{Z}) = 0$) and finite second moments.

Let $\bar{X}_{hj}$, $j = 1, 2$, be the sample mean for the hth variable, S_j the biased sample covariance matrix (with divisor n_j), and S the common covariance matrix, given by

$$S = (n_1 S_1 + n_2 S_2)/(n_1 + n_2 - 2).$$

Therefore the specific hypotheses may be expressed as

$$H_0 : \bigcap_{h=1}^{k} \{X_{h1} \stackrel{d}{=} X_{h2}\} \quad \text{vs.} \quad H_1 : \bigcup_{h}^{k} \{(X_{h1} + \delta) <\stackrel{d}{\neq}> X_{h2}\},$$

where $\stackrel{d}{=}$ stands for equality in distribution. Under homoscedasticity, this obviously implies

$$H_0 : \bigcap_{h=1}^{k} \{\mathbb{E}(X_{h1}) = \mathbb{E}(X_{h2})\} \quad \text{vs.} \quad H_1 : \bigcup_{h=1}^{k} \{\mathbb{E}(X_{h1}) <\neq> \mathbb{E}(X_{h2})\},$$

where $<\neq>$ means '$<$', '$\neq$' or '$>$'.

In the shape analysis context, when dealing with two independent random samples from independent populations with mean shapes $[\boldsymbol{\mu}_1]$ and $[\boldsymbol{\mu}_2]$, usually the hypotheses become:

$$H_0 : [\boldsymbol{\mu}_1] = [\boldsymbol{\mu}_2] \text{ versus } H_1 : [\boldsymbol{\mu}_1] \neq [\boldsymbol{\mu}_2] \,.$$

We recall that $[\boldsymbol{\mu}_1]$ (or $[\boldsymbol{\mu}_2]$) indicates the mean shape of population 1 (or 2, respectively), usually a $k \times m$ matrix, representing k landmarks in m dimensions.

Let $T_h^o(0)$ and $T_h^*(0)$ indicate respectively the observed and permutation values of T_h under H_0. We recall that the assumptions regarding the set of partial tests $\mathbf{T} = \{T_h, h = 1, \ldots, k\}$ sufficient for nonparametric combination are:

(1) All permutation partial tests T_h are marginally unbiased and significant for large values, so that they are stochastically larger in H_1 than in H_0.

(2) All permutation partial tests T_h are consistent, that is,

$$\Pr\{T_h \geq T_{h\alpha}|H_{1h}\} \to 1, \forall\alpha > 0, h = 1, \ldots, k,$$

as n tends to infinity, where $T_{h\alpha} < +\infty$ is the critical value of T_h at level α. In order to obtain traditional global consistency, it suffices that at least one partial test is consistent (Pesarin, 2001).

It is known that the set $\lambda_h, h = 1, \ldots, k$ of p-values associated with partial tests in $\mathbf{T}$ is positively dependent in the alternative, irrespective of dependence relations among component variables in X (Lehmann, 1986).

We define $T^{''}$, a permutation T^2-type counterpart, by

$$T^{''} = \sum_{h=1}^{k} \left(\frac{\bar{X}^*_{h1} - \bar{X}^*_{h2}}{S^*_{hh}} \right)^2,$$

where the symbol * indicates a permutation of the original data, $\bar{X}^*_{hj}$, $j = 1, 2$ are multivariate permutation sample means, and S^*_{hh} are the diagonal elements of S^*. If homoscedasticity in the alternative cannot be assumed, Hotelling's T^2 cannot be calculated correctly, whereas its permutation counterpart becomes $T^{''} = \sum \Phi^{-1}(1 - \lambda_h)$, provided that for each component variable there is stochastic dominance, i.e. $X_h(0) \overset{d}{=} X_h(\delta_h)$, where Φ is the standard normal c.d.f. We remark that the underlying dependence structure is nonparametrically and implicitly taken into consideration since the true permutation distribution of $T^{''}$ is generated by the procedure.

To illustrate, in the shape analysis field, the index $h = 1, \ldots, km$ represents the k landmarks in m dimensions. In order to apply the NPC methodology, the testing problem is broken down into two stages, considering both the coordinate and the landmark level (and, if present, the domain level too). Hence, we formulate partial test statistics for two-sided hypotheses and then we consider the global test T'' obtained after combining at the first stage with respect to m, then with respect to k.

For example, if we consider 4 landmarks, first of all one can derive a test for each coordinate (x and y coordinates in the 2D case) of each landmark. Once the aspects of interest are decided upon, one could focus on the coordinate level or on the landmark level, after combining coordinates, or on the domain level as well, and finally on the global test (see Figure 1).

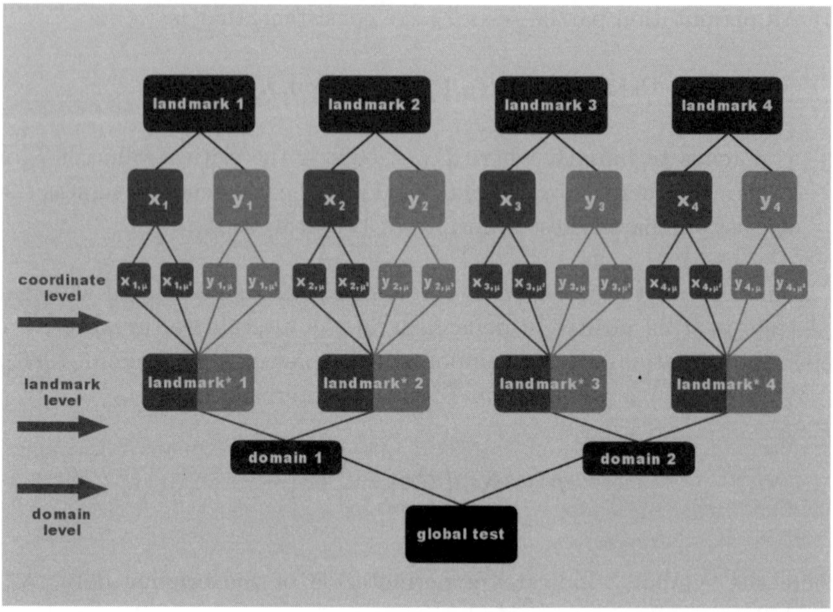

Fig. 1. Different levels of combination.

3. Motivation

Simulation results displayed in Tables 1–3 furnish the motivation for the comprehensive simulation study presented in the next section. In Table 1, T^2 stands for the standard parametric Hotelling T^2 test statistic.

Simulations have been performed in the normal homoscedastic situation, assuming $\delta_{hj} = \delta_j$, i.e., all variables have the same noncentrality parameter. We set the number, B, of permutations equal to 1000 and the number, MC, of Monte Carlo iterations equal to 1000.

With two samples of $n_1 = n_2 = 10$ individuals, it is possible to calculate the T^2 test statistic only if k, the number of variables, is less than 19 in order to guarantee positive degrees of freedom. The results obtained show that the power of T^2 increases for $k < 14$ and decreases for $k > 15$, whereas power of T'' increases when the number of processed variables increases (see Table 1) or the value of the noncentrality parameter δ increases, even when the number of variables k is larger than the permutation sample space (see Table 2–3). We point out that with $n_1 = n_2 = 3$, the cardinality of the permutation sample space is 20. These findings are very important since they show that it is possible to obtain powerful tests in a nonparametric framework by increasing the number of informative variables while the number of cases is fixed. By informative variables, we mean those variables that do not diminish the induced standardized noncentrality, thus producing non-decreasing signal-to-noise ratio. In this way, it is possible to solve a recurring problem in shape analysis where, very often, there are many informative landmarks and semi-landmarks (i.e., 'capturing' informative outlines) but few available specimens.

Table 1. Simulations under H_1 ($n_1 = n_2 = 10$, $\mu = 0$, $\delta = 0.40$, $B = 1000$, $MC = 1000$).

		α=0.01	α=0.05	α=0.10	α=0.20	α=0.30	α=0.50
$k = 15$	T^2	0.027	0.118	0.233	0.419	0.566	0.789
	T''	0.231	0.484	0.623	0.771	0.856	0.941
$k = 18$	T^2	0.013	0.067	0.132	0.269	0.414	0.642
	T''	0.253	0.543	0.667	0.816	0.874	0.956
$k = 19$	T^2	0	0	0	0	0	0
	T''	0.244	0.544	0.683	0.825	0.904	0.965
$k = 20$	T''	0.318	0.552	0.700	0.837	0.905	0.977
$k = 22$	T''	0.340	0.618	0.744	0.845	0.906	0.964
$k = 25$	T''	0.365	0.656	0.774	0.880	0.930	0.970

Table 2. Simulations under H_1 ($n_1 = n_2 = 3$, $\mu = 0$, $\delta = 0.40$, $MC = 1000$).

		α=0.10	α=0.20	α=0.30	α=0.50
$k = 3$	T''	0.059	0.194	0.337	0.554
$k = 20$	T''	0.090	0.264	0.390	0.611
$k = 40$	T''	0.089	0.277	0.442	0.667

4. A simulation study and results

Different multivariate distributions have been considered: normal $\mathcal{N}(0,1)$, Cauchy $\mathcal{C}y(0,1)$, Student's $\mathcal{S}t(2)$ with 2 degrees of freedom, and Pareto $\mathcal{P}a(1,1)$ distributions. We note that, because of the chosen parameters,

Table 3. Simulations under H_1 ($n_1 = n_2 = 3$, $\mu = 0$, $\delta = 1$, $MC = 1000$).

		α=0.10	α=0.20	α=0.30	α=0.50
$k = 3$	T''	0.187	0.454	0.629	0.800
$k = 20$	T''	0.324	0.779	0.902	0.977
$k = 50$	T''	0.434	0.907	0.985	0.999

here we deal with "particular" distributions. Actually, Cauchy $\mathcal{C}y(0,1)$ has no mean and infinite variance, Student's t with 2 d.f. has finite mean and infinite variance, and Pareto $\mathcal{P}a(1,1)$ has infinite mean and infinite variance.

The notion of unconditional finite sample consistency, defined for divergent fixed effects δ, is different from the traditional notion of (unconditional) consistency of a test, which considers the behavior of the rejection rate for a given δ when $\min(n_1, n_2)$ diverges. It is known that, in order to attain permutation (unconditional) consistency, it is required that random deviates Z have finite second moment (Lehmann, 1986; Romano, 1990). Here we only require that random deviates Z are not provided with finite moments of any integer order ≥ 2 (Pesarin and Salmaso, 2010).

We focus on the two independent sample case in the particular case in which only 3 specimens are available in each group. We have chosen this simulation setting because we are interested in investigating the behavior of the power of the test when the number of variables is larger than the cardinality of the permutation sample space. We remark that with $n_1 = n_2 = 3$, the cardinality of the permutation sample space is $\binom{6}{3} = 20$. Hence, we wish to analyze the behavior of the power using all the possible permutations and recalling that, in this case, the minimum attainable α level is 2/20 for two-sided tests (1/20 for one-sided tests). Figure 2 shows simulation results in the non-directional case, and Figure 3 shows those for the directional case, when the underlying distribution is multivariate normal. In the case of non-directional (two-sided) alternatives, we consider the test statistic given by $T^* = \sum_h (\bar{X}^*_{h1} - \bar{X}^*_{h2})^2$, where $\bar{X}^*_{hj}, j = 1, 2$ are multivariate permutation sample means, whereas with directional alternatives we simply consider the test statistic given by $T^* = \sum_h (\bar{X}^*_{h2})$.

Focusing on the directional case:

- In Figure 4 are shown simulation results for the case in which the underlying distribution is the multivariate Cauchy, $\mathcal{C}y(0,1)$.

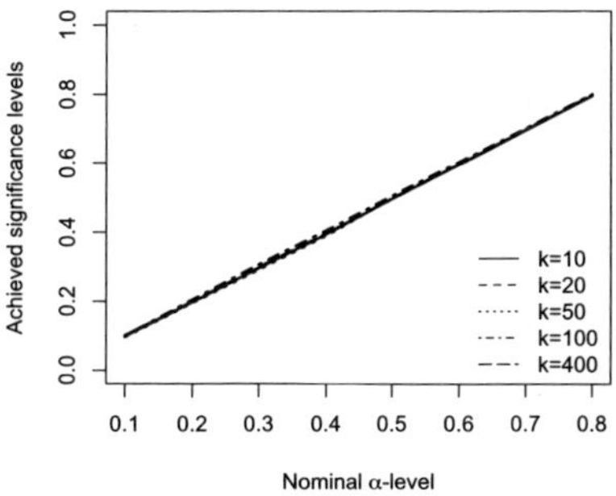

(a) Simulations under H_0, k variables

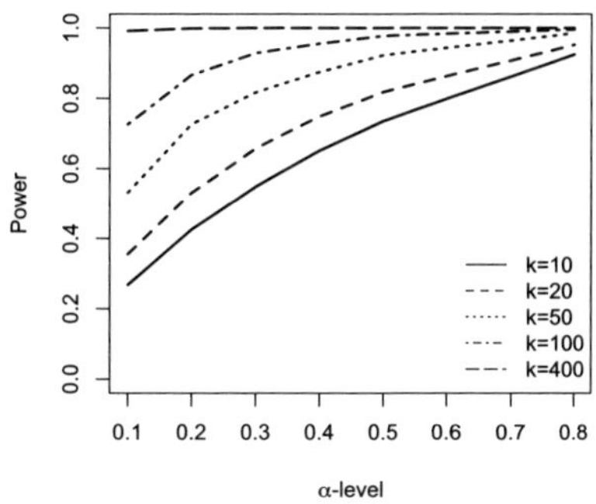

(b) Simulations under H_1,k variables, δ=0.5)

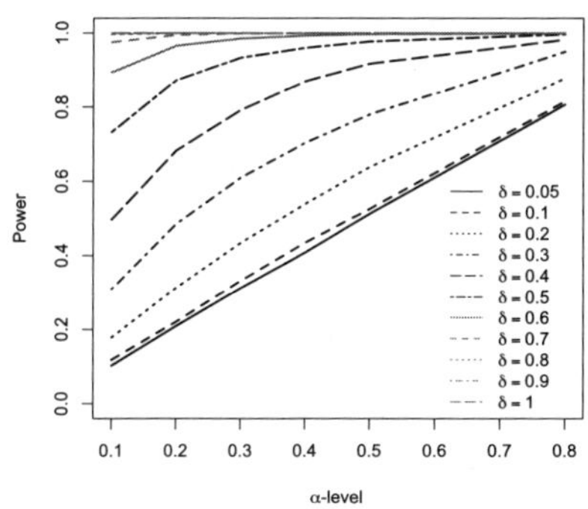

(c) Simulations under H_1, $k = 100$

Fig. 2. Simulation results under a multivariate normal $\mathcal{N}(0,1)$ distribution, non-directional alternatives

- In Figure 5 are displayed simulation results for the case in which data are distributed according to a Student t with 2 d.f., denoted by $\mathcal{S}t(2)$.
- In Figure 6 are shown simulation results for the case in which data follows a Pareto distribution, $\mathcal{P}a(1,1)$, with a shape parameter equal to 1.

In both the non-directional and directional cases, under the multivariate normal distribution, the power increases when the number of informative variables and/or the value of the noncentrality parameter δ increases. We

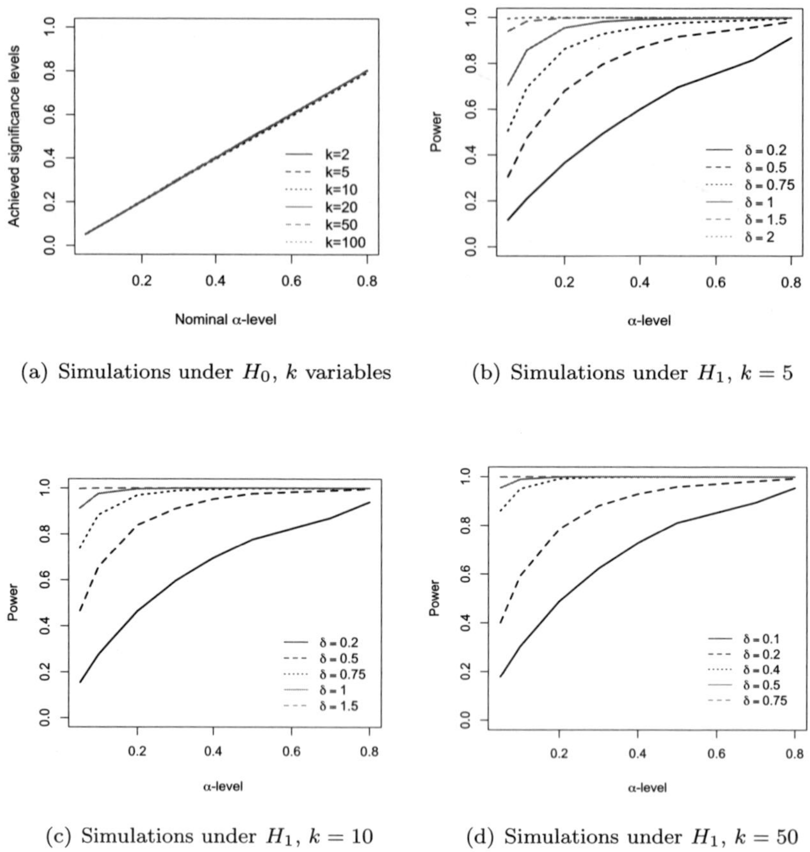

(a) Simulations under H_0, k variables

(b) Simulations under H_1, $k = 5$

(c) Simulations under H_1, $k = 10$

(d) Simulations under H_1, $k = 50$

Fig. 3. Simulation results under a multivariate normal distribution $\mathcal{N}(0, 1)$, directional alternatives

have also found that, under the Cauchy distribution, the power of T (a) is invariant for fixed δ when the number of variables increases, and (b) increases for fixed sample sizes when the standardized noncentrality increases, thus producing evidence for its conditional consistency.

Under the $\mathcal{S}t(2)$ distribution, it is possible to show that the test is consistent for a divergent number of variables even if the data distribution does not admit finite variance.

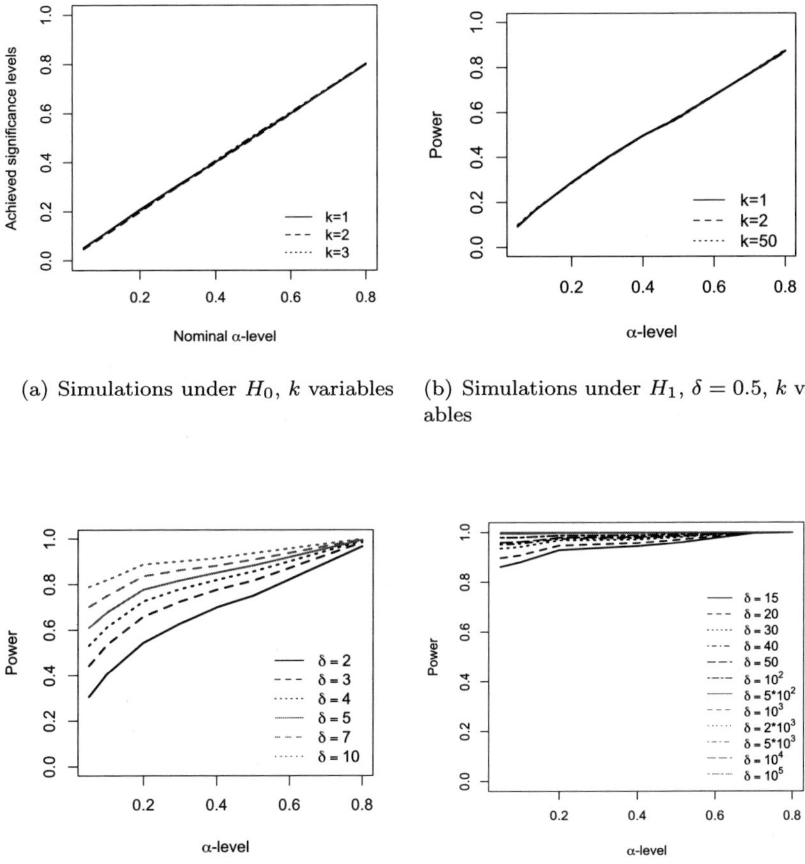

(a) Simulations under H_0, k variables

(b) Simulations under H_1, $\delta = 0.5$, k variables

(c) Simulations under H_1, $k = 2$

(d) Simulations under H_1, $k = 2$

Fig. 4. Simulation results under Cauchy $Cy(0, 1)$ distribution, directional alternatives

5. Application

We consider 12 Mediterranean monk seal (*Monachus Monachus*) skulls (Mo, 2005), of which 4 are considered young (i.e., the age class category is less or equal to 4) and 8 are adult (i.e., the age class category is greater than 4). For details on methods for determining age class category, we refer the reader to Mo (2005). In brief, age class category less or equal to 4 includes specimens between 0 and 4 months of age (from newborns to fat short pups), whereas

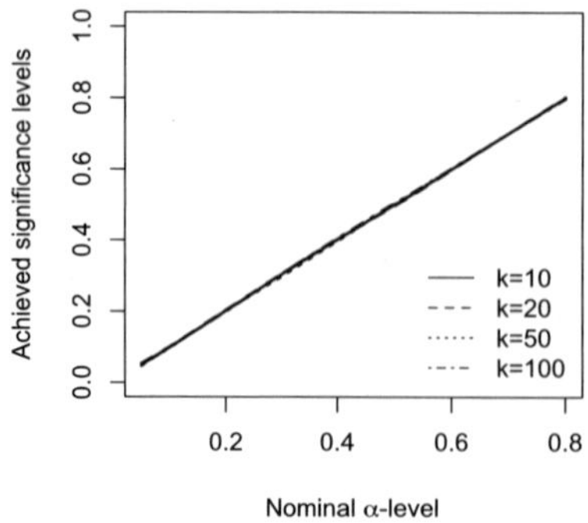

(a) Simulations under H_0, k variables

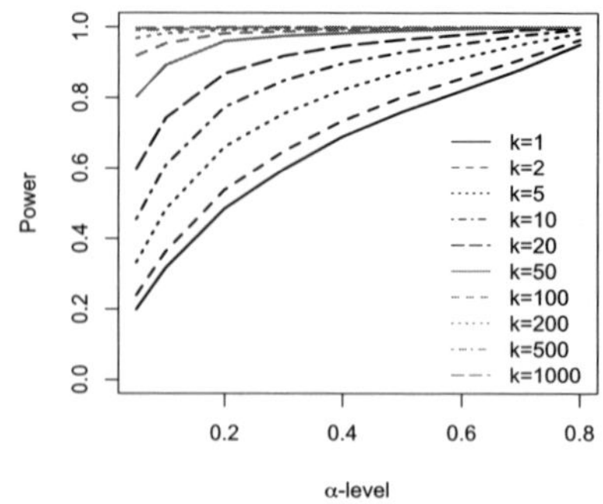

(b) Simulations under H_1, $\delta = 1$, k variables

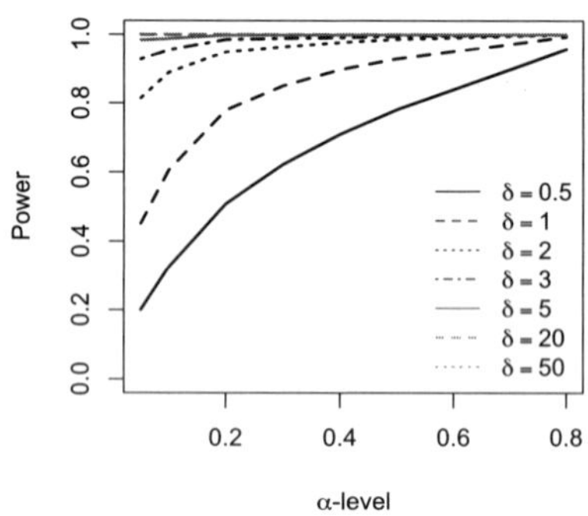

(c) Simulations under H_1, $k = 10$

Fig. 5. Simulation results under Student's $\mathcal{S}t(2)$ distribution, directional alternatives

age class categories greater than 4, i.e., age class categories 5, 6, 7, 8, and 9, includes specimens older than 5 months (from weaned youngsters to black adult males). It is of interest to assess whether or not there are differences in shape (in the 2D landmark coordinates) between different age categories. Given the small sample sizes, we have applied the nonparametric combination (NPC) methodology in the two independent sample case. We have chosen 4 anatomical landmarks (auriculare, rhinion, and two suture points between them) and 24 semi-landmarks. In particular we have digitized 6

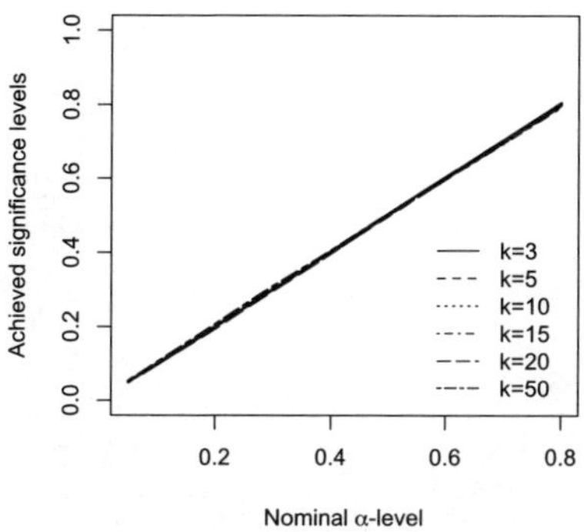

(a) Simulations under H_0, k variables

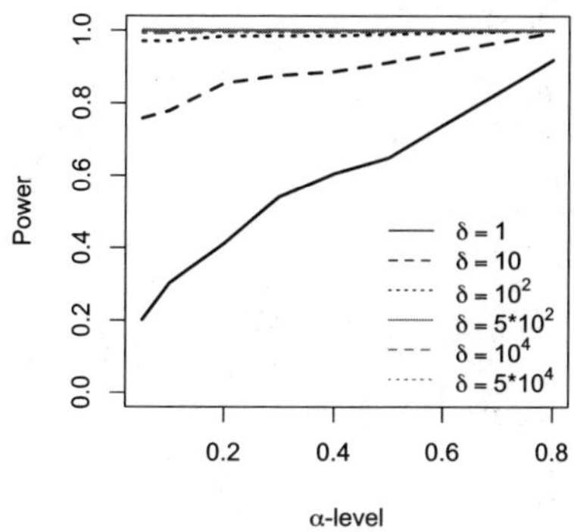

(b) Simulations under H_1, $k = 1$

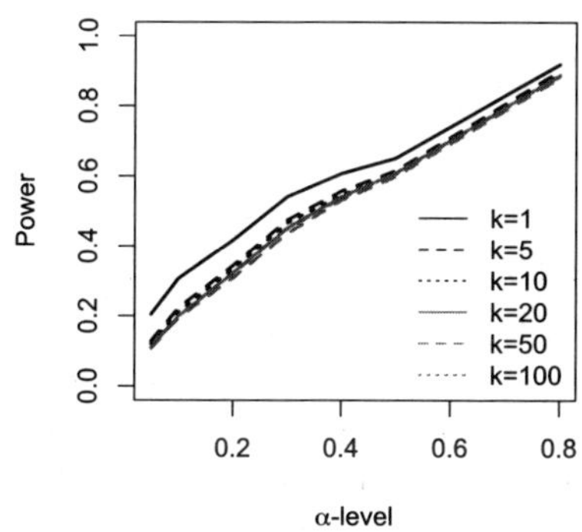

(c) Simulations under H_1, $\delta = 1$, k variables

Fig. 6. Simulation results under Pareto $\mathcal{P}a(1, 1)$ distribution, directional alternatives

semi-landmarks in the first curve from auriculare to the second point and 9 in each of the other two curves from point 2 to point 3 and from point 3 to rhinion. For semi-landmarks, we refer to points that are defined relative to other landmarks, for example 'midway between landmarks 1 and 2'. As we noted earlier, semi-landmark points do not enjoy the homology property and, to be precise, semi-landmarks lie on homologous curves yet

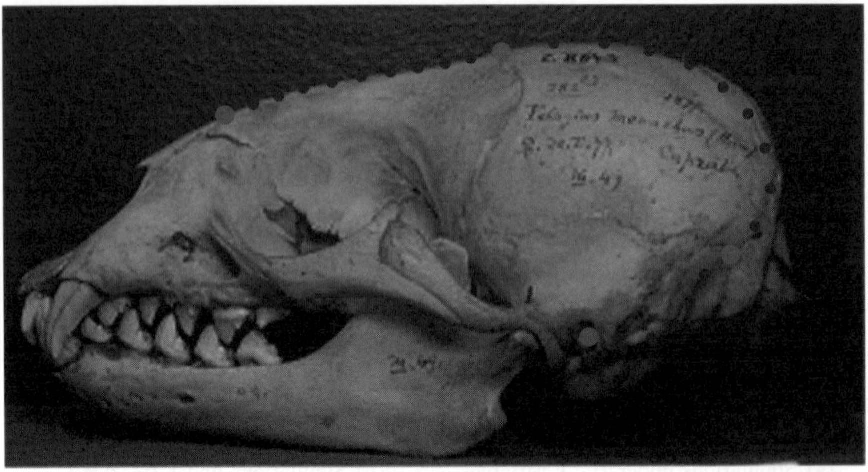

Fig. 7. Landmarks (4 large, lighter dots) and semilandmarks (24 small, darker dots).

their exact position along these (usually smooth) curves is unclear. As a part of the superimposition procedure, the semi-landmarks are allowed to slide along their curves in order to minimize the Procrustes distance from the actual landmark configuration to the sample average configuration. We have considered the data obtained after sliding semi-landmarks and after superimposition, minimizing the Procrustes distance.

As we wrote earlier, the NPC methodology is based on a decomposition of the hypotheses into k, $k > 1$, sub-hypotheses where, for each sub-hypothesis, a suitable partial permutation test statistic is available. In this case, x and y landmark and semi-landmark coordinates could be considered the sub-hypotheses of the problem, thus providing a set of partial tests. By combining these partial tests (thus obtaining a second-order statistic) we can acquire information (i.e., a p-value) for each 2D landmark and semi-landmark.

In particular, by combining x and y coordinates, we obtain p-values of 0.0026 for auriculare, 0.0045 for the second point, 0.0050 for the third point, and 0.0026 for rhinion. We can also compute a p-value for each curve, since a curve could be seen again as a combination of semi-landmarks. We obtain p-values of 0.0975 for the first curve, 0.0045 for the second curve, and 0.0026 for the third curve. We can finally consider a global combination of all the partial tests previously obtained. In this example, the global p-value is $p = 0.0026$.

A noteworthy comment is that this kind of analysis cannot be carried out in a parametric framework because the standard Hotelling T^2 statistic is approximately distributed according to an F_{M,n_1+n_2-M-1}, where $M = km - m - m(m-1)/2 - 1$ is the dimension of the tangent space. In this case, $k = 28$, $m = 2$ and $n_1 + n_2 = 12$; hence we would need to calculate $F_{52,-41}$, which is impossible.

6. Discussion

Due to their nonparametric nature, the suggested tests provide efficient solutions in multivariate small-sample-size problems. Moreover, these tests do not rely on a well-defined distributional model, they have the obvious advantage of not requiring the assumption of homogeneity of variance in the alternative, they deal with one-sided as well as two-sided tests of hypotheses, and they allow a more flexible analysis even in terms of the nature of the variables (continuous, categorical, or mixed) involved in the analysis. In the simulation study, we have evaluated the power of multivariate permutation tests based on Pesarin's nonparametric combining functions (Pesarin, 2001), showing that the power for the suggested tests increases when the number of processed variables increases, provided that the noncentrality parameter δ increases, even when the number of variables is larger than the permutation sample space. These findings are very important, since they demonstrate that it is possible to obtain powerful tests in a nonparametric framework by increasing the number of informative variables while leaving the number of cases fixed. Through the application to shape analysis, we have highlighted the power and the flexibility of the nonparametric permutation solution, showing how it is possible to deal with data sets containing many informative landmarks and few specimens.

Acknowledgements

Authors wish to thank the University of Padova (CPDA088513/08 and CPDA092350/09) and the Italian Ministry for University and Research (2008WKHJPK_002) for providing the financial support for this research.

References

Adams, D. C., Rohlf, F. J. and Slice, D. E. (2004). Geometric morphometrics: Ten years of progress following the 'revolution', *Italian Journal of Zoology* **71**, pp. 5–16.

Blair, R. C., Higgins, J. J., Karniski, W. and Kromrey, J. D. (1994). A study of multivariate permutation tests which may replace Hotelling's t^2 test in prescribed circumstances, *Multivariate Behavioral Research* **29**, pp. 141–163.

Bookstein, F. L. (1986). Size and shape spaces for landmark data in two dimensions, *Statistical Science* **1**, pp. 181–242.

Bookstein, F. L. (1997). Shape and the information in medical images: A decade of the morphometric synthesis, *Computer Vision and Image Understanding* **66**, pp. 97–118.

Celant, G., Pesarin, F. and Salmaso, L. (2000). Two sample permutation tests for repeated measures with missing values, *Journal of Applied Statistical Science* **9**, pp. 291–304.

Dryden, I. L. and Mardia, K. V. (1998). *Statistical Shape Analysis* (John Wiley & Sons, London).

Good, P. (2000). *Permutation tests: A practical guide to resampling methods for testing hypotheses* (Springer, New York).

Katina, S., Bookstein, F. L., Gunz, P. and Schaefer, K. (2007). What is worth digitizing all those curves? A worked example from craniofacial primatology, *American Journal of Physical Anthropology, Abstracts of American Academy of Physician Assistants (AAPA) poster and podium presentations* **132, S44**, p. 140.

Kent, J. T. (1994). The complex Bingham distribution and shape analysis, *Journal of the Royal Statistical Society, Series B* **56**, pp. 285–299.

Lehmann, E. L. (1986). *Testing Statistical Hypotheses* (Wiley, New York).

Mo, G. (2005). *Anatomy of the skull of mediterranean monk seal, Monachus monachus, functional morphology, densitometry, morphometrics and notes on natural history*, Ph.D. thesis, University of Padova.

Pesarin, F. (2001). *Multivariate Permutation Tests: With Application in Biostatistics* (John Wiley & Sons, Chichester-New York).

Pesarin, F. and Salmaso, L. (2010). *Permutation Tests for Complex Data: Theory, Applications and Software* (Wiley, Chichester).

Rohlf, F. J. (1999). Shape statistics: Procrustes superimpositions and tangent spaces, *Journal of Classification* **16**, pp. 197–225.

Rohlf, F. J. (2000). Statistical power comparisons among alternative morphometric methods, *American Journal of Physical Anthropology* **111**, pp. 463–478.

Romano, J. P. (1990). On the behavior of randomization tests without a group invariance assumption, *Journal of the American Statistical Association* **85**, pp. 686–692.

Slice, D. E., Bookstein, F. L., Marcus, L. F. and Rohlf, F. J. (1996). A glossary for geometric morphometrics, *Advances in Morphometrics* **284**, pp. 531–551.

Small, C. G. (1996). *The Statistical Theory of Shape* (Springer, New York).

Srivastava, A., Joshi, S., Mio, W. and Liu, X. (2005). Statistical shape analysis: Clustering, learning, and testing, *IEEE Transactions on Pattern Analysis and Machine Intelligence* **27**, pp. 590–602.

A Non-Parametric Cramér-von Mises Penalty Function Smoother

Bruce M. Brown

Department of Statistics, University of New South Wales
Sydney 2052, Australia
E-mail: Bruce.Brown@unsw.edu.au

Chenlei Leng

Department of Statistics and Applied Probability
National University of Singapore, Singapore 117546
E-mail: stalc@nus.edu.sg

Dedicated to Tom Hettmansperger, a lifetime colleague and friend

In many forms of penalty function smoothing, the choice of smoothing parameter is critical. Existing procedures for this choice include cross-validation or variations on likelihood methods. An alternative is introduced here, based on taking an intuitive non-parametric approach. For the linear spline form of penalty function, a Cramér-von Mises test statistic for the amount of smoothing is equal to the ratio of the least-squares and roughness components of the penalty formulation. Thus, constraining this test statistic to be equal to a central value of the Cramér-von Mises distribution provides a rational method for choosing the smoothing parameter. The necessary computations can be carried out by a convergent fixed-point iteration method, along with $O(n)$ equation solving techniques facilitated by a Cholesky-like decomposition. An example is used to illustrate the method, and simulations show that it compares favorably to the use of cross-validation.

Keywords: Cholesky decomposition; Functional iteration; Matrix spectral decomposition; Non-parametric smoothing; Smoothing coefficient

1. Introduction

The smooth function regression model is as follows. Consider an $(n+1) \times 1$ vector y of observations and a corresponding vector f of function values $f_i = f(x_i)$, generated by the model $y_i = f(x_i) + \varepsilon_i$ for $i = 0, 1, \ldots, n$, where the function $f(.)$ is smooth; the $\{\varepsilon_i\}$ are independent, identically distributed, variance σ^2 random variables; and $x_0 < x_1 < \cdots < x_n$ are known constants, either fixed in advance or realized values of a random explanatory variable.

The usual form of penalty function smoother chooses the function f to minimize a linear combination of a least-squares dispersion measure and a roughness penalty. A well-known choice for the roughness penalty is an integrated squared second derivative, leading to a cubic spline solution; see Wahba (1990) and Green and Silverman (1994). In this and other similar formulations, the coefficient of the roughness penalty is unspecified and estimated retrospectively. This could be by cross-validation or generalized cross-validation (Wahba, 1990) or, using likelihood methods, based on somewhat arbitrary assumptions about the probability mechanism generating the true f from a proposed class of functions; see Ruppert *et al.* (2003). In the present paper we introduce another method, based upon an intuitive non-parametric approach.

To illustrate the ideas, we discuss a simple penalty function criterion,

$$\begin{aligned} Q &= (y-f)^\top(y-f) \\ &\quad +\lambda\left[\int_{x_0}^{x_n}\{f'(x)\}^2dx - (x_n-x_0)^{-1}\left\{\int_{x_0}^{x_n} f'(x)dx\right\}^2\right], \end{aligned} \tag{1}$$

where f and y are $(n+1)\times 1$ vectors given by $(f)_i = f_i$ and $(y)_i = y_i$. The form of the roughness penalty is a variance of f' values, and, like the corresponding cubic spline penalty, is zero whenever f is linear. For fixed $\{f_i\}$ values it is minimized by choosing the function f to be linear on the intervals (x_{i-1}, x_i), allowing the objective function to be re-written as

$$Q = \sum_{i=0}^{n}(y_i - f_i)^2 + \lambda\left\{\sum_{i=1}^{n}\frac{(f_i - f_{i-1})^2}{h_i} - \frac{(f_n - f_0)^2}{\sum_1^n h_i}\right\}, \tag{2}$$

where $h_i = x_i - x_{i-1}$. This formulation is therefore suitable for smoothing of the observations $\{y_i\}$ into estimated $\{\hat{f}_i\}$ values, but not for the estimation of intermediate f values, because of the non-smooth piecewise linear nature of the full function $\hat{f}$.

Non-parametric methods often possess the attractive feature of being based upon sound, intuitively understandable criteria. Thus we keep this in mind in seeking an estimate of $\lambda \geq 0$, which controls the extent of smoothing. When λ is small, the fitted f_i are close to the observations y_i and little smoothing occurs, but when λ is high, the fitted f_i will be nearly linear, corresponding to over-smoothing. In the case of over-smoothing there will be patches of successive residuals $r_i = y_i - f_i$ having the same sign.

Such an effect is magnified by taking cumulative sums of residuals, and can be detected by high values of a Cramér-von Mises statistic of the form: sum of squares of cumulated residuals. In the case of undersmoothing, the intuitive effect is harder to disentangle, but if followed through it leads to a similar result: low values of the same Cramér-von Mises statistic. Thus, the smoothness coefficient λ can be chosen objectively by equating this Cramér-von Mises criterion to a central value of its null distribution.

What makes this suggestion more than just another *ad hoc* proposal is the following property. When normalized against an estimate of error variance σ^2, the statistic is the ratio of the two components, least squares and roughness penalty, in the penalty function formulation. Thus the proposed method appears to be intrinsic to the problem in a fundamental way. It is then not surprising that the proposed procedure turns out to have good computational properties.

The general formulation is spelled out in Section 2, the iterative computation scheme for calculating λ is outlined in Section 3, a Cholesky-type $O(n)$ method for updating estimates is in Section 4, the formula for setting the central Cramér-von Mises value is derived in Section 5, an example and simulations are presented in Section 6, and Section 7 contains various remarks about future directions.

2. The Cramér-von Mises formulation

The objective function in (1) and (2) can be re-written as

$$Q = (y - f)^\top (y - f) + \lambda f^\top D f, \tag{3}$$

where the symmetric $(n+1) \times (n+1)$ matrix D is specified by $D_{i,i+1} = D_{i+1,i} = -1/h_{i+1}$ for $i = 0, \ldots, n-1$, $D_{i,i} = h_i^{-1} + h_{i+1}^{-1}$ for $i = 1, \ldots, n-1$, $D_{0,0} = h_1^{-1} - (\sum h_i)^{-1}$, $D_{n,n} = h_n^{-1} - (\sum h_i)^{-1}$, $D_{0,n} = D_{n,0} = (\sum h_i)^{-1}$, and all other entries $= 0$.

Note that the labeling system for rows and columns of D runs from 0 to n, to align with the labels for $\{x_i\}$ and $\{y_i\}$. This labeling is followed throughout the rest of the paper.

It is easy to show that D is non-negative definite, from the observation that the term in brackets $\{\cdot\}$ in (2) is a variance, and hence always ≥ 0.

The minimum value of Q is given by

$$\hat{f} = (I + \lambda D)^{-1} y, \tag{4}$$

and the residual vector is

$$r = y - \hat{f} = \lambda D (I + \lambda D)^{-1} y. \tag{5}$$

The rows of D sum to zero, so that 1_{n+1}, the $(n+1)$ column of ones, is a zero-eigenvector of D. Then it is easy to show that $r^T 1_{n+1} = 0$, i.e. the residuals sum to zero. In fact, D has another zero-eigenvector, with entries proportional to $\{x_i\}$. The explanation for the existence of these two zero-eigenvectors is that the roughness penalty is unchanged by the addition of an arbitrary linear function to the observed $\{y_i\}$.

Now introduce an $n \times (n+1)$ lower triangular weight matrix A, determined by the spacings $\{h_i\}$: $A_{i,j} = \sqrt{h_i}$ for $j \leq i$, or $= 0$ for $j > i$. Note that the far right-hand column of A consists of zeroes. Then a Cramér-von Mises statistic, an interval-weighted sum of squares of cumulated residuals, is given by

$$C = (Ar)^T(Ar) = h_1 r_0^2 + h_2(r_0 + r_1)^2 + \cdots + h_n(r_0 + \cdots + r_{n-1})^2.$$

The correct attachment of h_j to the jth interval (x_{j-1}, x_j) can be seen by subtracting half the zero-sum of all residuals from each cumulated sum, so that, for example, $2(r_0 + \cdots + r_{j-1}) = (r_0 + \cdots + r_{j-1}) - (r_j + \cdots + r_n)$.

Denote $(\sum h_i)^{-1} = (x_n - x_0)^{-1}$ by s. Then the $n \times (n+1)$ matrix product AD becomes

$$\begin{pmatrix} h_1^{-1/2} - sh_1^{1/2} & -h_1^{-1/2} & 0 & 0 & \cdots & \cdots & \cdots & sh_1^{1/2} \\ -sh_2^{1/2} & h_2^{-1/2} & -h_2^{-1/2} & 0 & \cdots & \cdots & \cdots & sh_2^{1/2} \\ -sh_3^{1/2} & 0 & h_3^{-1/2} & -h_3^{-1/2} & 0 & \cdots & \cdots & sh_3^{1/2} \\ \cdots & \cdots & \cdots & \cdots & \cdots & \cdots & \cdots & \cdots \\ \cdots & \cdots & \cdots & \cdots & \cdots & \cdots & \cdots & \cdots \\ -sh_{n-1}^{1/2} & 0 & \cdots & \cdots & \cdots & h_{n-1}^{-1/2} & -h_{n-1}^{-1/2} & sh_{n-1}^{1/2} \\ -sh_n^{1/2} & 0 & \cdots & \cdots & \cdots & 0 & h_n^{-1/2} & -h_n^{-1/2} + sh_n^{1/2} \end{pmatrix}.$$

From this expression it follows that the $(n+1) \times (n+1)$ matrix $DA^\top AD = (AD)^\top(AD)$ is equal to D, i.e. $A^\top A$ is a generalized inverse of D. This in turn, with (4) and (5), shows that

$$\begin{aligned} C = r^\top A^\top Ar &= \lambda^2 y^\top (I + \lambda D)^{-1} DA^\top AD(I + \lambda D)^{-1} y \\ &= \lambda^2 y^\top (I + \lambda D)^{-1} D(I + \lambda D)^{-1} y = \lambda^2 \hat{f}^\top D \hat{f}. \end{aligned}$$

Thus the statistic C is proportional to the roughness term in the penalty function formulation (3). But, as a statistic of Cramér-von Mises type, when normalized by an estimate of error variance $\sigma^2 = \text{var}(\varepsilon_i)$, it will have an approximate Cramér-von Mises large-sample distribution. The natural estimate of σ^2 is $r^\top r/(n-1)$, the average squared residual, which is proportional to the least-squares term in (3).

Thus, an approximate Cramér-von Mises distributed statistic is the ratio $C/\hat{\sigma}^2 = (n-1)\lambda^2 \hat{f}^\top D\hat{f}/r^\top r$, proportional to the ratio of least-squares and roughness penalty terms in (3). This ratio is controlled by the degree of smoothing imposed by the smoothness coefficient λ. Then it is natural to define λ so that

$$\frac{C}{\hat{\sigma}^2} = \frac{(n-1)\lambda^2 \hat{f}^\top D\hat{f}}{r^\top r} = \eta, \tag{6}$$

a central value of the asymptotic Cramér-von Mises distribution. It is shown in the next section to be the basis of a convergent iterative scheme, requiring only $O(n)$ computational steps.

The divisor in the estimate of error variance σ^2 is $n-1$ and not the number of observations $n+1$; the loss of two degrees of freedom is attributable to the two zero eigenvalues of D. This and other aspects of the Cramér-von Mises statistic, including the choice of η, are discussed in Section 5.

3. A computation method

To construct a reliable and concise computation method, there are two aspects to be dealt with. The first is the evaluation of the smoothness coefficient λ through a convergent iterative scheme. At each stage, for each trial λ value, the calculation of $\hat{f}$ is then straightforward in principle, from (4). However, because this seems to require a matrix inversion and hence $O(n^3)$ calculations, the case of large n may be a concern. Thus the second aspect is finding a solution method for $\hat{f}$, from (4), in $O(n)$ steps. The first aspect is discussed in the present section, and the second aspect is dealt with in Section 4.

Equation (6) suggests the nth approximation to λ,

$$\lambda_n = \sqrt{\frac{\eta}{(n-1)} \frac{r_{n-1}^\top r_{n-1}}{\hat{f}_{n-1}^\top D\hat{f}_{n-1}}}, \tag{7}$$

where $\hat{f}_{n-1}$ is the version of $\hat{f}$ from (4) using $\lambda = \lambda_{n-1}$ and $r_{n-1} = y - \hat{f}_{n-1}$.

This has the form $\lambda_n = g(\lambda_{n-1})$, i.e., it is a fixed point or functional iteration method of solving the equation $\lambda = g(\lambda)$. To ensure convergence to a unique solution, it is enough to show that g is non-decreasing and that $0 < g'(\lambda) < 1$ at any solution λ.

Theorem 3.1. *Let the function g be given by*

$$g(x) = \sqrt{\frac{\eta}{(n-1)} \frac{r_x^\top r_x}{f_x^\top D f_x}} \tag{8}$$

where, from (4), $f_x = (I+xD)^{-1}y$, and $r_x = y - f_x$. Then g is an increasing function and $0 < g'(x) < 1$ at any solution to $x = g(x)$.

Remark 3.1. Equation (7) comes from (6) after taking square roots. This step enables the convergence result of Theorem 1 to hold for any chosen central value η.

Proof. Techniques used in the proof to follow, but omitted now, show that $r_x^\top r_x \uparrow$ and $f_x^\top D f_x \downarrow$ as $x \uparrow$, so g is an increasing function.

Write $x = \lambda + \delta$, where $\lambda = g(\lambda)$ and δ is small. Then

$$\begin{aligned} f_x &= (I + \lambda D + \delta D)^{-1} y \\ &= (I + \lambda D)^{-1} y - \delta (I + \lambda D)^{-1} D (I + \lambda D)^{-1} y + O(\delta^2) \\ &= \hat{f} - \delta \lambda^{-1} (I + \lambda D)^{-1} r + O(\delta^2) \end{aligned}$$

and $r_x = y - f_x = r + \delta\lambda^{-1}(I + \lambda D)^{-1} r + O(\delta^2)$. Thus

$$\begin{aligned} r_x^\top r_x &= r^\top r + 2\delta\lambda^{-1} r^\top (I + \lambda D)^{-1} r + O(\delta^2) \\ \text{and } f_x^\top D f_x &= \hat{f}^\top D \hat{f} - 2\delta\lambda^{-1} \hat{f}^\top D (I + \lambda D)^{-1} r + O(\delta^2). \end{aligned}$$

Substituting into (8) and applying first-order calculations in δ gives

$$g(x) = g(\lambda + \delta) = \lambda + \delta \left\{ \frac{r^\top (I + \lambda D)^{-1} r}{r^\top r} + \frac{\hat{f}^\top D (I + \lambda D)^{-1} r}{\hat{f}^\top D \hat{f}} \right\}. \tag{9}$$

Thus the factor in brackets $\{\cdot\}$ is $g'(\lambda)$. Call this the *bracketed factor*. Clearly it is > 0 from (4) and (5) since D is non-negative definite, so all that remains to show is that it is < 1.

Use the spectral decomposition of D:

$$D = \sum_{j=1}^{n-1} \theta_j \xi_j \xi_j^\top, \tag{10}$$

where non-negative eigenvalues are $\theta_1 \le \theta_2 \le \cdots \le \theta_{n-1}$, with corresponding eigenvectors $\xi_1, \xi_2, \ldots, \xi_{n-1}$. The two zero eigenvalues are omitted. Let $c_j = \xi_j^\top y$, so that $y = I_{n+1} y = (\sum \xi_j \xi_j^\top) y = \sum c_j \xi_j$. Thus from (4) and (5) we have

$$\begin{aligned} r^\top r &= \lambda^2 \sum \frac{\theta_j^2 c_j^2}{(1 + \lambda\theta_j)^2}, \\ f^\top D f &= \sum \frac{\theta_j c_j^2}{(1 + \lambda\theta_j)^2}, \quad \text{and} \\ r^\top (1 + \lambda D)^{-1} r &= \lambda^2 \sum \frac{\theta_j^2 c_j^2}{(1 + \lambda\theta_j)^3} \quad = \quad \lambda f^\top D (I + \lambda D)^{-1} r. \end{aligned}$$

The bracketed factor is

$$\left\{\sum \theta_j^2(1+\lambda\theta_j)^{-3}c_j^2\right\}\left\{\frac{1}{\sum \theta_j^2(1+\lambda\theta_j)^{-2}c_j^2}+\frac{\lambda}{\sum \theta_j(1+\lambda\theta_j)^{-2}c_j^2}\right\},$$

which simplifies to

$$\frac{\{\sum \theta_j(1+\lambda\theta_j)^{-1}c_j^2\}\{\sum \theta_j^2(1+\lambda\theta_j)^{-3}c_j^2\}}{\{\sum \theta_j^2(1+\lambda\theta_j)^{-2}c_j^2\}\{\sum \theta_j(1+\lambda\theta_j)^{-2}c_j^2\}}.$$

This expression can be written as $E(XY)/\{E(X)E(Y)\}$, where the random variables $(X,Y)=\{(1+\lambda\theta_j)^{-1},\theta_j(1+\lambda\theta_j)^{-1}\}$ with probability proportional to $\theta_j(1+\lambda\theta_j)^{-1}c_j^2$. But this ratio and hence the bracketed factor must be < 1 since $X+\lambda Y = 1$ implies $\mathrm{cov}(X,Y) < 0$, and hence $E(XY) < E(X)E(Y)$. □

Thus the proposed functional iteration, or fixed point method for evaluating λ, has guaranteed, reliable convergence, starting from any trial value $\lambda > 0$ and all values of η.

4. Updating the functional estimates

In the iterative solution method for λ described in Section 3, a key input into each step is an updated value of the estimated $\hat{f}$ from (4) for each new trial value of λ. Because this evaluation is expressed in terms of the inverse matrix $(I+\lambda D)^{-1}$, which appears to imply $O(n^3)$ computational steps, it is desirable to find more efficient methods of solving the equation $(I+\lambda D)f = y$. If D were a banded matrix, then a Cholesky decomposition expressing $I+\lambda D = L\Lambda L^\top$ could be constructed, where L is a lower triangular banded matrix with unit diagonal entries and Λ is diagonal. Then it would be possible to solve for f by solving $Lu = y$ for u, $\Lambda v = u$ for v, and $L^\top f = v$ for f, all three solutions requiring only $O(n)$ steps. But the matrix $I+\lambda D$ is not banded because of the non-zero entries $D_{0,n} = D_{n,0}$.

However, the Cholesky-type solution strategy will still work. All that happens is that the lower triangular matrix L is almost but not quite banded. It has two diagonal bands and a bottom row of non-zero entries. This hardly interferes at all with the usual successive solution method based on lower triangular matrices. The computational details are as follows.

The diagonal elements of Λ are defined recursively by $\Lambda_0 = D_{00}$ and $\Lambda_i = D_{ii} - D_{i,i-1}^2/\Lambda_{i-1}$ for $i = 1,2,\ldots,n$.

The diagonal elements of L are all one, and the adjacent diagonal elements are given by $L_{i+1,i} = D_{i+1,i}/\Lambda_i$ for $i = 0,1,\ldots,n-2$. The last of these diagonal elements, $L_{n,n-1}$, is excluded because the bottom row of L has a special formulation. This is as follows: $L_{n,0} = D_{n,0}/\Lambda_0, L_{n,n} = 1, L_{n,i} = \rho_i$ for $i = 1, \ldots, n-2$, and $L_{n,n-1} = \rho_{n-1} + D_{n,n-1}/\Lambda_{n-1}$. The "correction factors" ρ_i are given by

$$\rho_1 = -\frac{D_{1,0}D_{n,0}}{\Lambda_1\Lambda_0} \quad \text{and} \quad \rho_i = -\frac{D_{i,i-1}}{\Lambda_i}\rho_{i-1}.$$

5. Choosing the central Cramér-von Mises value

The proposal is to choose the smoothness coefficient λ so that a Cramér-von Mises measure, the ratio of least-squares and roughness criteria in the penalty function formulation, is equal to a central value of its approximate distribution. This will ensure that the degree of smoothness is appropriate, in the sense that the serial behavior of the resulting residuals matches what would be expected from cumulated independent random errors. Because the aim of the present paper is to propose a simple method, we suggest the most convenient choice of a central value η, namely, the null expected value, which can be computed directly. Thus it is not necessary to know the specific form of the asymptotic Cramér-von Mises distribution.

It is easy to see that adding a linear function to all the observations has the effect of adding the same linear function to the estimated f, while having no effect on the roughness penalty. Equivalently, the errors $\{\varepsilon_i\}$ have a simple linear regression extracted from them. Thus, when calculating the expected Cramér-von Mises value $E(\varepsilon^\top A^\top A\varepsilon)$, where ε is the $(n+1) \times 1$ vector of independent zero-mean variance σ^2 random errors, replace ε by $(I_{n+1} - P)\varepsilon$, where P is the projection matrix projecting the observations $\{y_i\}$ onto the line of best fit. In effect, the errors $\{\varepsilon_i\}$ are replaced by the corresponding residuals after fitting the simple linear regression.

Thus we seek to calculate the central value η through

$$\sigma^2\eta = E\Big(\varepsilon^\top(I_{n+1} - P)A^\top A(I_{n+1} - P)\varepsilon\Big),$$

giving $\eta = \mathrm{tr}\{(I_{n+1} - P)A^\top A(I_{n+1} - P)\}$. From the spectral decomposition of D in (10), let B be a generalized inverse of D given by $B = \sum_{j=1}^{n-1}\theta_j^{-1}\xi_j\xi_j^\top$. Because D and $I_{n+1} - P$ have the same zero-eigenvectors, we can write $I_{n+1} - P = BD$. Since $DA^\top AD = D$, the expected value we need is $\eta = \mathrm{tr}(BDB) = \sum_{j=1}^{n-1}\theta_j^{-1}$.

This is a compact representation that can be found from an eigen-analysis of the matrix D. However, when n is large, it is desirable to have an alternative formulation that can be calculated in $O(n)$ steps. The following result provides the necessary formula.

Theorem 5.1. *The value of* η *is given by*

$$\eta = \mathrm{tr}\{(I_{n+1} - P)A^\top A(I_{n+1} - P)\}$$
$$= \frac{1}{(n+1)} \sum_{j=0}^{n} (2j - n)x_j - \frac{\sum_{j=1}^{n} (x_j - x_{j-1}) \{ \sum_{i=0}^{j-1} (x_i - \bar{x}) \}^2}{S_{xx}},$$

where $S_{xx} = \sum_{j=0}^{n} (x_j - \bar{x})^2$.

Corollary 5.1. *When the* $\{x_i\}$ *are equally spaced integers,* $\eta/(n-1) = (n+3)/15$.

Proof. The required trace expression is the sum of squares of elements of the matrix $(I_{n+1} - P)A^\top$, whose ith column consists of the residuals after fitting a simple linear regression based on the "observed" column c_i, where $c_i^\top = (\sqrt{h_i}, \ldots, \sqrt{h_i}, 0, \ldots, 0)$, there being exactly i non-zero elements. If the sums of squares of these residuals is denoted by $S_{ee}^{(i)}$, then the trace is $\sum_i S_{ee}^{(i)}$. A routine calculation using the elementary expression for simple linear regression residuals gives the result. □

6. An example and simulations

As an illustration, we apply the proposed method to the Nitrogen oxides data of Gu (2002, Section 3.7.1), where the $\{y_i\}$ are log-ten-transformed NO_x and the $\{x_i\}$ are corresponding values of equivalence ratio. With a starting value of 1, successive values of λ are 0.245, 0.108, 0.0823, 0.075, 0.073, 0.072, and 0.072, and after the sixth step, λ remains virtually unchanged. Figure 1 produces the fitted line together with the fit obtained by using the generalized cross validation (GCV) method (Wahba, 1990; Gu, 2002) to choose λ in (1). As a reminder, the GCV method chooses λ to minimize

$$\frac{(y-f)^\top (y-f)}{\{n - \mathrm{tr}(H)\}^2},$$

where $H = (I + \lambda D)^{-1}$.

The Cramér-Von Mises smoother produces visually smoother estimates than the linear smoothing spline estimate with GCV. This is to be expected, because cross-validation makes no attempt to penalize roughness in its assessments of fit using mean squared error, and hence will generally choose smaller values of the smoothing parameter λ than those chosen by the proposed method. This is also confirmed by the results of some simulations, described now.

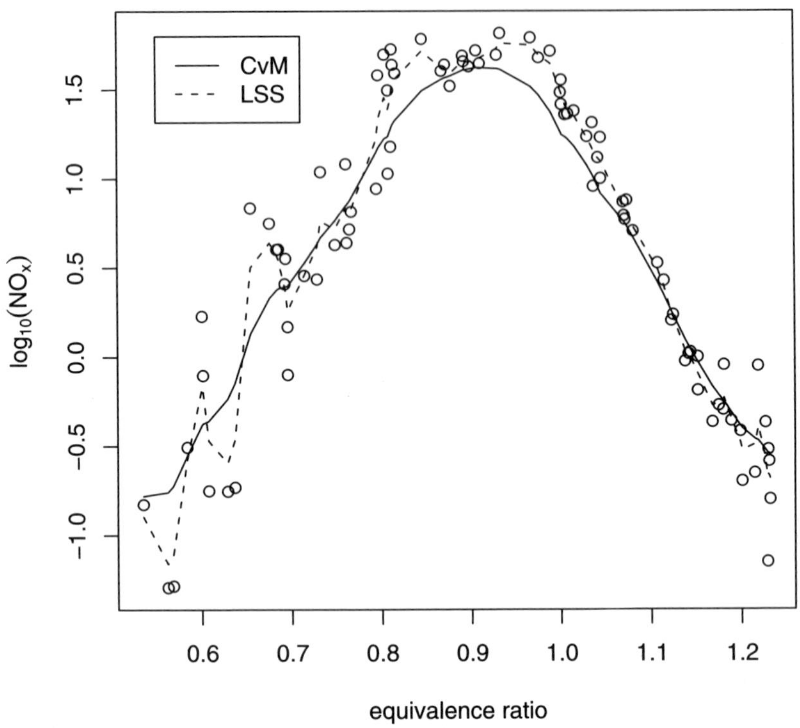

Fig. 1. Two linear smoothing spline fits to NO_x data. CvM: Cramér-von Mises smoother; LSS: Using GCV.

For these experiments, data are simulated from the following four models:

$$y = f_j(x) + \sigma_j \varepsilon, \ j = 1, 2, 3, 4,$$

where x follows a uniform distribution on $[-2, 2]$ and ε follows a standard normal distribution. The functions f_j, taken from Fan and Gijbels (1996,

Section 4.2), are $f_1(x) = x + 2e^{-16x^2}$, $f_2(x) = \sin(2x) + 2e^{-16x^2}$, $f_3(x) = 0.3e^{-4(x+1)^2} + 0.6e^{-16(x-1)^2}$, and $f_4(x) = 0.4x + 1$. The σ_j's are taken to be 0.4, 0.3, 0.1, and 0.15, respectively. The proposed smoother is compared to the linear smoothing spline with GCV in a way that acknowledges the implicit importance of smoothness.

For each f, two hundred data sets, each with 100 data points, are generated. Using the value of λ obtained from the Cramér-von Mises method, an objective method of choice, the roughness penalty terms (PEN) from the penalized criterion and the mean squared errors (MSE), defined as $n\int_{x_0}^{x_n}\{f(x)-\hat{f}(x)\}^2dx$, are calculated for each of the two methods, Cramér-von Mises (CvM) and generalized cross-validation (GCV). The factor n is included in the MSE formula to make the combined criterion, the total of MSE and PEN, compatible with the original penalty formulation, where the least-squares term is a sum, not an expected value. Details are shown in Table 1. The listed values are those for CvM, with the corresponding values for GCV shown in parentheses.

Table 1. Comparison of mean squared errors and penalty terms for the CvM (and GCV) methods.

Example	f_1	f_2	f_3	f_4
MSE	0.0508 (0.0293)	0.0277 (0.0195)	0.00470 (0.00258)	0.000416 (0.00114)
PEN	0.0510 (0.2462)	0.0623 (0.1224)	0.00544 (0.0202)	0.000170 (5.3136)
total	0.1089 (0.2755)	0.0901 (0.1418)	0.0101 (0.0228)	0.000586 (5.3147)

As is expected, cross-validation tends to undersmooth, using smaller values of λ than the Cramér-von Mises method. The effect is well-illustrated in Figure 1. On the other hand, cross-validation leads to smaller values of mean squared error but larger values of the combined criterion. The case of f_4 is interesting; here, for consistent estimation of the completely smooth linear function, it is necessary that $\hat{\lambda} \to \infty$, but the GCV method is not concerned with smoothness, leading to dramatically worse roughness penalty values than for the Cramér-von Mises method.

7. Remarks

The proposed Cramér-von Mises method is non-parametric; it is based on seeking an intuitive measure to detect under-smoothing or over-smoothing, not dependent on any extra assumptions about the form of the error distribution or the underlying smooth function f. In reducing to the ratio of roughness and least-squares terms in the penalty function formulation, it appears to be the natural way of choosing the smoothing parameter λ on statistical grounds, a claim reinforced by regular computational properties and the results of some simulations.

There are some possibilities for further investigations, listed as follows.

(1) A simpler formulation is to replace the roughness measure by $\int_{x_0}^{x_n} \{f'(x)\}^2 dx$, a second moment, not a variance, of f' values. Linear functions are penalized unless they are flat, i.e., the constant function. Thus this approach is suitable for some situations involving stationary processes. The treatment becomes easier; for example, the D matrix is banded, with only one zero-eigenvector, having constant entries. The details correspond closely to those of the present method.

(2) The existing estimation method can be adapted to statistical testing. To test $H_0 : f = f_0$, where $(f_0)_i = f_0(x_i)$, it is natural to use the Cramér-von Mises statistic $r_0^\top A^\top A r_0/\hat{\sigma}^2$, with $\hat{\sigma}^2 = r_0^\top r_0/(n-1)$ and $r_0 = y - f_0$. It is reasonable to suppose that the hypothesized f_0 is chosen so $(I_{n+1} - P)r_0 = 0$, i.e., so that a linear fit to the data has already been taken care of. The test statistic is $(n-1)r_0^\top A^\top A r_0/r_0^\top r_0$. This would be referred to the large-sample distribution of a Cramér-von Mises statistic with a linear component extracted. The details of the asymptotics are not completely straightforward, because of the pattern of spacings of the $\{x_i\}$. The present paper has been able to avoid the topic of Cramér-von Mises asymptotics, but if testing procedures were to be developed, this topic would have to be addressed.

(3) The existing method is suitable for smoothing of the observations $\{y_i\}$, replacing them by $\{\hat{f}_i\}$, but not for estimating the function f at intermediate values, because $\hat{f}$ is piecewise linear, and not smooth, at the nodes $\{(x_i, \hat{f}_i)\}$. To obtain a fully smooth function, it would be necessary to use a roughness criterion like $\int_{x_0}^{x_n} \{f''(x)\}^2 dx$, leading to a cubic spline solution, as is well-known; see Green and Silverman (1994). Derivation of a Cramér-von Mises approach in this context is more complex than in the present setting, and is a topic for later development.

(4) When data is generated from errors that are dependent, with substantial negative correlations, it can happen that no value of λ is large enough to solve (6) for the recommended central value of η. Thus, an extension of the method to deal with correlated errors is another worthwhile topic for later development.

References

Fan, J. and Gijbels, I. (1996). *Local Polynomial Modelling and Its Applications* (Chapman & Hall, London).

Green, P. J. and Silverman, B. W. (1994). *Nonparametric Regression and Generalized Linear Models* (Chapman & Hall, London).

Gu, C. (2002). *Smoothing Spline ANOVA Models* (Springer, New York).

Ruppert, D., Wand, M. P. and Carroll, R. J. (2003). *Semi-parametric Regression* (Cambridge University Press, Cambridge).

Wahba, G. (1990). *Spline Models for Observational Data*, CBMS-NSF Regional Conference Series in Applied Mathematics (SIAM, Philadelphia).

Statistical Models for Globular Cluster Luminosity Distribution

Max-Louis G. Buot
Department of Mathematics & Computer Science, Xavier University
Cincinnati, OH 45207, USA
E-mail: buotm@xavier.edu

Donald St. P. Richards
Department of Statistics, Pennsylvania State University
University Park, PA 16802, USA
E-mail: richards@stat.psu.edu

We consider statistical models that have been proposed for luminosity distributions for the globular clusters in the Milky Way and M31. Although earlier research showed that the cluster luminosity functions in those two galaxies were well fit by Gaussian distributions, subsequent investigations suggested that their luminosities were better fit by t-, rather than Gaussian, distributions. By applying the Bayesian Information Criterion, we do not find overwhelming statistical evidence that the t-distribution is superior to the Gaussian distribution as a model of luminosity distribution for the Milky Way. In the case of M31, we find moderate evidence that the Gaussian distribution is superior to the t-distribution. In neither case do we find strong evidence to support the use of one distribution over the other as a statistical model for the luminosities of the globular clusters in the Milky Way and M31. Consequently, we recommend that the Gaussian be retained as the statistical model for luminosity distribution. Moreover, we urge caution in the use of the Kolmogorov-Smirnov statistic to justify the choice of statistical models for globular cluster luminosity functions.

Keywords: Bayesian Information Criterion; Galaxies; The Galaxy; Gaussian distribution; Globular cluster systems; Kolmogorov-Smirnov statistic; Luminosity distribution; M31; The Milky Way; Maximum likelihood estimation; Nonparametric statistics; Schwarz criterion; t-distribution

1. Introduction

Globular clusters are some of the oldest objects in the Universe, they are innately luminous, and there is strong evidence that they are formed during periods of major star formation (Larsen and Richtler, 1999). For these and other reasons, globular clusters have played an important role in research on the formation of galaxies (Harris, 2001; van den Bergh, 2000).

The study of the globular cluster luminosity function (GCLF) of a galaxy is motivated by the distribution of luminosities of the globular clusters within the galaxy. Precisely, the GCLF of a galaxy is the relative number of its globular clusters at a given luminosity. In the Milky Way, empirical evidence suggests that the corresponding GCLF is usually unimodal and nearly symmetric, and the peak of the luminosity function occurs at a magnitude which varies little from galaxy to galaxy (Harris, 1998). For these reasons, it has been found that the GCLF peak in the Milky Way is sometimes appropriate as a standard candle for distance measurement (Whitmore, 1997).

In early work on luminosity functions, much attention was paid to plausible analytical forms of the GCLF. In several instances, including globular clusters in the Milky Way and in M31, it was shown that a Gaussian distribution was a good analytical fit to the empirically observed distribution of luminosities (Racine and Shara, 1979; van den Bergh, 1985; Harris *et al.*, 1991). A subsequent analysis (Secker, 1992) argued that a t-distribution provided a better fit, and that claim has led to wide acceptance of the t-distribution in subsequent research on luminosity functions (Barmby *et al.*, 2001).

As measured by the maximum likelihood procedure, the t-distribution has been shown to fit the empirical data more closely than the Gaussian distribution (Secker, 1992). However, this comparison is complicated by the fact that the analytical form of the Gaussian distribution is based on only two parameters, the mean and standard deviation, whereas the t-distribution has an additional index, or shape, parameter. This raises the issue of whether or not the increase in the likelihood function for the t-distribution over the Gaussian may be caused by the presence of an additional parameter.

The issue of the goodness-of-fit of a hypothesized distribution *vis-a-vis* the number of parameters in the underlying analytical form of that distribution is precisely the *raison d'être* of the Bayesian Information Criterion (BIC) or Schwarz criterion (Schwarz, 1978). Simply put, the BIC is designed to ascertain the extent to which an improved fit is due to an increase in the number of parameters in the analytical form of the distribution.

In the sequel, we apply the BIC to the data provided by Secker (1992). We do not find strong statistical evidence in support of the t-distribution over the Gaussian distribution, or vice versa, as a statistical model for the luminosity distribution of globular clusters in the Milky Way or M31.

Consequently, given the smaller number of parameters in the Gaussian distribution and the many well-known attractive features of that distribution, we recommend that the Gaussian be utilized as a statistical model for the luminosity distribution. Moreover, we urge caution in the use of the classical Kolmogorov-Smirnov statistic as justification for the choice of statistical models for globular cluster luminosity distributions.

2. The Bayesian Information Criterion

Suppose that two statistical distributions are plausible models for an observed data set. These models may be fit to the data using a variety of statistical procedures, e.g., residual sums of squares, the method of moments, or the method of maximum likelihood.

In polynomial regression, for example, the residual sum of squares can be reduced simply by increasing the degree of the polynomial regression function. In general, the more complex the mathematical form of a model, the better a model will be seen to fit the data. Therefore, it is clear that the choice of a statistical model should not be based entirely on measures such as residual sums of squares or likelihood function values, for, by increasing the number of parameters in the hypothesized model, we can obtain a relentless reduction in the residual sums of squares or an increase in the values of the likelihood function.

The Bayesian Information Criterion (BIC) constitutes a standard approach to assessing the relative plausibility of two competing statistical models that are being fit to data with *large* sample sizes (Schwarz, 1978). To balance any difference in the number of parameters between two statistical models, the BIC penalizes a model that has a larger number of free parameters. To illustrate this approach, consider the situation of two competing models, $f_1(x; \theta_1, \ldots, \theta_{m_1})$ and $f_2(x; \phi_1, \ldots, \phi_{m_2})$. Here, $\theta_1, \ldots, \theta_{m_1}$ and $\phi_1, \ldots, \phi_{m_2}$ are parameters of the corresponding density functions f_1 and f_2, respectively. On being given a random sample $X_1, \ldots, X_n$, we construct the likelihood functions

$$L_1(\theta_1, \ldots, \theta_{m_1}) = \prod_{i=1}^{n} f_1(x_i; \theta_1, \ldots, \theta_{m_1})$$

and

$$L_2(\phi_1, \ldots, \phi_{m_2}) = \prod_{i=1}^{n} f_2(x_i; \phi_1, \ldots, \phi_{m_2}),$$

and the corresponding BIC are, respectively,

$$\mathrm{BIC}_1 = -2 \ln L_1(\theta_1, \ldots, \theta_{m_1}) + m_1 \ln n$$

and

$$\mathrm{BIC}_2 = -2 \ln L_2(\phi_1, \ldots, \phi_{m_2})] + m_2 \ln n.$$

Given the parameters $\theta_1, \ldots, \theta_{m_1}$ and $\phi_1, \ldots, \phi_{m_2}$, and explicit formulas for the density functions f_1 and f_2, the relative superiority of the model $f_1(x; \theta_1, \ldots, \theta_{m_1})$ over the model $f_2(x; \phi_1, \ldots, \phi_{m_2})$ is measured by the difference in BICs, *viz.*,

$$\begin{aligned}\Delta_{\mathrm{BIC}} &= \mathrm{BIC}_2 - \mathrm{BIC}_1 \\ &= 2[\ln L_1(\theta_1, \ldots, \theta_{m_1}) - \ln L_2(\phi_1, \ldots, \phi_{m_2})] - (m_1 - m_2) \ln n.\end{aligned}$$

The first term in this expression is a measure of the increase in the likelihood function values of the first model over the second, and the second term is a penalty term reflecting the difference in the numbers of parameters in the models. Thus, Δ_{BIC} assesses any increase in the likelihood in light of the additional number of parameters necessary to achieve such an increase.

In practice, the values of $\theta_1, \ldots, \theta_{m_1}$ and $\phi_1, \ldots, \phi_{m_2}$ are unknown and need to be estimated from the data. Thus, we calculate the corresponding maximum likelihood estimates $\widehat{\theta}_1, \ldots, \widehat{\theta}_{m_1}$ and $\widehat{\phi}_1, \ldots, \widehat{\phi}_{m_2}$ and use those results to calculate

$$\widehat{\mathrm{BIC}}_1 = -2 \ln L_1(\widehat{\theta}_1, \ldots, \widehat{\theta}_{m_1}) + m_1 \ln n$$

and

$$\widehat{\mathrm{BIC}}_2 = -2 \ln L_2(\widehat{\phi}_1, \ldots, \widehat{\phi}_{m_2}) + m_2 \ln n,$$

the corresponding maximum likelihood estimates of BIC_1 and BIC_2, respectively. Finally, we calculate

$$\begin{aligned}\widehat{\Delta}_{\mathrm{BIC}} &= \widehat{\mathrm{BIC}}_2 - \widehat{\mathrm{BIC}}_1 \\ &= 2[\ln L_1(\hat{\theta}_1, \ldots, \hat{\theta}_{m_1}) - \ln L_2(\hat{\phi}_1, \ldots, \hat{\phi}_{m_2})] - (m_1 - m_2) \ln n,\end{aligned} \tag{1}$$

the maximum likelihood estimate of Δ_{BIC}.

General rules for using $\widehat{\Delta}_{\mathrm{BIC}}$ to assess the relative goodness-of-fit of the models f_1 and f_2 are as follows (Kass and Raftery, 1995; Mukherjee *et al.*, 1998; Jeffreys, 1961, Appendix B):

- $0 \le \widehat{\Delta}_{\mathrm{BIC}} < 2$: Weak evidence that Model 1 is superior to Model 2
- $2 \le \widehat{\Delta}_{\mathrm{BIC}} \le 6$: Moderate evidence that Model 1 is superior to Model 2
- $6 < \widehat{\Delta}_{\mathrm{BIC}} \le 10$: Strong evidence that Model 1 is superior to Model 2

- $\widehat{\Delta}_{\mathrm{BIC}} > 10$: Very strong evidence that Model 1 is superior to Model 2

For the case in which $\widehat{\Delta}_{\mathrm{BIC}} < 0$, one simply reverses the roles of L_1 and L_2 in order to apply these general rules.

We now apply these procedures to compare three statistical models for GCLF data.

3. Models for GCLF Distribution in the Milky Way and in M31

Consider the following competing models for GCLF in the Milky Way: A Gaussian model (van den Bergh, 1985),

$$f_1(x;\mu,\sigma) = \frac{1}{\sqrt{2\pi}\sigma} \exp\left[-\frac{(x-\mu)^2}{2\sigma^2}\right], \tag{2}$$

and a t-distribution model (Secker, 1992),

$$f_2(x;\mu,\sigma,\delta) = \frac{\Gamma\big((\delta+1)/2\big)}{\sqrt{\pi\delta}\,\sigma\,\Gamma(\delta/2)}\left[1+\frac{(x-\mu)^2}{\delta\sigma^2}\right]^{-(\delta+1)/2}, \tag{3}$$

where, in each model, $-\infty < x < \infty$ and the permissible ranges of the parameters are $-\infty < \mu < \infty$, $\sigma > 0$, and $\delta > 0$. In each model, μ represents the population mean; σ is a measure of variability; and, in the case of the t-distribution model, δ is a shape parameter.

Under the Gaussian model (2), the likelihood function corresponding to a random sample $X_1,\ldots,X_n$ is

$$\begin{aligned} L_1(\mu,\sigma) &= \prod_{i=1}^{n} f(X_i;\mu,\sigma) \\ &= \frac{1}{(2\pi\sigma^2)^{n/2}} \exp\left[-\frac{1}{2\sigma^2}\sum_{i=1}^{n}(X_i-\mu)^2\right]. \end{aligned}$$

It is well-known that the maximum likelihood estimator of the parameter μ is $\widehat{\mu} = \bar{X}$, the sample mean. Further, denoting by S the sample standard deviation, it is well-known that the maximum likelihood estimator of σ^2 is $\widehat{\sigma}^2 = (n-1)S^2/n$. Because $(n-1)/n \approx 1$ for large values of n and because the estimator S^2 has numerous desirable statistical properties, including unbiasedness, it is common practice to estimate σ^2 by S^2.

Using the 100 observations comprising the Milky Way data from Secker (1992), Table 1, it is found (Secker, 1992, p. 1475) that the likelihood function L_1 for the Gaussian model is maximized at $(\widehat{\mu},\widehat{\sigma}) = (-7.14, 1.41)$, and then we obtain (Secker, 1992, p. 1476)

$$\ln L_1(-7.14, 1.41) = -176.4, \tag{4}$$

consequently, $\widehat{\mathrm{BIC}}_1 = 362.0$.

In the case of the t-distribution model (3), the likelihood function is

$$\begin{aligned} L_2(\mu, \sigma, \delta) &= \prod_{i=1}^{n} f_2(X_i; \mu, \sigma, \delta) \\ &= \prod_{i=1}^{n} \frac{\Gamma(\frac{\delta+1}{2})}{\sqrt{\pi\delta}\,\sigma\,\Gamma(\frac{\delta}{2})} \left[1 + \frac{(X_i - \mu)^2}{\delta\sigma^2}\right]^{-(\delta+1)/2}. \end{aligned}$$

Unlike the Gaussian case, no algebraic formulas for $\widehat{\mu}$, $\widehat{\sigma}$, or $\widehat{\delta}$ are available for this model, and therefore the maximization of L_2 and the corresponding likelihood estimates are obtained numerically. Calculating again from the Milky Way data (Secker, 1992, p. 1476) leads to the maximum likelihood estimates $(\widehat{\mu}, \widehat{\sigma}, \widehat{\delta}) = (-7.31, 1.03, 3.55)$.* Then, the corresponding value of L_2 is given by

$$\ln L_2(-7.31, 1.03, 3.55) = -173.0, \tag{5}$$

consequently, $\widehat{\mathrm{BIC}}_2 = 359.8$. The maximum likelihood calculations in (4) and (5) suggest that the Gaussian model (2) is inferior to the t-model as a distribution for GCLF data in the Milky Way. However, there remains the issue of whether the increase in likelihood in (5) over (4) is due to the larger number of parameters in (3).

On applying the formula for $\widehat{\Delta}_{\mathrm{BIC}}$ in (1) with the values derived from (4) and (5), we obtain

$$\widehat{\Delta}_{\mathrm{BIC}} = \widehat{\mathrm{BIC}}_2 - \widehat{\mathrm{BIC}}_2 = -2.2\,.$$

By the general rules on application of $\widehat{\Delta}_{\mathrm{BIC}}$ for model comparison, we have only moderate evidence that the t-distribution is superior to the Gaussian distribution as a model for the Galactic data. In particular, we do not have overwhelming evidence in support of one model over the other.

In the case of 82 globular clusters from M31 (Secker, 1992, Table 2), similar calculations (Secker, 1992, p. 1478) for the Gaussian model lead to the

*We remark that optimization routines in the statistical package R (R Development Core Team, 2009) returned the estimates $(\widehat{\mu}, \widehat{\sigma}, \widehat{\delta}) = (-7.30, 1.07, 4.28)$, which are different from those reported in Secker (1992), *loc. cit.* However, to afford direct comparison with the results of Secker (1992), and because our overall conclusions are the same for both reported estimates, we opt to work with the values given in Secker (1992).

maximum likelihood estimates $(\widehat{\mu}, \widehat{\sigma}) = (16.98, 0.99)$ and the corresponding likelihood function value is given by

$$\ln L_1(16.98, 0.99) = -115.4, \tag{6}$$

hence, $\widehat{\text{BIC}}_1 = 239.6$. In the case of the t-distribution model, the parameters are found (Secker, 1992, p. 1478) to have maximum likelihood estimates $(\widehat{\mu}, \widehat{\sigma}, \widehat{\delta}) = (17.0, 0.90, 11.02)$,† and the corresponding likelihood function value is

$$\ln L_2(17.0, 0.90, 11.02) = -115.1, \tag{7}$$

hence, $\widehat{\text{BIC}}_2 = 243.4$. As before, the maximum likelihood calculations in (6) and (7) suggest that the Gaussian model (2) is inferior to the t-model as a distribution for the GCLF data in M31.

On applying the formula for $\widehat{\Delta}_{\text{BIC}}$ in (1) with the values derived from (6) and (7), we obtain

$$\widehat{\Delta}_{\text{BIC}} = \widehat{\text{BIC}}_2 - \widehat{\text{BIC}}_2 = 3.8\,.$$

Consequently, we have moderate evidence that the Gaussian distribution is superior to the t-distribution as a statistical model for the GCLF of M31. Again, we find no overwhelming evidence in support of one model over the other.

To infer whether the observed values of $\widehat{\Delta}_{\text{BIC}}$, *viz.*, -2.2 (in the case of the Milky Way data) and 3.8 (in the case of the M31 data), are significantly small, we estimated the standard error of $\widehat{\Delta}_{\text{BIC}}$ using bootstrap methods, with all calculations being done with the statistical package R (R Development Core Team, 2009). In each instance, we simulated from the respective models a set of observations equal in number to the sample sizes of the actual data sets in Secker (1992). From those observations, we then computed the corresponding values of $\widehat{\Delta}_{\text{BIC}}$, and then calculated the estimated standard error of $\widehat{\Delta}_{\text{BIC}}$ from its simulated values based on 20,000 bootstrap samples.

To fit a parametric bootstrap, we used the t-distribution to model the Milky Way data. The estimated standard error of $\widehat{\Delta}_{\text{BIC}}$ is 21.9, which is extremely large. Hence, the conclusion that the observed value of -2.2 indicates moderate evidence in favor of the t- over the Gaussian distribution should be accepted cautiously.

---

†In this case, optimization routines in the package R (R Development Core Team, 2009) returned the estimates $(\widehat{\mu}, \widehat{\sigma}, \widehat{\delta}) = (16.99, 0.91, 12.53)$.

In the case of the M31 data, we used the Gaussian distribution to model the data. The estimated standard error of $\widehat{\Delta}_{\mathrm{BIC}}$ is 0.9, a small value. Consequently, we infer moderate evidence in favor of the Gaussian over the t-distribution.

A Cauchy model was also examined (Secker, 1992, p. 1474) as a possible statistical model for GCLF in the Milky Way and M31. In this case, the analytical form of the density function is

$$f_3(x; \mu, \sigma) = \frac{\sigma}{\pi[\sigma^2 + (x - \mu)^2]}, \tag{8}$$

where $-\infty < x < \infty$, and the parameter ranges are $-\infty < \mu < \infty$ and $\sigma > 0$. The parameter μ represents the median and σ a measure of the spread of the distribution. Here again, we apply maximum likelihood calculations (Secker, 1992, pp. 1476–1478) to compare the Cauchy with the Gaussian or t-distribution models. When the Gaussian or t- models are compared to the Cauchy model as a fit to the Milky Way GCLF data, we obtain $\widehat{\Delta}_{\mathrm{BIC}} > 20$ in both cases. When the Gaussian or t- models are compared to the Cauchy model for the M31 data, we obtain $\widehat{\Delta}_{\mathrm{BIC}} = 17.4$ in the Gaussian case and 24.2 in the t- case. In all instances, there is strong or very strong evidence that the Gaussian and t- distributions each are superior to the Cauchy distribution as a fit to the GCLF data from the Milky Way or from M31.

In comparing the Gaussian or the t- distributions to the Cauchy, we again calculated bootstrap estimates of the standard errors of the statistic $\widehat{\Delta}_{\mathrm{BIC}}$. To construct bootstrap samples, we utilized the same simulation procedures as before, sampling from the Gaussian model because of the prior evidence that it is a superior fit to the data. The estimated standard errors for the data sets are 7.0 and 6.4 for the Milky Way and M31 data, respectively, so we have even stronger evidence in favor of the Gaussian over the Cauchy distribution as a fit for either data set.

Similarly, in comparing the t-distribution model to the Cauchy in both data sets, we calculated bootstrap estimates of the standard errors of $\widehat{\Delta}_{\mathrm{BIC}}$, using the t-distribution as the model. In this case, the estimated standard errors of $\widehat{\Delta}_{\mathrm{BIC}}$ for the data sets are 6.4 and 5.9 for the Milky Way and M31 data, respectively. Consequently, we have further evidence that the t-distribution is a better fit for the data than the Cauchy distribution.

As regards the fitting of more complex models, with higher numbers of parameters, to GCLF data, it is now clear that such models must be compared to the Gaussian model not only on the basis of the likelihood function but also with regard to the values of the BIC. As a final comment in support of the simpler Gaussian model, we point out that Harris (2001,

p. 293) also commented that although models with greater numbers of free parameters will fit the data more accurately, such models do not appear to provide "any immediate insight about the astrophysical processes governing the cluster luminosities and masses ..."

4. Concluding Remarks

The BIC is only one of many procedures for testing the goodness-of-fit of a statistical model and, as with any procedure, it should be used carefully. In particular, it is not a panacea and, in general, it should be applied in conjunction with other information criteria.

There are also drawbacks of the BIC. Findley (1991) notes that under specific circumstances, the BIC will tend to select the model that has fewer parameters. Precisely, Findley proves that "if the log-likelihood-ratio sequence of two models with different numbers of estimated parameters is bounded in probability then the BIC will, with asymptotic probability 1, select the model having fewer parameters." At least heuristically, this result can be deduced via (1), as follows: Suppose that $m_1 < m_2$ and suppose also that the first term in (1) is bounded above and below by universal constants, i.e., constants which do not depend on n. Then $\widehat{\Delta}_{\text{BIC}}$ in (1) is dominated by $(m_2 - m_1)\ln n$ as n increases; consequently, $\widehat{\Delta}_{\text{BIC}}$ will become positive. By the general rules for application of the BIC, we would then infer strong evidence in favor of $f_1(x;\theta_1,\ldots,\theta_{m_1})$, the model having the fewer number of parameters.

In the context of astrophysical applications of comparing two statistical models, $f_1(x;\theta_1,\ldots,\theta_{m_1})$ and $f_2(x;\phi_1,\ldots,\phi_{m_2})$, for globular cluster luminosity functions, an application of Findley's theorem requires a sequential calculation of the corresponding log-likelihood ratios

$$\ln \frac{L_1(\hat{\theta}_1,\ldots,\hat{\theta}_{m_1})}{L_2(\hat{\phi}_1,\ldots,\hat{\phi}_{m_2})}$$

for successively increasing sample sizes n. If it is *believed* that the observed sequence of log-likelihood ratios will be bounded between two universal constants, and such a belief necessarily will have to be justified on intrinsic astrophysical arguments, then it be necessary to look for alternatives to the BIC in measuring the relative plausibility of a statistical model.

In fitting models to GCLF data, it is important to bear in mind that the BIC is *consistent*: As $n \to \infty$, the probability that the BIC will determine the correct statistical model converges to 1 (Jeffreys, 1961). The property of consistency does not hold, in general, for model-fitting information criteria.

There is also the issue of utilizing the Kolmogorov-Smirnov statistic to measure the goodness-of-fit of a statistical model. In the case of the Milky Way and M31 data considered in this paper, the Kolmogorov-Smirnov statistic was also applied (Secker, 1992, p. 1467 *ff.*) to support the choice of the t-distribution over the Gaussian and Cauchy models. Here again, though, there is cause for concern. The importance of statistics based on empirical distribution functions, such as the Kolmogorov-Smirnov statistic, stems from the fact that they are distribution-free in the case of continuous data, such as luminosity measurements. However, *these statistics are no longer distribution-free when the model parameters need to be estimated from the data.* Consequently, in model fitting contexts, goodness-of-fit levels of significance derived from the Kolmogorov-Smirnov statistic are usually incorrect when applied with estimated parameters (Babu and Rao, 2004).

We urge that caution be exercised in using the Kolmogorov-Smirnov statistic to justify the fitting of the models (2), (3), or (8) to GCLF data. The basic problem with the Kolmogorov-Smirnov statistic is that it is not penalized; it is possible that a penalized form of that statistic may work well, and indeed a main point of this paper is that penalized statistics generally are better for fitting models to luminosity data. Moreover, any parameter estimation in a penalized Kolmogorov-Smirnov statistic can be dealt with by bootstrap methods.

Acknowledgments

We are grateful to Tom Hettmansperger and Mercedes Richards for their comments on earlier drafts of the manuscript. The second author is supported in part by the National Science Foundation, grant DMS-0705210.

References

Babu, G. J. and Rao, C. R. (2004). Goodness-of-fit tests when parameters are estimated, *Sankhyā* **66**, pp. 63–74.

Barmby, P., Huchra, J. P. and Brodie, J. P. (2001). The m31 globular cluster luminosity function, *The Astronomical Journal* **121**, pp. 1482–1496.

Findley, D. F. (1991). Counterexamples to parsimony and bic, *Annals of the Institute of Statistical Mathematics* **43**, pp. 505–514.

Harris, W. E. (1998). Globular cluster systems as distance indicators, in C. Pritchet and S. van den Bergh (eds.), *The Extragalactic Distance Scale*, ASP Conf. Ser. 4 (ASP, San Francisco), pp. 231–254.

Harris, W. E. (2001). Globular cluster systems, in L. Labhardt and B. Binggeli (eds.), *Star Clusters: Saas-Fee Advanced Course 28*, Swiss Society for Astrophysics and Astronomy Lecture Notes 1998 (Springer-Verlag, Berlin), pp. 223–408.

Harris, W. E., Allwright, J. W. B., Pritchet, C. J. and van den Bergh, S. (1991). The luminosity distribution of globular clusters in three giant virgo ellipticals, *The Astrophysical Journal Supplement* **76**, pp. 115–151.

Jeffreys, H. (1961). *Theory of Probability, 3rd ed.* (Clarendon Press, Oxford).

Kass, R. E. and Raftery, A. E. (1995). Bayes factors, *Journal of the American Statistical Association* **90**, pp. 773–795.

Larsen, S. S. and Richtler, T. (1999). Young massive star clusters in nearby galaxies, i. identification and general properties of the cluster systems. *Astronomy and Astrophysics* **345**, pp. 59–72.

Mukherjee, S., Feigelson, E. D., Babu, G. J., Murtagh, F., Fraley, C. and Raftery, A. (1998). Three types of gamma-ray bursts, *The Astrophysical Journal* **508**, pp. 314–327.

R Development Core Team (2009). *R: A Language and Environment for Statistical Computing*, R Foundation for Statistical Computing, Vienna, Austria, URL `http://www.R-project.org`, ISBN 3-900051-07-0.

Racine, R. and Shara, M. (1979). The luminosity distribution of globular clusters in m31, *The Astronomical Journal* **84**, pp. 1694–1696.

Schwarz, G. E. (1978). Estimating the dimension of a model, *Annals of Statistics* **6**, pp. 461–464.

Secker, J. (1992). A statistical investigation into the shape of the globular cluster luminosity distribution, *The Astronomical Journal* **104**, pp. 1472–1481.

van den Bergh, S. (1985). The luminosity function of globular clusters, *The Astrophysical Journal* **297**, pp. 361–364.

van den Bergh, S. (2000). Some musings on globular cluster systems, *Publications of the Astronomical Society of the Pacific* **112**, pp. 932–941.

Whitmore, B. C. (1997). Globular clusters as distance indicators, in M. D. M. Livio and N. Panagia (eds.), *The Extragalactic Distance Scale* (Cambridge University Press, Cambridge), pp. 254–272.

A Likelihood-Tuned Density Estimator Via a Nonparametric Mixture Model

Yeojin Chung† & Bruce G. Lindsay‡

Department of Statistics, Pennsylvania State University
University Park, PA 16802, USA
† *E-mail: ychung@psu.edu*
‡ *E-mail: bgl@psu.edu*

We consider an improved density estimator which arises from treating the kernel density estimator as an element of the model that consists of all mixtures of the kernel, continuous or discrete. One can obtain the kernel density estimator with "likelihood-tuning" by using the uniform density as the starting value in an EM algorithm. The second tuning leads to a fitted density with higher likelihood than the kernel density estimator. The two step EM estimator can be written explicitly with a Gaussian kernel, and its bias is one order of magnitude smaller than the kernel estimator. In addition, the order of magnitude of the variance stays of the same order, so that the asymptotic mean square error can be reduced significantly. Compared with other modified density estimators, the simulation results show that the two-step likelihood-tuned density estimator performs robustly against different types of true density.

Keywords: Nonparametric density estimation; Nonparametric mixture model; Nonparametric maximum likelihood

1. Introduction

Statistical inference using finite mixture models has largely been treated as a special, and challenging, type of parametric inference. However, these models also have great potential within the area of nonparametric inference, as illustrated beautifully by Thomas and Hettmansperger (2001). In this paper, we show how one can use the theory of nonparametric mixture estimation fruitfully in the area of nonparametric density estimation.

The kernel density estimator is a widely used nonparametric density estimator. Let $X_1, \cdots, X_n$ be a random sample from $f(x)$. Then the kernel density estimator of $f(x)$, denoted $\hat{f}_{\mathrm{KER}}(x)$ in this paper, is defined by

$$\hat{f}_{\mathrm{KER}}(x) = n^{-1}\sum_{i=1}^{n} K_h(x, X_i), \tag{1}$$

where $K_h(\cdot,\cdot)$ is a known kernel function with a bandwidth h. Many authors have modified it to reduce its bias, a key element of its mean squared error. In this paper, we propose a new method that is based on applying one step of the EM algorithm to $\hat{f}_{\mathrm{KER}}$ in a class of mixture models. This also reduces bias, and it has an advantage over other methods in that it is based on a likelihood device that can easily be generalized to other smoothing problems.

Breiman *et al.* (1977) proposed to replace h in (1) by a variable bandwidth $h(X_i)$ depending on an observation. Abramson (1982) suggested taking $h(X_i)$ proportional to $f(X_i)^{-1/2}$ by showing that this reduces the bias to $O(h^4)$. Although this procedure requires one to know the true $f(x)$, Silverman (1986) mentioned that the basic kernel estimator in (1) works well as a pilot estimator of $f(X_i)$ for the variable bandwidth. In addition, he defined *the adaptive bandwidth density estimator*, denoted by $\hat{f}_{\mathrm{ABW}}$ throughout this paper, as

$$\hat{f}_{\mathrm{ABW}}(x) = n^{-1}\sum_{i=1}^{n} K_{h(\hat{f}_{\mathrm{KER}}(X_i)/g)^{-1/2}}(x, X_i), \tag{2}$$

where g is the geometric mean of the $\hat{f}_{\mathrm{KER}}(X_1), \cdots, \hat{f}_{\mathrm{KER}}(X_n)$.

Jones *et al.* (1995) considered a multiplicative bias correction that results in the density estimator given by

$$\hat{f}_{\mathrm{MBC}}(x) = \hat{f}_{\mathrm{KER}}(x)\cdot n^{-1}\sum_{i=1}^{n} \hat{f}_{\mathrm{KER}}(X_i)^{-1} K_h(x, X_i). \tag{3}$$

The leading bias term of $\hat{f}_{\mathrm{KER}}$ in the numerator is canceled by the leading bias of $\hat{f}_{\mathrm{KER}}$ in the denominator so that $\hat{f}_{\mathrm{MBC}}$ attains a bias of order $O(h^4)$ (Jones and Signorini, 1997). The estimator in (3) is referred as the *multiplicative bias correction density estimator* in this paper, following Jones and Signorini (1997).

Recently, Di Marzio and Taylor (2004) applied boosting to kernel density estimation. They adopted a goodness-of-fit measure that compares the kernel density estimator with the leave-one-out estimator (Silverman, 1986). Based on this measure, the boosting step updates the weight for each $K(X_i,\cdot)$ and fits the weighted kernel density estimator. Starting from the uniform initial weight, n^{-1}, the first boosting step provides $\hat{f}_{\mathrm{MBC}}$. While a further boosting step reduces the average ISE for the Gaussian distribution, the simulation result shows that it does not clearly improve the performance in the case of non-Gaussian distributions.

In this paper, we will compare our new method with $\hat{f}_{\mathrm{KER}}$, $\hat{f}_{\mathrm{ABW}}$ and $\hat{f}_{\mathrm{MBC}}$. We have chosen $\hat{f}_{\mathrm{ABW}}$ because this seems to be the most commonly used modified density estimator. In addition, Jones *et al.* (1995) compared six higher-order bias density estimators, including $\hat{f}_{\mathrm{ABW}}$ and $\hat{f}_{\mathrm{MBC}}$, and found that these two estimators were the most competitive. Adopting another tuning parameter of the bandwidth for $\hat{f}_{\mathrm{KER}}$ in (2) or in (3) can give better performance for some densities. However, for a fair comparison, we will use the common bandwidth for $\hat{f}_{\mathrm{KER}}$ and the kernel function in (2) and (3) because it is most natural to compare our estimator with other estimators in their basic form.

This article proposes a new density estimator that reduces the bias. To create it, we look at the kernel density estimator as an estimator via the nonparametric mixture model and use the EM algorithm to improve its likelihood. In Section 2.1, we describe the connection between the nonparametric mixture model and nonparametric density estimation, two areas that have been mostly treated in separate literatures to date. In Section 2.2, the likelihood-tuning procedure and the resulting density estimators are introduced. In Section 3, we investigate the asymptotic properties of the new density estimator and compare them with existing estimators such as the basic kernel, adaptive bandwidth and multiplicative bias correction density estimator. The simulation comparisons are given in Section 4.

2. Methodology

2.1. *Background*

Consider the nonparametric mixture model with a mixing (latent) distribution $\Pi(\phi)$, given by

$$f(x;\Pi) = \int K(x,\phi)d\Pi(\phi), \tag{4}$$

where $K(\cdot,\cdot)$ is a known density function, called a component density, and ϕ represents a component parameter. Here the distribution function $\Pi(\phi)$ is allowed to be discrete, continuous or from any specific family of distributions. If Π is discrete with point masses π_i at ϕ_i, $i = 1, \cdots, m$, then (4) becomes the m-component finite mixture model, written as

$$f(x; \phi_1, \cdots \phi_m, \pi_1, \cdots, \pi_m) = \sum_{i=1}^{m} \pi_i K(x, \phi_i). \tag{5}$$

Let $X_1, \cdots, X_n$ be a random sample from $f(x; \Pi)$. If Π is estimated by $\hat{\Pi}(\phi) = n^{-1}$ at $\phi = X_i$, $i = 1, \cdots, n$ and 0 elsewhere, (4) can be written as

$$f(x; \hat{\Pi}) = n^{-1} \sum_{i=1}^{n} K(x, X_i),$$

and this is the same as the basic kernel density estimator in (1). This view of the kernel density estimator in the context of nonparametric mixture models leads to the idea that an improved estimator of Π would provide a good estimator of $f(x)$.

There is an extensive literature concerning the nonparametric maximum likelihood estimator of the mixing distribution Π. Consider the likelihood function of an observation, given by

$$L_i(\Pi) = \int K(X_i, \phi) d\Pi(\phi),$$

and the log-likelihood function with multiple observations $n(i)$ of a single L_i, given by

$$l(\Pi) = \sum_{i=1}^{D} n(i) \ln(L_i(\Pi)), \tag{6}$$

where $\sum_{i=1}^{D} n(i) = n$. Here we do not specify any parametric form of Π, allowing it to be either discrete or continuous. The estimator $\hat{\Pi}$ that maximizes (6) is called the nonparametric maximum likelihood estimator (NPMLE) of Π (Lindsay, 1995). Lindsay (1995) proves that if the L_i's are all distinct and all $n(i) > 0$, then there exists an NPMLE, $\hat{\Pi}$, that is a discrete distribution with no more than D distinct points of support.

Since the NPMLE is discrete on finitely many support points, we can apply the EM algorithm by restricting $\Pi(\phi)$ to be a discrete distribution on a large number of support points (Laird, 1978; Vardi *et al.*, 1985). On the other hand, we could consider the generalized EM algorithm that allows the initial $\Pi(\phi)$ to be continuous with a density function $\pi(\phi)$, even though we know it converges to the discrete NPMLE. Vardi and Lee (1993) and Lindsay (1995) described *the continuous EM algorithm for NPMLE*, given by

$$\hat{\pi}_{(k+1)}(\phi) = \hat{\pi}_{(k)}(\phi) \cdot n^{-1} \sum_{i=1}^{n} \frac{K(X_i, \phi)}{L_i(\hat{\Pi}_{(k)})}. \tag{7}$$

Notice that this is a generalization of the standard EM algorithm for updating the mixing proportion π_j in the discrete mixture model in (5), written as

$$\pi_{j,(k+1)} = \pi_{j,(k)} \cdot n^{-1} \sum_{i=1}^{n} \frac{K(X_i, \phi_j)}{f(x_i)}.$$

In this paper, we consider the consequences of using a continuous uniform density, $\pi(\phi) = 1$, as the initial estimate of $\hat{\pi}$. Each update by (7) will produce a spikier estimate of π that assigns more weights to the support points of the discrete NPMLE. A single update by the EM gives the basic kernel density estimator and the second update generates our new and improved density estimator. Although the second-step estimator of π is still not the NPMLE, it does provide a density estimator that is asymptotically superior to the kernel density estimator in many cases. It also shows better overall performance in our simulation study than other modified kernel density estimators.

Similar ideas have appeared in the mixture literature, but the focus has been on the estimator of Π, not on the resulting estimator for the density of x. Laird and Louis (1991) viewed $\Pi(\phi)$ as a prior distribution of a parameter ϕ and x_i's as the realizations from $x_i|\phi \sim N(\phi, 1)$. They employed a uniform prior and updated Π with an empirical Bayes estimate. This procedure of updating a prior distribution Π is identical to the continuous EM algorithm of updating the mixing distribution Π in (7).

There have been other attempts to smooth the NPMLE for Π that do not depend on the EM. Goutis (1997)'s method is motivated by the kernel density estimator of the unobserved data ϕ_i's in the incomplete data (X_i, ϕ_i) problem. Since we cannot observe the ϕ_i's, he proposed to iteratively estimate the conditional expectation of the kernel density estimate of ϕ_i. With the Gaussian kernel with a bandwidth h, this procedure turns out to be the same as the generalized EM in (7) except that the update of π is generated by kernels with a larger bandwidth than the update of π by the EM. Silverman *et al.* (1990) proposed the EMS algorithm, which adds a smoothing step to the EM for smoothing the estimate in each iteration. These authors have focused on reducing the spiky feature of the NPMLE of $\hat{\Pi}$. On the contrary, we are utilizing our methodology to improve the kernel density estimator for x.

2.2. *Likelihood-tuning Procedure*

Prior to introducing the likelihood-tuning procedure, we define a gradient function of the log-likelihood function $l(\Pi)$ in (6) at Π_0 toward the direction of ϕ by

$$D_{\Pi_0}(\phi) = \sum_{i=1}^{D} n(i) \left(\frac{K(X_i, \phi)}{L_i(\Pi_0)} - 1 \right)$$

(Lindsay, 1995). This is derived from the directional derivative of $l(\Pi)$ at Π_0 along the path between Π_0 and Π_1, that is,

$$D_{\Pi_0}(\Pi_1) = \frac{\partial}{\partial \alpha} l(\alpha \Pi_0 + (1-\alpha)\Pi_1)|_{\alpha=0}.$$

When Π_1 is a degenerate distribution at ϕ, $D_{\Pi_0}(\Pi_1)$ becomes $D_{\Pi_0}(\phi)$. Thus if $D_{\Pi_0}(\phi)$ is positive, $l(\Pi_0)$ will be increased by adding more probability at ϕ in the mixing distribution Π_0. On the contrary, if $D_{\Pi_0}(\phi)$ is negative for ϕ, $l(\Pi_0)$ will be increased by shrinking the mass at ϕ in Π_0.

Using the gradient function, the continuous EM algorithm in (7) can be written as

$$\hat{\pi}_{(k+1)}(\phi) = \hat{\pi}_{(k)}(\phi) \left(1 + n^{-1} D_{\hat{\Pi}_{(k)}}(\phi) \right).$$

This equation implies that the EM algorithm increases the density at ϕ where the gradient function is positive and reduces the density where the gradient function is negative. We note that $\hat{\pi}_{(k+1)}(\phi)$ integrates to 1 if $\hat{\pi}_{(k)}(\phi)$ does.

The likelihood-tuning procedure that updates a density estimator of x includes two steps: updating the mixing density π and updating the density estimator of x. Given an initial estimate $\pi_0(\phi)$, the *likelihood-tuning procedure* is as follows.

(1) Update the estimator of the mixing density π by

$$\hat{\pi}_{(k+1)}(\phi) = \hat{\pi}_{(k)}(\phi) \Delta_{(k)}(\phi),$$

where $\Delta_{(k)}(\phi) = 1 + n^{-1} D_{\Pi_{(k)}}(\phi)$.

(2) Update the density estimator of x by

$$\hat{f}_{(k+1)}(x) = \int K(x, \phi) \hat{\pi}_{(k+1)}(\phi)\, d\phi.$$

Notice that it is desirable to avoid numerical integration in step 2. We will take the uniform density $\pi_0(\phi) = 1$ to be the initial estimate of π given a lack of any prior information about the location of peaks. There are sound theoretical reasons for this choice, as it gives a gradient function that tends to identify the needed points of support in the NPMLE. In each step thereafter, $\Delta_{(k)}(\phi)$ indicates the deviation of the latent density $\pi_{(k+1)}(\phi)$ from $\pi_{(k)}(\phi)$, which therefore increases mass at the highest gradient values. In fact, one can show that repeated application of the continuous EM algorithm converges to the NPMLE without the parameter space searches ("gradient checks") that are required when one uses discrete Π estimators (Lindsay, 1995).

Let $K_h(\cdot,\cdot)$ be a kernel function with a bandwidth h. If we consider a diffusion kernel function, which satisfies the diffusion equation

$$\int K_h(x,\phi)K_h(\phi,y)d\phi = K_{\sqrt{2}h}(x,y),$$

then the estimators of $\pi(\phi)$ and $f(x)$ by the likelihood-tuning procedure are reduced to explicit forms. The Gaussian kernel function, which is such a diffusion kernel, is defined by $K_h(x,\phi) = h^{-1}\varphi\left\{h^{-1}(x-\phi)\right\}$, where φ is the standard normal density function. Then the first likelihood-tuning iteration provides

$$\hat{\pi}_{(1)}(\phi) = n^{-1}\sum_{i=1}^{n} K_h(\phi, X_i)$$

and

$$\begin{aligned}\hat{f}_{(1)}(x) &= \int K_h(x,\phi)\cdot n^{-1}\sum_{i=1}^{n} K_h(\phi, X_i)\, d\phi \\ &= n^{-1}\sum_{i=1}^{n} K_{h\sqrt{2}}(x, X_i). \qquad (8)\end{aligned}$$

Notice that the estimate $\hat{\pi}_{(1)}(\phi)$ from a single likelihood-tuning step turns out to be the basic kernel density estimator. The resulting fitted density for x, $\hat{f}_{(1)}$, is also the basic kernel density estimator but with a wider bandwidth than the bandwidth in $\hat{\pi}_{(1)}$.

If we apply the likelihood-tuning procedure once more, we obtain

$$\hat{\pi}_{(2)}(\phi) = n^{-2}\sum_{i=1}^{n}\sum_{j=1}^{n} w_{ij}K_{h/\sqrt{2}}(\phi, \bar{X}_{ij}),$$

where $w_{ij} = K_{h\sqrt{2}}(X_i, X_j)/\hat{f}_1(X_i)$ and $\bar{X}_{ij} = (X_i + X_j)/2$. The second step density estimator is given by

$$\hat{f}_{(2)}(x) = n^{-2}\sum_i\sum_j w_{ij}\int K_h(x,\phi)K_{h/\sqrt{2}}(\phi,\bar{X}_{ij})d\phi$$
$$= n^{-2}\sum_i\sum_j w_{ij}K_{h\sqrt{\frac{3}{2}}}(x,\bar{X}_{ij}). \tag{9}$$

We define this second step density estimator in (9) as the *two-step likelihood-tuned density estimator* and denote it by $\hat{f}_{\mathrm{EM2}}(x)$.

Proceeding in the same manner, one can take more EM steps to move $\hat{\pi}$ even closer to the NPMLE than $\hat{\pi}_{(2)}$. For any positive integer k, $\pi_{(k)}$ is generalized by

$$\hat{\pi}_{(k)}(\phi) = \frac{1}{n^k}\sum_{i_1=1}^{n}\cdots\sum_{i_k=1}^{n} w_{i_1,\cdots,i_{k-1}}K_{h/\sqrt{k}}(\phi,\bar{X}_{i_1,\cdots,i_k}),$$

where

$$w_{i_1,\cdots,i_{k-1}} = c(h)\cdot\frac{K_{h\sqrt{k}}(X_{i_1},X_{i_2})\cdots K_{h\sqrt{k}}(X_{i_{k-1}},X_{i_k})}{\hat{f}_{(1)}(X_{i_1})\cdots\hat{f}_{(k-1)}(X_{i_{k-1}})}$$

with known constant $c(h)$ and $\bar{X}_{i_1,\cdots,i_k}$ is the mean of $(X_{i_1},\cdots,X_{i_k})$. The updated density estimator can be expressed in the generalized form

$$\hat{f}_{(k)}(x) = \frac{1}{n^k}\sum_{i_1=1}^{n}\cdots\sum_{i_k=1}^{n} w_{i_1,\cdots,i_{k-1}}K_{h\sqrt{\frac{k+1}{k}}}(x,\bar{X}_{i_1,\cdots,i_k}). \tag{10}$$

Although the density estimator in (10) with $k \geq 3$ might have more desirable asymptotic properties than the second step estimator, it requires intensive computations. In fact, when one moves from the $(k-1)$th step estimator to the kth step, the number of summands is increased from n^{k-1} to n^k. Without proceeding further, therefore, this paper focuses on the second step estimator, $\hat{f}_{\mathrm{EM2}}(x)$, and will show that it has desirable asymptotic properties and promising simulation results in Sections 3 and 4.

3. Asymptotic Properties

In this section, we show the asymptotic properties of the likelihood-tuned density estimator and compare them with other improved density estimators.

Theorem 3.1. *Suppose that $f(x)$ is four times continuously differentiable* *and* $K_h(x,\phi) = h^{-1}\varphi\left(h^{-1}(x-\phi)\right)$. *Then, when* $h \to 0$ *and* $nh \to \infty$,

$$E\left[\hat{f}_{\mathrm{EM2}}(x)\right] = f(x)\left(-\frac{f^{(4)}(x)}{f(x)} + \frac{f^{(3)}(x)f'(x) + f''(x)^2}{f(x)^2} - \frac{f''(x)f'(x)^2}{f(x)^3}\right)h^4 \\ + f(x) + o(h^4)$$

and

$$Var\left[\hat{f}_{\mathrm{EM2}}(x)\right] = (nh)^{-1}f(x)\frac{1}{\sqrt{\pi}}\left(\frac{1}{4} + \sqrt{2} - \frac{2}{\sqrt{3}}\right) + o\left((nh)^{-1}\right).$$

See the Appendix for an outline proof.

Theorem 3.1 reveals that the likelihood-tuned density estimator has bias of order $O(h^4)$ and variance of order $O((nh)^{-1})$, as do the other modified density estimators. Before comparing it with other density estimators, one should note that we had a bandwidth of $h\sqrt{2}$ instead of h for the one-step kernel density estimator in Section 2.2. Thus it is reasonable to compare the one-step and two-step estimators to see the effect of likelihood tuning. When we make comparisons with the adaptive bandwidth and the multiplicative bias correction estimator, we also need to account for bandwidth effects.

Jones *et al.* (1995) provided bias and variance formulas for the adaptive bandwidth density estimator in (2). With the Gaussian kernel defined in Theorem 1, the asymptotic bias reduces to

$$\left(-\frac{3f^{(4)}(x)}{2f(x)} + \frac{6f^{(3)}(x)f'(x) + 4f''(x)^2}{f(x)^2} \right. \\ \left. -20\frac{f''(x)f'(x)^2}{f(x)^3} + 12\frac{f'(x)^4}{f(x)^4}\right)h^4 + o(h^4)$$

and the asymptotic variance to

$$(nh)^{-1}f(x)\frac{1}{\sqrt{\pi}}\left(\frac{1}{2\sqrt{2}} + \frac{1}{2\sqrt{3}} + \frac{1}{16}\left(1 + \frac{x^2}{4}\right)\right).$$

Since $\hat{f}_{\mathrm{MBC}}(x)$ given in (3) does not integrate to 1, Jones *et al.* (1995) suggested rescaling it by dividing $\hat{f}_{\mathrm{MBC}}(x)$ by its integral. They provided the bias of the rescaled version. With the Gaussian kernel function, the rescaled bias becomes

$$f(x)\left(-\frac{f^{(4)}(x)}{f(x)} + \frac{2f^{(3)}(x)f'(x) + f''(x)^2}{f(x)^2} \right. \\ \left. -\frac{2f''(x)f'(x)^2}{f(x)^3} + \int \frac{f''(z)^2}{f(z)}dz\right)h^4 + o(h^4).$$

The rescaled $\hat{f}_{\text{MBC}}(x)$ has exactly the same asymptotic variance as the likelihood-tuned density estimator.

To compare the theoretical MSE of $\hat{f}_{\text{ABW}}$, $\hat{f}_{\text{MBC}}$, and $\hat{f}_{\text{EM2}}$, we replace h of $\hat{f}_{\text{ABW}}$ by $h \cdot AVar(\hat{f}_{\text{ABW}})/AVar(\hat{f}_{\text{EM2}})$, as Jones *et al.* (1995) proposed. Then $\hat{f}_{\text{ABW}}$ also has the same asymptotic variance as $\hat{f}_{\text{EM2}}$, so that MSE comparison can be based only on the rescaled bias.

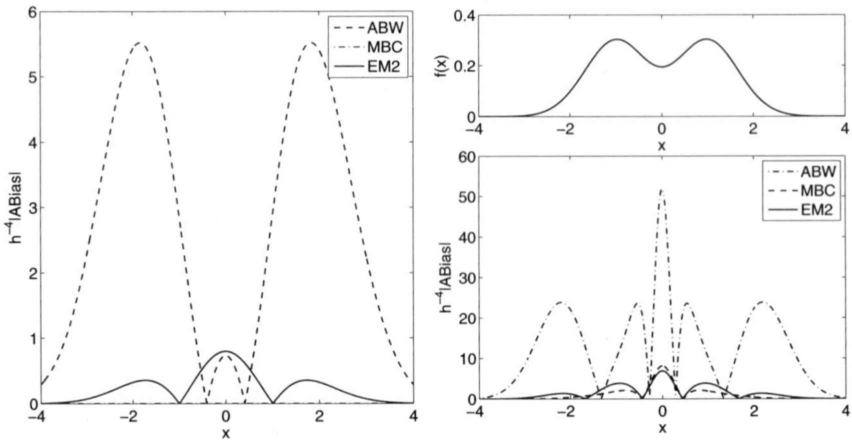

(a) Normal(0,1)

(b) Bimodal (with density shown at top)

Fig. 1. Asymptotic biases after rescaling for $\hat{f}_{\text{ABW}}$.

Figure 1(a) displays the absolute values of asymptotic biases after rescaling the bias of $\hat{f}_{\text{ABW}}$ for the standard normal distribution (after removing h^4). Since $\hat{f}_{\text{MBC}}$ has zero bias for $N(0,1)$ with order $o(h^4)$, the dash-dotted line is flat on the horizontal axis. Around the mode, $\hat{f}_{\text{ABW}}$ is superior to $\hat{f}_{\text{EM2}}$. However, in most regions, $\hat{f}_{\text{ABW}}$ has much larger bias than $\hat{f}_{\text{EM2}}$ and so $\hat{f}_{\text{EM2}}$ beats $\hat{f}_{\text{ABW}}$ in terms of the asymptotic MSE.

Figure 1(b) illustrates the absolute values of asymptotic biases for the bimodal distribution defined in Marron and Wand (1992), also after rescaling the bias of $\hat{f}_{\text{ABW}}$. The upper plot in Figure 1 (b) is the true density $f(x)$. The adaptive bandwidth density estimator, $\hat{f}_{\text{ABW}}$, again has larger bias than the others except for a small region around the shoulders of both modes. The likelihood-tuned estimator, $\hat{f}_{\text{EM2}}$, beats both $\hat{f}_{\text{ABW}}$ and $\hat{f}_{\text{MBC}}$ around the "valley" of f, while it is more biased around the two modes and the tails.

In both densities, the adaptive bandwidth density estimator has larger MSE than the others in most regions. Although the likelihood-tuned estimator is asymptotically worse than the multiplicative bias correction estimator in some places, it is known that asymptotic results may differ significantly from actual finite sample performance (Bowman and Foster, 1993) and so we need to study the methods further by simulation.

4. Simulation Comparison

In this section, we use simulated data to compare the performances of $\hat{f}_{\text{KER}}(x)$, $\hat{f}_{\text{ABW}}(x)$, $\hat{f}_{\text{MBC}}(x)$ and $\hat{f}_{\text{EM2}}(x)$. We consider the first eight Marron-Wand distributions (Marron and Wand, 1992), which are mixtures of Gaussian densities with various shapes, named 'Gaussian,' 'skewed unimodal,' 'strongly skewed,' 'kurtotic unimodal,' 'outlier,' 'bimodal,' 'separated bimodal,' and 'skewed bimodal.' In addition, four non-Gaussian distributions, namely Gamma$(2,1)$, Beta$(2,5)$, Beta$(2,2)$ and Beta$(1,3)$, were considered. Each density estimate was obtained on a grid of 301 points on [-3,3] for Gaussian mixtures, 350 points on [0,7] for Gamma$(2,5)$, and 100 points on [0,1] for beta distributions. For each distribution and each estimator, we calculated an *optimal bandwidth* by minimizing the average of integrated square errors (ISE), given by

$$ISE(\hat{f}) = \int \left\{ \hat{f}(x) - f(x) \right\}^2 dx.$$

The ISE was calculated by numerical integration except that the ISE values for Gaussian mixture distributions with $n = 100$ were analytically calculated.

To investigate local performance, 500 random samples of size 500 were generated and the density was estimated by the four density estimators using the corresponding optimal bandwidths. Then, at each grid point of x, we found the mean of the squared errors (SEs) over 500 replicates. Figure 2 displays the bimodal density and the square root of mean SEs for this density. Here the mean SE of $\hat{f}_{\text{EM2}}$ (thick solid line) was uniformly smaller than that of $\hat{f}_{\text{KER}}$ (thin solid line). Although $\hat{f}_{\text{ABW}}$ (dashed line) behaved slightly better in the left and right shoulder, it clearly had higher mean SEs in both the peaks and tails. The most competitive estimator in this case was $\hat{f}_{\text{MBC}}$ (dash-dotted line), which performed better in the tails and peaks. However, we can see that $\hat{f}_{\text{EM2}}$ was still superior in the valley, where it also beat $\hat{f}_{\text{MBC}}$ in asymptotic bias, as shown in Section 3.

Figure 3 shows the square root of the mean SE for Beta$(2,5)$, one of the non-Gaussian distributions. The squared errors were calculated on

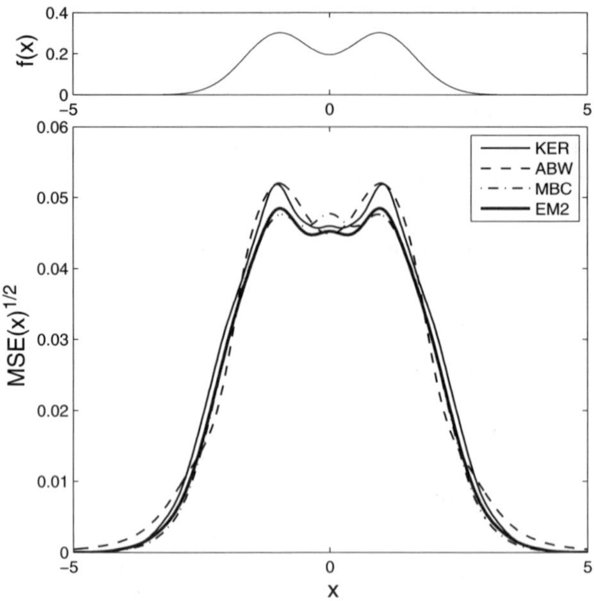

Fig. 2. Bimodal density (top) and symmetrized mean squared errors for Bimodal with n=500 and replicates=500 (bottom).

$[-0.5, 1.5]$ to observe the behavior beyond the support of the true density, $[0, 1]$. In this case, $\hat{f}_{\text{KER}}$ worked much better than in the bimodal case. In Figure 3, the thin solid line for $\hat{f}_{\text{KER}}$ is found significantly below the others around the left shoulder. Though $\hat{f}_{\text{EM2}}$ is located slightly above $\hat{f}_{\text{KER}}$ in the left shoulder, it beats all of three competitors in the peak. In addition, $\hat{f}_{\text{EM2}}$ was superior to $\hat{f}_{\text{ABW}}$ and $\hat{f}_{\text{MBC}}$ in most areas except where x was negative. Outside of the support of f, $\hat{f}_{\text{MBC}}$ had the smallest mean SE. However, it clearly lost to other estimators inside the support. For both the bimodal and Beta$(2, 5)$, we cannot conclude that the likelihood-tuned estimator was the best. However, it still performed very well for both, while the other estimators tended to work well for one or the other.

Table 1 provides a summary of simulation results over all our trial densities. For the comparison of global performances, 1000 random samples were generated by each f and we calculated density estimates by $\hat{f}_{\text{KER}}$, $\hat{f}_{\text{ABW}}$, $\hat{f}_{\text{MBC}}$, and $\hat{f}_{\text{EM2}}$ with corresponding optimal bandwidths. The ISE was calculated exactly or approximately as in the bandwidth selection, then we found the mean ISE for each density esitmator. The simulation was conducted with sample sizes $n = 100$ and $n = 500$, separately. Table 1 contains

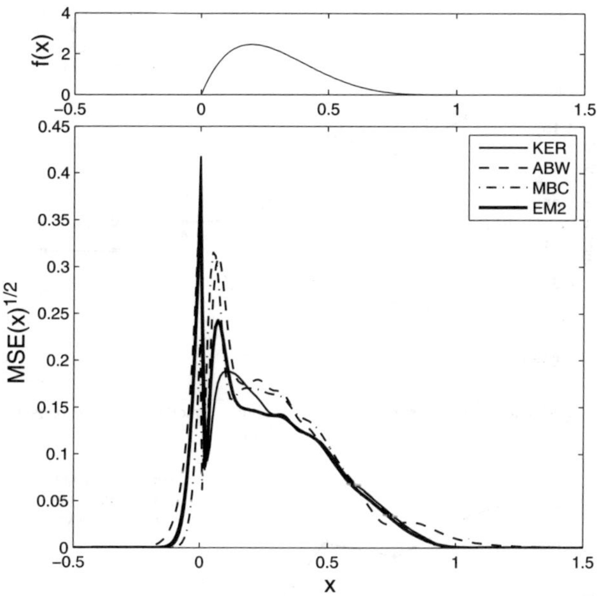

Fig. 3. Beta(2, 5) density (top) and mean squared errors for Beta(2, 5) with $n = 500$ and 500 replicates (bottom).

the ratio of the mean ISE for the basic kernel, the adaptive bandwidth and the multiplicative bias correction estimator versus the likelihood-tuned

Table 1. Ratio of the mean ISE of $\hat{f}_{\text{KER}}$, $\hat{f}_{\text{ABW}}$ and $\hat{f}_{\text{MBC}}$ against $\hat{f}_{\text{EM2}}$ for sample sizes $n = 100$ and $n = 500$ from eight Gaussian mixture densities and four non-Gaussian densities over 1000 simulations.

	$\hat{f}_{\text{KER}}$ vs. $\hat{f}_{\text{EM2}}$		$\hat{f}_{\text{ABW}}$ vs. $\hat{f}_{\text{EM2}}$		$\hat{f}_{\text{MBC}}$ vs. $\hat{f}_{\text{EM2}}$	
	$n = 100$	$n = 500$	$n = 100$	$n = 500$	$n = 100$	$n = 500$
Gaussian	1.4259	1.6010	1.1716	1.2504	0.7327	0.6638
skewed unimodal	1.3019	1.4359	1.0245	1.0843	0.8976	0.8913
strongly skewed	0.9819	1.0316	0.8776	0.8397	1.0476	1.0247
kurtotic unimodal	1.0644	1.1581	0.7853	0.7731	1.0105	0.9806
outlier	1.3536	1.5135	0.9885	1.0501	0.8076	0.7724
bimodal	1.0543	1.1900	1.1227	1.1181	1.0231	0.9924
separated bimodal	1.2366	1.3925	1.1675	1.2230	0.8600	0.7827
skewed bimodal	0.9799	1.0952	1.0551	1.0258	1.0358	1.0008
Gamma(2, 1)	0.9836	0.9575	1.0944	1.1963	1.1966	1.2704
Beta(2, 5)	1.0686	1.0210	1.2214	1.3511	1.1548	1.2159
Beta(2, 2)	1.1649	1.0685	1.4961	1.6988	0.9035	1.3418
Beta(1, 3)	0.8735	0.8515	1.0057	1.0073	1.2134	1.2404

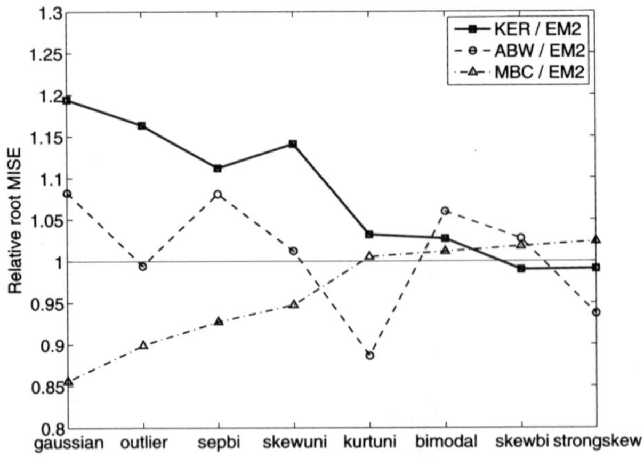

(a) $n = 100$

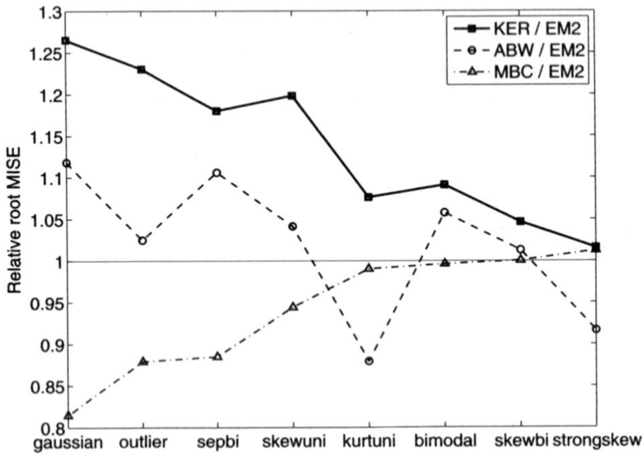

(b) $n = 500$

Fig. 4. Relative root mean ISE for $\hat{f}_{\mathrm{KER}}$, $\hat{f}_{\mathrm{ABW}}$ and $\hat{f}_{\mathrm{MBC}}$ vs. $\hat{f}_{\mathrm{EM2}}$ for Gaussian mixture distributions. Dotted line represents the ratio 1.

estimator. Thus, a ratio greater than 1 implies that the likelihood-tuned density estimator was superior to the corresponding density estimator.

When $n = 100$, $\hat{f}_{\mathrm{EM2}}$ beat $\hat{f}_{\mathrm{KER}}$ for eight out of 12 densities, it beat $\hat{f}_{\mathrm{ABW}}$ for nine densities, and it beat $\hat{f}_{\mathrm{MBC}}$ for seven densities. When $n = 500$, for 10 of 12 densities, $\hat{f}_{\mathrm{EM2}}$ had the smaller mean ISE than $\hat{f}_{\mathrm{KER}}$ and $\hat{f}_{\mathrm{ABW}}$. However, $\hat{f}_{\mathrm{MBC}}$ outperformed $\hat{f}_{\mathrm{EM2}}$ for six out of 12 densities.

Figure 4 shows the relative root mean ISE of $\hat{f}_{\mathrm{KER}}$, $\hat{f}_{\mathrm{ABW}}$ and $\hat{f}_{\mathrm{MBC}}$ compared to $\hat{f}_{\mathrm{EM2}}$ for eight Gaussian mixture distributions at two sample sizes, $n = 100$ in plot (a) and $n = 500$ in plot (b). Since the mean ISE of $\hat{f}_{\text{EM2}}$ is in the denominator, a relative root mean ISE greater than one implies that $\hat{f}_{\text{EM2}}$ was superior to the competitor for the corresponding distribution in terms of the mean ISE. The thin solid line represents the ratio one. The horizontal axis represents eight Gaussian mixture distributions ordered so that the relative root mean ISE of $\hat{f}_{\text{MBC}}$ was increasing. The $\hat{f}_{\text{MBC}}$ estimator turned out to be the most competitive density estimator for Gaussian mixture distributions.

For the Gaussian mixture densities, $\hat{f}_{\text{EM2}}$ was superior to $\hat{f}_{\text{KER}}$ at both sample sizes while $\hat{f}_{\text{MBC}}$ performed somewhat better than $\hat{f}_{\text{EM2}}$ at both. When $n = 100$ in Figure 4 (a), one can observe that six out of eight points of $\hat{f}_{\text{KER}}$ are above the thin solid line. In the case of $n = 500$ in Figure 4 (b), all points of $\hat{f}_{\text{KER}}$ are greater than one. On the other hand, a half of relative root mean ISEs of $\hat{f}_{\text{MBC}}$ were less than one when $n = 100$ and six out of eight were less than one when $n = 500$. Thus we can conclude that $\hat{f}_{\text{MBC}}$ worked better than $\hat{f}_{\text{EM2}}$ for Gaussian mixture distributions. Based on the trend, we expect that, as the sample size increases, the performance of $\hat{f}_{\text{EM2}}$ and $\hat{f}_{\text{MBC}}$ would be enhanced relative to $\hat{f}_{\text{KER}}$ and $\hat{f}_{\text{EM2}}$, respectively. The performance of $\hat{f}_{\text{ABW}}$ relative to $\hat{f}_{\text{EM2}}$ was mixed, but $\hat{f}_{\text{EM2}}$ was superior in more cases.

Figure 5 includes relative root mean ISE for non-Gaussian distributions. In these plots, the horizontal axis is in increasing order of the relative root mean ISE of $\hat{f}_{\text{KER}}$, which performed best for the non-Gaussian distributions. For the non-Gaussian distributions, the result was opposite to the one for Gaussian mixture distributions. $\hat{f}_{\text{KER}}$ performed better than $\hat{f}_{\text{EM2}}$ while $\hat{f}_{\text{MBC}}$ performed worse than $\hat{f}_{\text{EM2}}$, and this pattern is clearest when the sample size is large. In Figures 5 (a) and (b), a half of relative root mean ISEs of $\hat{f}_{\text{KER}}$ are below the thin solid line whereas all points of $\hat{f}_{\text{MBC}}$ are above the line. For non-Gaussian distributions, $\hat{f}_{\text{ABW}}$ was inferior to $\hat{f}_{\text{EM2}}$. In Figure 5, all points are greater than one regardless of the sample size.

In conclusion, the two-step likelihood-tuned density estimator was not the best for the Gaussian mixture densities or for the non-Gaussian densities, but it was more robust than the other estimators. The adaptive bandwidth density estimator was worse than the likelihood-tuned for both the Gaussian mixtures and the non-Gaussians. The kernel density estimator performed well for the non-Gaussians while it worked poorly for the Gaussian mixtures. On the other hand, the multiplicative bias correction estimator beat all the others for the Gaussian mixtures while it did not perform well for the non-Gaussians. However, the likelihood-tuned density

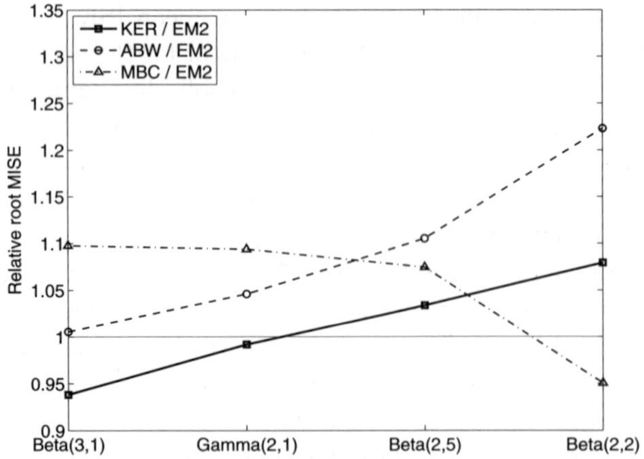

(a) $n = 100$

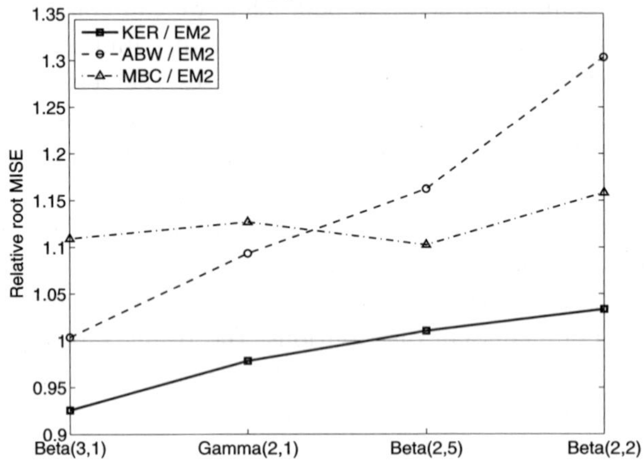

(b) $n = 500$

Fig. 5. Relative root mean ISE for $\hat{f}_{\rm KER}$, $\hat{f}_{\rm ABW}$ and $\hat{f}_{\rm MBC}$ vs. $\hat{f}_{\rm EM2}$ for non-Gaussian distributions. Dotted line represents the ratio 1.

estimator showed a robust performance for both types of density.

5. Discussion

The likelihood-tuned density estimator is a two-step EM estimator, taking the uniform density as a prior for the mixing density $\pi(\phi)$. However, this procedure can be modified by applying a different prior density or proceeding with further tuning steps. By experiment, we found that using $N(\bar{X}, S^2 - h^2)$ as the initial Π_0 results in an estimator that is more robust against outliers than the two-step likelihood-tuned density estimator proposed here. We also suspect it would further debias the estimator in the Gaussian case. As mentioned in Section 2.2, further tuning steps could possibly improve asymptotic properties, but at a heavy computational cost.

The likelihood-tuning procedure and the resulting density estimator can be naturally extended to the multi-dimensional case. Compared to the adaptive bandwidth density estimator, the performance of the likelihood-tuned density estimator is expected to be even better than in the univariate case. This is because the adaptive bandwidth density estimator has a disadvantage in sparse areas, a situation made worse in multiple dimensions. Compared to the multiplicative bias correction estimator, we expect the likelihood-tuned estimator to be still more robust against the type of the true density.

Appendix A. Outline proof of Theorem 1

The likelihood-tuned density estimator in (9) can be expressed as

$$\hat{f}_{\mathrm{EM2}}(x) = n^{-2}\sum_{i=1}^{n}\sum_{j=1}^{n}\hat{f}_{(1)}(X_j)^{-1}\int K_h(X_i,y)K_h(X_j,y)K_h(y,x)dy$$

if we do not expand terms in the second tuning step. Then the expectation of $\hat{f}_{\mathrm{EM2}}(x)$ is

$$\int\int \frac{\int K_h(x_i,y)K_h(x_j,y)K_h(y,x)dy}{\hat{f}_{(1)}(x_j)}dF(x_i)dF(x_j), \qquad \text{(A.1)}$$

where F is a distribution function for the true density f. The numerator of the integrand in (A.1) is regarded as a conditional density of (x_i, x_j) given y. This conditional density can be factored into two terms:

$$K(x_i|x_j,x) \;=\; N\left(\frac{x+x_j}{2},\frac{3}{2}h^2\right) \;\text{ and }\; K(x_j|x) \;=\; N(x,2h^2).$$

Using these terms, we rewrite (A.1) as

$$\int\left\{\int K(x_i|x_j,x)f(x_i)dx_i\right\}\frac{K(x_j|x)f(x_j)dx_j}{\int K_{h\sqrt{2}}(x_k,x_j)f(x_k)dx_k}, \qquad \text{(A.2)}$$

where $K_h(x,y) = h^{-1}\varphi(h^{-1}(x-y))$ with φ denoting the standard normal density. Define $B_1(x)$, $B_2(x)$, and $B_3(x)$ as two numerators and a denominator as follows and change variables x_i, x_j, and x_k into z_1, z_2, and z_3 that follow the standard normal density $\varphi(\cdot)$:

$$\begin{aligned} B_1(x) &= \int K(x_i|x_j,x)f(x_i)dx_i = \int \varphi(z_1)\, f\left(\frac{x+x_j}{2} + z_1 h\sqrt{\frac{3}{2}}\right) dz_1; \\ B_2(x) &= K(x_j|x)f(x_j)dx_j = \varphi(z_2)f(x+z_2h\sqrt{2})dz_2; \\ B_3(x) &= \int K_{h\sqrt{2}}(x_k,x_j)f(x_k)dx_k = \int \varphi(z_3)f(x_j+z_3h\sqrt{2})dz_3. \end{aligned}$$

After we can expand $f(\cdot)$ in B_1, B_2, and B_3 around $h=0$, $B_1(x)$ and $B_3(x)$ are integrated over z_1 and z_3, respectively. Now there remains only one random variable, z_2, in three terms. By multiplying $B_2(x)/B_3(x)$ and $B_1(x)$ and integrating over z_2, the density estimator (A.2) is written as

$$\int \varphi(z_2)\{f(x)+a_1(z_2)h+a_2(z_2)h^2+a_3(z_2)h^3+a_4(z_2)h^4+O(h^5)\}dz_3$$

for appropriate coefficients $a_1(z_3),\cdots,a_4(z_3)$. With some calculation, we find that

$$\int a_1(z_2)dz_2 = \int a_2(z_2)dz_2 = \int a_3(z_2)dz_2 = 0$$

and

$$\int a_4(z_2)dz_2 = -f^{(4)}(x) + \frac{1}{f(x)}\left(f'(x)f^{(3)}(x) + f''^2(x)\right) - \frac{1}{f^2(x)}f''(x)f'^2(x).$$

Thus we attain

$$\begin{aligned} E\left[\hat{f}_{\mathrm{EM2}}(x)\right] &= f(x)\left(-\frac{f^{(4)}(x)}{f(x)} + \frac{f^{(3)}(x)f'(x)+f''(x)^2}{f(x)^2} - \frac{f''(x)f'(x)^2}{f(x)^3}\right)h^4 \\ &\quad + f(x) + o(h^4). \end{aligned}$$

For the asymptotic variance, rewrite (A.2) as a functional on the distribution F as

$$T(F) = \int\int K_{h\sqrt{3/2}}\left(\frac{x+x_j}{2}, x_i\right) dF(x_i) \frac{K_{h\sqrt{2}}(x_j,x)dF(x_j)}{\int K_{h\sqrt{2}}(x_k,x_j)dF(x_k)}.$$

We will find the first von Mises derivative $T'(y)$, and then use the result that

$$T(\hat{F}) - T(F) \approx \int T'(y)d(\hat{F}-F)$$

so that the asymptotic variance for $T(\hat{F})$ is

$$\mathrm{asyvar}(T(\hat{F})) = \mathrm{Var}_F(T'(y))/n.$$

The first von Mises derivative $T'(y)$ is a sum of three terms:

$$\int \frac{\int K_{h\sqrt{3/2}}(\frac{x+x_j}{2}, x_i)dF(x_i)}{\int K_{h\sqrt{2}}(x_k, x_j)dF(x_k)} K_{h\sqrt{2}}(x_j, x)d\Delta(x_j),$$

$$\int \frac{\int K_{h\sqrt{3/2}}(\frac{x+x_j}{2}, x_i)d\Delta(x_i)}{\int K_{h\sqrt{2}}(x_k, x_j)dF(x_k)} K_{h\sqrt{2}}(x_j, x)dF(x_j), \text{ and}$$

$$-\int \frac{\int K_{h\sqrt{3/2}}(\frac{x+x_j}{2}, x_i)dF(x_i) \int K_{h\sqrt{2}}(x_k, x_j)d\Delta(x_k)}{\left(\int K_{h\sqrt{2}}(x_k, x_j)dF(x_k)\right)^2} K_{h\sqrt{2}}(x_j, x)dF(x_j),$$

where the measure $d\Delta(x) = d\delta_y(x) - dF(x)$ and δ_y is the distribution degenerate at y. We can rewrite by letting

$$C_1(y) = \frac{\int K_{h\sqrt{3/2}}(\frac{x+y}{2}, x_i)dF(x_i)}{\int K_{h\sqrt{2}}(x_k, y)dF(x_k)} K_{h\sqrt{2}}(y, x),$$

$$C_2(y) = \int \frac{K_{h\sqrt{3/2}}(\frac{x+x_j}{2}, y)}{\int K_{h\sqrt{2}}(x_k, x_j)dF(x_k)} K_{h\sqrt{2}}(x_j, x)dF(x_j),$$

$$C_3(y) = -\int \frac{\int K_{h\sqrt{3/2}}(\frac{x+x_j}{2}, x_i)dF(x_i) \cdot K_{h\sqrt{2}}(y, x_j)}{\left(\int K_{h\sqrt{2}}(x_k, x_j)dF(x_k)\right)^2} K_{h\sqrt{2}}(x_j, x)dF(x_j),$$

and noting that $T'(y) = C_1(y) + C_2(y) + C_3(y) - E[C_1(y) + C_2(y) + C_3(y)]$.

Further, $E[C_1(y) + C_2(y) + C_3(y)]$ is just the asymptotic mean calculated above. Thus we seek $E[C_1(y) + C_2(y) + C_3(y)]^2$. By expanding and integrating as in calculating the asymptotic mean, we have the following limiting results:

$$h\int C_1(y)^2 f(y)dy \to f(x)\frac{1}{\sqrt{2}}\int \varphi(z)^2 dz,$$

$$h\int C_2(y)^2 f(y)dy \to f(x)\frac{1}{\sqrt{2}}\int \varphi^2(z)dz,$$

$$h\int C_3(y)^2 f(y)dy \to f(x)\frac{1}{2}\int \varphi^2(z)dz,$$

$$h\int C_1(y)C_2(y)f(y)dy \to f(x)\frac{1}{\sqrt{2}}\int \varphi^2(z)dz,$$

$$h\int C_1(y)C_3(y)f(y)dy \to -f(x)\frac{1}{\sqrt{3}}\int \varphi^2(z)dz, \text{ and}$$

$$h\int C_2(y)C_3(y)f(y)dy \to -f(x)\frac{1}{\sqrt{3}}\int \varphi^2(z)dz.$$

By adding three square terms and three cross terms twice, we have the asymptotic variance,

$$\mathrm{Var}\left[\hat{f}_{\mathrm{EM2}}(x)\right] = (nh)^{-1}f(x)\left(\frac{1}{2}+\frac{4}{\sqrt{2}}-\frac{4}{\sqrt{3}}\right)\int\varphi^2(z)dz + o\left((nh)^{-1}\right).$$

References

Abramson, I. S. (1982). On bandwidth variation in kernel estimates-a square root law, *The Annals of Statistics* **10**, 4, pp. 1217–1223.

Bowman, A. W. and Foster, P. J. (1993). Adaptive smoothing and density-based tests of multivariate normality, *Journal of the American Statistical Association* **88**, 422, pp. 529–537.

Breiman, L., Meisel, W. and Purcell, E. (1977). Variable kernel estimates of multivariate densities, *Technometrics* **19**, 2, pp. 135–144.

Di Marzio, M. and Taylor, C. C. (2004). Boosting kernel density estimates: A bias reduction technique? *Biometrika* **91**, 1, pp. 226–233.

Goutis, C. (1997). Nonparametric Estimation of a Mixing Density Via the Kernel Method. *Journal of the American Statistical Association* **92**, 440, pp. 1445–1450.

Jones, M., Linton, O. and Nielsen, J. (1995). A simple bias reduction method for density estimation, *Biometrika* **82**, 2, pp. 327–338.

Jones, M. and Signorini, D. F. (1997). A comparison of higher-order bias kernel density estimators. *Journal of the American Statistical Association* **92**, 439, pp. 1063–1073.

Laird, N. M. (1978). Nonparametric maximum likelihood estimation of a mixing distribution, *Journal of the American Statistical Association* **73**, 1978, pp. 805–811.

Laird, N. M. and Louis, T. A. (1991). Smoothing the non-parametric estimate of a prior distribution by roughening: A computational study, *Computational Statistics and Data Analysis* **12**, pp. 27–37.

Lindsay, B. G. (1995). *Mixture Models: Theory, Geometry, and Applications* (Institute of Mathematical Statistics).

Marron, J. S. and Wand, M. P. (1992). Exact mean integrated squared error, *The Annals of Statistics* **20**, 2, pp. 712–736.

Silverman, B. W. (1986). *Density Estimation for Statistics and Data Analysis* (Chapman & Hall/CRC).

Silverman, B. W., Jones, M. C., Wilson, J. D. and Nychka, D. W. (1990). A smoothed EM approach to indirect estimation problems, with particular reference to stereology and emission tomography, *Journal of the Royal Statistical Society, Series B* **52**, 2, pp. 271–324.

Thomas, H. and Hettmansperger, T. P. (2001). Modelling change in cognitive understanding with finite mixtures, *Journal of the Royal Statistical Society, Series C* **50**, 4, pp. 435–448.

Vardi, Y. and Lee, D. (1993). From image deblurring to optimal investments: Maximum likelihood solutions for positive linear inverse problems, *Journal of the Royal Statistical Society, Series B* **55**, 3, pp. 569–612.

Vardi, Y., Shepp, L. A. and Kaufman, L. (1985). A statistical model for positron emission tomography (with discussion), *Journal of the American Statistical Association* **80**, pp. 8–37.

Shock Models for Defaults: Parametric and Nonparametric Approaches

Pasquale Cirillo[†] & Jürg Hüsler[‡]

Institut für Mathematische Statistik und Versicherungslehre
University of Bern, Bern, Switzerland
[†] *E-mail: pasquale.cirillo@stat.unibe.ch*
[‡] *E-mail: huesler@stat.unibe.ch*

We review shock models for defaults, both in the standard and in the urn-based approach. Shock models are motivated by engineering problems where a material can break because of stress, but they can also be efficiently used in other fields, such as economics and biology.

Standard shock models are parametric models, whereas the urn-based shock models are nonparametric models assuming as little as possible for the prediction of defaults. First, we mention some results for the two models, in particular describing the finite and asymptotic behavior of time to failure or of the chance for defaults. Finally, we also present an application of the second model to defaults of Italian firms, comparing our results with a standard prediction model of economics, the Z-score. We note that our model predicts the default behavior of these firms better.

Keywords: Convergence; Default; Extremes; Firm's failure; First passage times; Moments; Renewal theory; Shock; Stopped random walk

1. Introduction

The setup in extreme shock models is a family $\{(X_k, Y_k),\ k \geq 0\}$ of independent, identically distributed (i.i.d.) two-dimensional random vectors, with X_k the magnitude of the k th shock and Y_k the time between the $(k-1)$ th and the k th shock. The main object of interest is the lifetime/failure time of the system, assuming certain schemes for the failure. These models are motivated by the possible breakdown of a material or by a system subject to random shocks of random magnitude, as occur in engineering. It is easy to see useful applications also in other fields, such as economics, medicine and biology.

Cumulative shock models and extreme shock models, as well as mixtures of both models, are discussed by Gut and Hüsler (1999, 2005) and the references therein.

In the cumulative shock model, we consider

$$T_n = \sum_{k \le n} Y_k \quad \text{and} \quad S_n = \sum_{k \le n} X_k,$$

for $n \ge 1$, with $T_0 = S_0 = 0$. The failure of the system occurs if $S_n > \alpha$ for some n and α. Here α denotes the critical level of the system. The time until the system fails the first time or the failure time T_τ with $\tau = \min\{n : S_n > \alpha\}$ are then of interest. For results, see, e.g., Gut (1990).

In the simple extreme shock model, one large or extreme shock, larger than a given threshold γ, may cause the failure of the system. The lifetime of the system is in this case defined as T_ν, where

$$\nu = \min\{n : X_n > \alpha\}. \tag{1}$$

This model was dealt with in Gut and Hüsler (1999) and Gut (2001). Gut and Hüsler (2005) extended this extreme shock model to more realistic shock models assuming that the threshold is not a constant, but may vary with time depending on the experienced shocks. They assumed that the threshold is decreasing in time. This is quite sensible but in some applications the threshold might even increase, say in particular in the beginning or break-in period of some equipment. We present in the following section this more realistic model with some theoretical results, which are based on parametric assumptions.

In Section 3, we derive some exact and some asymptotic univariate and multivariate distributions of the parametric model.

Shock models can even be studied using nonparametric approaches. In Section 4, we discuss an urn-based version of the generalized extreme shock model of Section 2 to investigate defaults. This kind of models is more nonparametric, assuming as little as possible for the prediction of defaults of firms. Such models are rather simple but generate quite general results, which can be conveniently used to study failures and, in particular, economical phenomena. Standard shock models are possibly less appropriate for application to economics, because too many parameters should be known.

In Section 5, we present an application of the urn-based model for the prediction of some Italian firms' defaults, comparing our results with a standard benchmark used by practitioners.

2. Extreme shock models with varying threshold

With the critical threshold α, the breakdown or failure of a system, a firm, or a piece of technical equipment occurs because of one large load on the system which is larger than α. But with age, the system can be partially damaged or worn out because of obsolescence, such that the critical load is no longer the designed load. This is realistic in particular if a large load is damaging the structure, e.g., producing some cracks in the structure which weakens the system. This means that the critical threshold for breakdown is decreasing with the number of large but not fatal shocks. Such models were proposed and theoretically discussed in Gut and Hüsler (2005). In some applications, a system has a break-in period during which the critical load can increase and the structure is strengthened because at the beginning the loads or shocks are large but non-fatal. This may happen in particular until the first damage or crack. After such an event, the system can only be weakened. Such a pattern can be modeled as follows.

We let the arrival times T_i of the shocks X_i be, as mentioned, a partial sum of i.i.d. inter-arrival times Y_j with distribution G. The loads $X_i, i \geq 1$, are an i.i.d. sequence of r.v.'s with distribution F. A shock or stroke X_i strengthens the material if $X_i \in [\gamma, \beta)$. At the beginning, the material supports a maximal load α, the critical threshold. After a strengthening stroke, the maximal load becomes larger, say $\alpha_1 = \alpha + b_1$ with $b_1 > 0$. This boundary increases with each strengthening stroke, inducing boundaries $\alpha_j = \alpha + b_j$, $j \geq 1$ with $b_j \uparrow$. After the first harmful stroke larger than β but smaller than the critical level at this time point, the load boundary decreases because of possible cracks or some weakening of the material. If the boundary has reached the level α_k (because of k strengthening strokes before the first harmful stroke), the critical level now becomes $\alpha_k - c_1$, and it decreases further with the next harmful, nonfatal strokes to $\alpha_k - c_2$, $\alpha_k - c_3$, ..., with $c_j \uparrow (\geq 0)$. This is shown in Figure 1. There might be an upper load limit α^* for the α_k, as well a lower load limit α_* for the $\alpha_k - c_l$, where we take $\alpha_* \geq \beta$. It is convenient to set $b_0 = c_0 = 0$.

For notational reasons, we define the number $N_-(n)$ of weakening shocks X_i, $i < n$, before the nth shock,

$$N_-(n) = \sum_{i<n} 1(X_i \in [\beta, \alpha + b_{N_+(i)} - c_{N_-(i)}),$$

with $N_-(0) = 0$, and the number $N_+(n)$ of strengthening strokes X_i, $i < n$, before the nth shock and before the first weakening or fatal shock,

$$N_+(n) = \sum_{i<n} 1(X_i \in [\gamma, \beta), N_-(i) = 0),$$

with $N_+(0) = 0$. Note that the critical boundary for X_i is $\alpha_i = \alpha + b_{N_+(i)} - c_{N_-(i)}$. In addition, let W be the index of the first harmful shock larger than β. It could even indicate a fatal shock being larger than the critical boundary at this time point. Hence, if such a shock occurs,

$$W = \inf\{i : X_i \geq \beta\} \leq \infty.$$

If the set is empty, we set $W = \infty$. Hence, the shock X_i has no impact if $X_i \leq \gamma$, it induces a strengthening of the material if $i < W$ and $X_i \in [\gamma, \beta)$, and it is fatal if $X_i \geq \alpha_i = \alpha + b_{N_+(i)} - c_{N_-(i)}$. Note also that $N_+(k) = N_+(W)$ for all $k \geq W(< \infty)$.

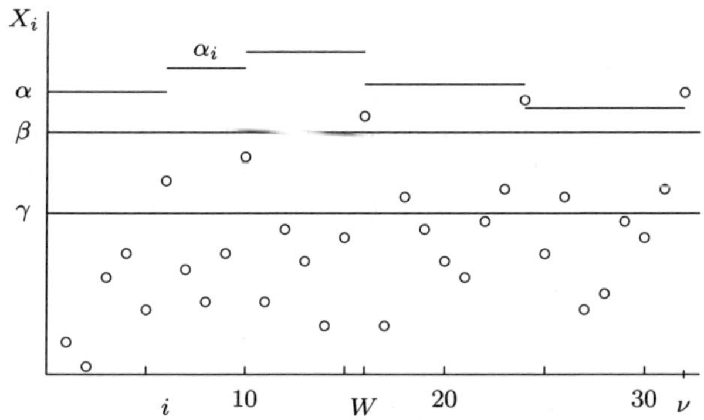

Fig. 1. Realization of a sequence of shocks with strengthening and weakening load limits, depending on the values X_i. Here we have $\nu = 32, W = 16, N_+(\nu) = 2, N_-(\nu) = 2$.

First, we analyze in Section 3 the distribution of the number ν of shocks until the first fatal shock:

$$\nu = \min\{i : X_i \geq \alpha + b_{N_+(i)} - c_{N_-(i)}\},$$

where we take $b_0 = 0$ and $c_0 = 0$. The time until the fatal shock is thus T_ν. Its distribution depends on the distribution of ν and G.

We consider the asymptotic behavior of these random variables by letting the parameters $\alpha = \alpha(t), \beta = \beta(t)$ and $\gamma = \gamma(t)$ tend to $x_F \leq \infty$ as $t \to \infty$, where x_F denotes the upper endpoint of the distribution F. We assume that x_F is a continuity point of F. For the limit distributions, certain additional restrictions will be imposed on the b_k and c_k, which are also dependent on t. Hence $\nu = \nu(t)$ will tend to ∞ in general, depending on the underlying distribution F.

3. Distribution for ν

The distribution of ν can be derived in this more general model as in the simpler model dealt with in Gut and Hüsler (2005). First, we derive the exact distribution and then we analyze the asymptotic distributions that depend on the behavior of the sequences b_k and c_k. For the derivation, we use the notation

$$\alpha_{k,l} = \alpha + b_k - c_l$$

for any $k, l \geq 0$.

3.1. *Exact distribution*

To derive $P\{\nu > m\}$, we have to condition on the other random variables. If $N_-(m+1) = 0$, then $W > m$ and $P\{\nu > m, N_-(m+1) = 0\}$ is simply $F^m(\beta)$. If $N_-(m+1) = l > 0$ with $k < j$ and $m \geq j + l - 1$, we consider the joint distribution

$$\begin{aligned} P\{\nu > m, N_+(m) = k, N_-(m+1) = l, W = j\} \\ = \binom{j-1}{k} F^{j-1-k}(\gamma)\,[\bar{F}(\gamma) - \bar{F}(\beta)]^k \\ \times \binom{m-j}{l-1} F^{m-j-l+1}(\beta) \prod_{h=1}^{l} [\bar{F}(\beta) - \bar{F}(\alpha_{k,h-1})] \end{aligned} \tag{2}$$

or, for $m \geq j + l$ and $k < j$,

$$\begin{aligned} P\{\nu = m, N_+(m) = k, N_-(m) = l, W = j\} \\ = \binom{j-1}{k} F^{j-1-k}(\gamma)\,[\bar{F}(\gamma) - \bar{F}(\beta)]^k \\ \times \binom{m-j-1}{l-1} F^{m-j-l}(\beta) \prod_{h=1}^{l} [\bar{F}(\beta) - \bar{F}(\alpha_{k,h-1})]\bar{F}(\alpha_{k,l})\,. \end{aligned} \tag{3}$$

If $k \geq j$ or $m \leq j + l - 1$, the latter probabilities are 0.

By summing the appropriate terms, we get the exact univariate and multivariate distributions for ν, $N_+(m)$, $N_-(l)$, and W, as well as for $N_+(\nu)$ and $N_-(\nu)$. For instance, the joint distribution of $N_+(\nu)$ and $N_-(\nu)$ with

$l \geq 1$ is

$$P\{N_+(\nu) = k, N_-(\nu) = l\} = \sum_m P\{\nu = m, N_+(m) = k, N_-(m) = l\}$$

$$= \sum_{m,j} \binom{j-1}{k} \binom{m-j-1}{l-1} F^{j-k-1}(\gamma)[\bar{F}(\gamma) - \bar{F}(\beta)]^k F^{m-j-l}(\beta)$$

$$\times \prod_{h=0}^{l-1} \{1 - \bar{F}(\alpha_{k,h})/\bar{F}(\beta)\} \bar{F}^l(\beta) \bar{F}(\alpha_{k,l})$$

$$= \sum_{j \geq k+1} \binom{j-1}{k} F^{j-k-1}(\gamma) \sum_{m \geq j+l} \binom{m-j-1}{l-1} F^{m-j-l}(\beta) \bar{F}^l(\beta)$$

$$\times [\bar{F}(\gamma) - \bar{F}(\beta)]^k \prod_{h=0}^{l-1} \{1 - \bar{F}(\alpha_{k,h})/\bar{F}(\beta)\} \bar{F}(\alpha_{k,l}). \quad (4)$$

Note that the sums are summing all the probabilities of a negative binomial distribution, hence

$$P\{N_+(\nu) = k, N_-(\nu) = l\} = \sum_{h \geq 0} \binom{h+k}{k} F^h(\gamma)$$

$$\times [\bar{F}(\gamma) - \bar{F}(\beta)]^k \prod_{h=0}^{l-1} \{1 - \bar{F}(\alpha_{k,h})/\bar{F}(\beta)\} \bar{F}(\alpha_{k,l})$$

$$= \bar{F}^{-k-1}(\gamma)[\bar{F}(\gamma) - \bar{F}(\beta)]^k \prod_{h=0}^{l-1} \{1 - \bar{F}(\alpha_{k,h})/\bar{F}(\beta)\} \bar{F}(\alpha_{k,l})$$

$$= [1 - \bar{F}(\beta)/\bar{F}(\gamma)]^k \prod_{h=0}^{l-1} \{1 - \bar{F}(\alpha_{k,h})/\bar{F}(\beta)\} [\bar{F}(\alpha_{k,l})/\bar{F}(\gamma)]. \quad (5)$$

For the case $l = 0$, we get in the same way

$$P\{N_+(\nu) = k, N_-(\nu) = 0\}$$

$$= \sum_{m > k} P\{\nu = m, N_+(m) = k, N_-(m) = 0, W = m\}$$

$$= \sum_{m > k} \binom{m-1}{k} F^{m-k-1}(\gamma)[\bar{F}(\gamma) - \bar{F}(\beta)]^k \bar{F}(\alpha_{k,0})$$

$$= [1 - \bar{F}(\beta)/\bar{F}(\gamma)]^k [\bar{F}(\alpha_{k,0})/\bar{F}(\gamma)]. \quad (6)$$

Other exact distributions can be derived in the same way by appropriate summation. For the derivation of the asymptotic distributions, we sometimes have to approximate these sums to simplify the formulas. We note for later use the following exact distribution:

$$
\begin{aligned}
P\{\nu > m, N_-(\nu) > 0\} \\
&= \sum_{k\geq 0, l>0, j\geq 1} P\{\nu > m, N_+(m) = k, N_-(m+1) = l, W = j\} \\
&= \sum_{k\geq 0, l>0, j\geq 1} \binom{j-1}{k} F^{j-1-k}(\gamma)\,[\bar{F}(\gamma) - \bar{F}(\beta)]^k \\
&\quad \times \binom{m-j}{l-1} F^{m-j-l+1}(\beta) \prod_{h=1}^{l} [\bar{F}(\beta) - \bar{F}(\alpha_{k,h-1})]. \qquad (7)
\end{aligned}
$$

3.2. *Asymptotic distribution*

For the asymptotic behavior, let $\alpha(t) \to \infty, \beta(t) \to \infty$, and $\gamma(t) \to \infty$ as $t \to \infty$. The asymptotic behavior also depends on the assumptions of the sequences b_k and c_k, which may also depend on the parameter t. But mostly, we do not indicate the dependence on the parameter t. From the above finite distributions, it is reasonable to use the conditions $\bar{F}(\beta(t)) \to 0$ as $t \to \infty$,

$$\lim_t \frac{\bar{F}(\beta(t))}{\bar{F}(\gamma(t))} = g \in [0, 1], \qquad (8)$$

and

$$\lim_t \frac{\bar{F}(\alpha(t) + b_k(t) - c_l(t))}{\bar{F}(\beta(t))} = \lim_t \frac{\bar{F}(\alpha_{k,l}(t))}{\bar{F}(\beta(t))} = a_{k,l} \in [0, 1]. \qquad (9)$$

Obviously, the $a_{k,l}$ are monotone by the assumed monotonicity of the sequences b_k and c_k, i.e., monotone decreasing in k with l fixed, and monotone increasing in l with k fixed. We consider only the interesting cases with $g, a_{k,l} \in (0, 1)$. The simplest case occurs if $a_{k,l} = a \in (0, 1)$ for all k, l. This implies that

$$\prod_{h=1}^{l} [\bar{F}(\beta) - \bar{F}(\alpha + b_k - c_{h-1})] \sim [\bar{F}(\beta)(1-a)]^l$$

for each $l \geq 1$, where here and throughout this article, "$\sim$" means "asymptotically equal to" in the sense that the ratio converges to 1 as $t \to \infty$. But in general, we approximate

$$\prod_{h=1}^{l} [\bar{F}(\beta) - \bar{F}(\alpha + b_k - c_{h-1})] \sim \bar{F}^l(\beta) \prod_{h=1}^{l} (1 - a_{k,h})$$

for each $l \geq 1$. For notational reasons, we also define $\prod_{h=1}^{0}(1 - a_{k,h}) = 1$. To simplify the notation, we do not indicate the dependence on t in the following, e.g., we write ν instead of $\nu(t)$.

Theorem 3.1. *If (8) and (9) hold with $g, a_{k,l} \in [0, 1]$, then for any $k \geq 0$ and $l \geq 0$,*

$$P\{N_+(\nu) = k, N_-(\nu) = l\} \to g(1-g)^k \prod_{h=0}^{l-1}(1 - a_{k,h})a_{k,l}$$

as $t \to \infty$.

Proof. Use (5) for $l \geq 1$ and (6) for $l = 0$, with assumptions (8) and (9), to derive the claim immediately. □

For the particular simple case that $a_{k,h} = a$ for all k and h, the limit distribution above is the product of two geometric distributions:

$$P\{N_+(\nu) = k, N_-(\nu) = l\} \to g(1-g)^k(1-a)^l a.$$

Because of the assumptions, the number of strengthening strokes does not have an influence on the number of weakening strokes asymptotically, which shows the asymptotic independence of $N_+(\nu)$ and $N_-(\nu)$ in this case.

Theorem 3.2. *If (8) and (9) hold with $a_{k,l} \in (0, 1]$ for each k, l, then:*

(i) For $k \geq 0 = l$,

$$\lim_{t\to\infty} P\{\nu \geq z/\bar{F}(\beta), N_+(\nu) = k, N_-(\nu) = 0\} = \int_{z/g}^{\infty} v^k e^{-v} dv g(1-g)^k a_{k,0}/k!\,.$$

(ii) For $l \geq 1$ and $k \geq 0$,

$$\begin{aligned}
&\lim_{t\to\infty} P\{\nu \geq z/\bar{F}(\beta), N_+(\nu) = k, N_-(\nu) = l\} \\
&\quad = \int_z^{\infty}\int_0^1 y^k(1-y)^{l-1}\exp\{-yu(g^{-1}-1)\}dy\ \exp\{-u\}\,u^{k+l}du \\
&\qquad \times ((1-g)/g)^k \prod_{h=0}^{l-1}(1-a_{k,h})a_{k,l}/(k!(l-1)!).
\end{aligned}$$

Proof. Let $k \geq 0 = l$ and $z_\beta = z/\bar{F}(\beta)$. Then

$$
\begin{aligned}
&P\{\nu \geq z_\beta, N_+(\nu) = k, N_-(\nu) = 0\} = \\
&= \sum_{m \geq z_\beta} P\{\nu = m, N_+(m) = k, N_-(m) = 0\} \\
&= \sum_{m=z_\beta}^{\infty} \binom{m-1}{k} F^{m-k-1}(\gamma)[\bar{F}(\gamma) - \bar{F}(\beta)]^k \bar{F}(\alpha_{k,0}) \\
&\sim \sum_{m=z_\beta}^{\infty} \frac{m^k}{k!} F^m(\gamma)[\bar{F}(\gamma) - \bar{F}(\beta)]^k \bar{F}(\alpha_{k,0}) \\
&\sim \frac{1}{k!} \int_z^{\infty} v^k \exp\{-v\bar{F}(\gamma)/\bar{F}(\beta)\}\, dv\, (\bar{F}(\beta))^{-k-1} [\bar{F}(\gamma) - \bar{F}(\beta)]^k \bar{F}(\alpha_{k,0}) \\
&\to \frac{1}{k!} \int_{z/g}^{\infty} y^k e^{-y} dy\, (1-g)^k g\, a_{k,0}
\end{aligned}
$$

as $t \to \infty$. If $k = 0$, then the integral is simply $e^{-z/g}$.

Now let $l > 0$. By (3), we have

$$
\begin{aligned}
&P\{\nu \geq z_\beta, N_+(\nu) = k, N_-(\nu) = l\} \\
&= \sum_{m \geq z_\beta} P\{\nu = m, N_+(m) = k, N_-(m) = l\} = \\
&= \sum_{m=z_\beta}^{\infty} \sum_{j=k+1}^{m-l} \binom{j-1}{k} \binom{m-j-1}{l-1} F^{j-k-1}(\gamma) F^{m-j-l}(\beta) \\
&\qquad \times \bar{F}^k(\gamma)(1-g+o(1))^k \bar{F}^{l+1}(\beta) \prod_{h=0}^{l-1} (1 - a_{k,h} + o(1))(a_{k,l} + o(1)).
\end{aligned}
$$

Split the inner sum into three parts, with $j \leq \epsilon m$, $\epsilon m < j < (1-\epsilon)m$ and $j \geq (1-\epsilon)m$ where we used $W = j$ in the summands. The first and the third sum are asymptotically negligible as $\epsilon \to 0$, which can be shown to be similar to the following derivations for the second sum. For the second sum, we get

$$
\begin{aligned}
&P\{\nu \geq z_\beta, N_+(\nu) = k, N_-(\nu) = l, \epsilon m < W < (1-\epsilon)m\} = \\
&= \sum_{m=z_\beta}^{\infty} \sum_{j>\epsilon m}^{(1-\epsilon)m} \binom{j-1}{k} \binom{m-j-1}{l-1} F^{j-k-1}(\gamma) F^{m-j-l}(\beta) \\
&\qquad \times \bar{F}^k(\gamma) \bar{F}^{l+1}(\beta)(1-g+o(1))^k \prod_{h=0}^{l-1} (1 - a_{k,h} + o(1))(a_{k,l} + o(1))
\end{aligned}
$$

$$\sim \sum_{m=z_\beta}^{\infty} \sum_{j>\epsilon m}^{[(1-\epsilon)m]} \frac{(j-1)!(m-j-1)!}{k!(j-1-k)!(l-1)!(m-j-l)!} F^{j-k-1}(\gamma) F^{m-j-l}(\beta)$$

$$\times \bar{F}^k(\gamma)\bar{F}^{l+1}(\beta)(1-g)^k \prod_{h=0}^{l-1}(1-a_{k,h})a_{k,l}$$

$$\sim \sum_{m=z_\beta}^{\infty} \sum_{j>\epsilon m}^{[(1-\epsilon)m]} \frac{j^k(m-j)^{l-1}}{k!(l-1)!} F^{j-k-1}(\gamma) F^{m-j-l}(\beta)$$

$$\times \bar{F}^k(\gamma)\bar{F}^{l+1}(\beta)(1-g)^k \prod_{h=0}^{l-1}(1-a_{k,h})a_{k,l} =: P_t.$$

Now we approximate the sums by integrals:

$$P_t \sim \sum_{m=z_\beta}^{\infty} \int_{\epsilon}^{1-\epsilon} (ym)^k (m(1-y))^{l-1} F^{ym}(\gamma) F^{m(1-y)}(\beta) dy\, (m/k!(l-1)!)$$

$$\times \bar{F}^k(\gamma)\bar{F}^{l+1}(\beta)(1-g)^k \prod_{h=0}^{l-1}(1-a_{k,h})a_{k,l}$$

$$= \sum_{m=z_\beta}^{\infty} \int_{\epsilon}^{1-\epsilon} y^k (1-y)^{l-1} F^{ym}(\gamma) F^{m(1-y)}(\beta) dy\, m^{k+l}$$

$$\times \bar{F}^k(\gamma)\bar{F}^{l+1}(\beta)(1-g)^k \prod_{h=0}^{l-1}(1-a_{k,h})a_{k,l}/(k!(l-1)!)$$

$$\sim \int_{z}^{\infty} \int_{\epsilon}^{1-\epsilon} y^k\, (1-y)^{l-1} F^{yu/\bar{F}(\beta)}(\gamma) F^{(1-y)u/\bar{F}(\beta)}(\beta) dy\, u^{k+l} du$$

$$\times \bar{F}^k(\gamma)\bar{F}^{-k}(\beta)(1-g)^k \prod_{h=0}^{l-1}(1-a_{k,h})a_{k,l}/(k!(l-1)!)$$

$$\sim \int_{z}^{\infty} \int_{\epsilon}^{1-\epsilon} y^k (1-y)^{l-1} F^{yu/\bar{F}(\beta)}(\gamma) F^{(1-y)u/\bar{F}(\beta)}(\beta) dy\ u^{k+l} du$$

$$\times g^{-k}(1-g)^k \prod_{h=0}^{l-1}(1-a_{k,h})a_{k,l}/(k!(l-1)!).$$

The factor $F^{yu/\bar{F}(\beta)}(\gamma)F^{(1-y)u/\bar{F}(\beta)}(\beta)$ can be approximated for large t by

$$\exp(-(1+o(1))yu\bar{F}(\gamma)/\bar{F}(\beta)-(1+o(1))(1-y)u) \sim \exp(-[yu(g^{-1}-1)-u]),$$

uniformly for u bounded. By (1) and the dominated convergence theorem, we get

$$P_t \sim \int_z^\infty \int_\epsilon^{1-\epsilon} y^k(1-y)^{l-1} \exp\{-yu(g^{-1}-1)\}dy\ \exp\{-u\}\, u^{k+l}du$$
$$\times ((1-g)/g)^k \prod_{h=0}^{l-1}(1-a_{k,h})a_{k,l}/(k!(l-1)!)$$
$$\to \int_z^\infty \int_0^1 y^k(1-y)^{l-1} \exp\{-yu(g^{-1}-1)\}dy\ \exp\{-u\}\, u^{k+l}du$$
$$\times ((1-g)/g)^k \prod_{h=0}^{l-1}(1-a_{k,h})a_{k,l}/(k!(l-1)!)$$

as $\epsilon \to 0$. □

If we set $z = 0$, the integrals of both statements can be determined explicitly, which implies the result of Theorem 3.1 in both cases $l = 0$ and $l > 0$.

Other limit distributions are determined by summing the terms $P\{\nu = m, N_+(m) = k, N_-(\nu) = l, W = j\}$. Only under additional assumptions can these distributions be simplified. We deal with such a particular case that generalizes Theorem 5.1 in Gut and Hüsler (2005). The formulas and the sums or integrals can be simplified, for instance, if the impact of the strengthening strokes is asymptotically negligible, i.e., if $a_{k,h} = a_h$ for all k, h. In this case, we consider the limit distribution of ν and T_ν.

Theorem 3.3. *If (8) and (9) hold with $a_{k,h} = a_h \in [0,1]$ for each k and $h \geq 1$, then for $z_\beta = z/\bar{F}(\beta)$ with $z > 0$,*

$$P\{\nu > z_\beta\} \to \sum_{l=0}^{\infty} \frac{z^l}{l!} e^{-z} \prod_{h=0}^{l-1}(1-a_h)$$
$$= 1 - H(z)$$

and

$$P\{T_\nu > z_\beta\} \to 1 - H(z/\mu)$$

as $t \to \infty$, where $\mu = E(Y_1) < \infty$.

Proof. The assumptions imply that $\prod_{h=0}^{l-1}(1-a_{k,h}) = \prod_{h=0}^{l-1}(1-a_h) =: \tilde{a}_l$. Let $\tilde{a}_0 = 1$.

(i) For the first statement, we start with (7) and apply assumptions (8) and (9) to derive the limiting distribution.

$$\begin{aligned}
&P\{\nu > m, N_-(m) > 0\} \\
&= \sum_{j=1}^{m} \sum_{l=1}^{m-j+1} \sum_{k=0}^{j-1} \binom{j-1}{k} F^{j-1-k}(\gamma)\,[\bar{F}(\gamma) - \bar{F}(\beta)]^k \\
&\qquad \times \binom{m-j}{l-1} F^{m-j-l+1}(\beta) \prod_{h=1}^{l} [\bar{F}(\beta) - \bar{F}(\alpha_{k,h-1})] \\
&\sim \sum_{j=1}^{m} \sum_{l=1}^{m-j+1} \sum_{k=0}^{j-1} \binom{j-1}{k} F^{j-1-k}(\gamma)\,[\bar{F}(\gamma) - \bar{F}(\beta)]^k \\
&\qquad \times \binom{m-j}{l-1} F^{m-j-l+1}(\beta)\, \bar{F}^l(\beta)\tilde{a}_l.
\end{aligned}$$

Now, as the sum on k can be simplified since it is a binomial sum, we get for this sum $(F(\gamma) + \bar{F}(\gamma) - \bar{F}(\beta))^{j-1} = (F(\beta))^{j-1}$. Thus,

$$\begin{aligned}
&P\{\nu > m, N_-(m) > 0\} \\
&= \sum_{j=1}^{m} \sum_{l=1}^{m-j+1} F^{j-1}(\beta) \binom{m-j}{l-1} F^{m-j-l+1}(\beta)\bar{F}^l(\beta)\tilde{a}_l \\
&= \sum_{l=1}^{m} \sum_{j=1}^{m-l+1} \binom{m-j}{l-1} F^{m-l}(\beta)\bar{F}^l(\beta)\tilde{a}_l \\
&= \sum_{l=1}^{m} \binom{m}{l} F^{m-l}(\beta)\bar{F}^l(\beta)\tilde{a}_l \\
&\to \sum_{l=1}^{\infty} \frac{z^l}{l!} \exp(-z)\tilde{a}_l
\end{aligned}$$

using the normalization $m = z/\bar{F}(\beta)$, which tends to ∞.

(ii) The second statement is immediate by applying the weak law of large numbers to show that $T_\nu/\nu \to \mu$ in probability as $t \to \infty$. □

In other more general cases, we have to sum the terms of Theorem 3.2 to get the limit distribution of ν and T_ν. Notice also that the dependence between X_k and Y_k has no influence on the limit distribution of T_ν.

4. A nonparametric urn-based generalized extreme shock model

We now introduce a nonparametric version of the discussed generalized extreme shock model without defining the distribution of the shock X_i and without defining the changes of the critical levels, i.e., without defining the sequences b_k and c_k. We propose a more general, nonparametric model that can be easily applied, for example, to the failure of firms, but also in survival analysis and mortality studies.

Instead of the fatal domain, the non-fatal but harmful domain, the eventually improving domain, and the no-impact domain, we present an intuitive approach to the discussed generalized shock models by using a triangular urn process that evolves according to the sampling of balls of three different colors.

The idea is indeed to create three different risk areas for the system—no risk or safe, risky, and default, to link every area to a particular color, and to work with the probability for the process to enter each area. If every time the process enters the risky area, the probability of failing increases, and this can be obtained with a triangular reinforcement matrix, we can consider such a modelization a sort of intuitive approach to generalized extreme shock models, getting around the definition of the moving threshold. In some sense, reinforcing the probability for the system to fail is like making the risky threshold move down, and vice-versa.

Urn processes (or urn models or urn schemes) constitute a very large family of probabilistic models in which the probability of certain events is represented in terms of sampling, replacing and adding balls to one or more urns or boxes. Urns are particularly suitable, thanks to their efficiency, for studying chance experiments, especially when these are characterized by countable spaces. Moreover, they have powerful, elegant combinatorial properties that allow for general, complex results in a rather concise form. In particular, using the many relationships and isomorphisms between urn models and other well-known mathematical objects, the researcher can switch from one approach to the other at her/his convenience, as in the case of the analytic urns used here.

In this work, we make use of Polya-like urns, which is to say urns that are reinforced over time. This kind of urn is consistent with the Bayesian paradigm of prior specification, knowledge update thanks to empirical observations, and posterior calculation. In particular, prior knowledge is given by the urn composition at time 0. Then every time a ball is sampled and its color observed, the beliefs about the possibility of sampling another ball of the same color change and, specifically, increase, thanks to the reinforcement mechanism of the urn. Furthermore, given the initial composition and the updates, it is possible at every stage to perform a prediction about the possibility of picking a given ball (the posterior).

As stated above, in our model, we consider an urn containing balls of three different colors: x, y, and w. The x-balls are related to the safe state, y-balls to the risky state, and w-balls to the default state. The process evolves as follows.

(1) At time n a ball is sampled from the urn. The probability of sampling a particular ball depends on the urn composition after time $n-1$;
(2) According to the color of the sampled ball, the process enters (or remains in) one of the three states of risk. For example, if the sampled ball is of type x, the process is in a safe state, while it fails if the chosen ball is w;
(3) The urn is then reinforced according to its reinforcement matrix. It means that if the sampled ball is of type x, then θ x-balls are added to the urn; if the sampled ball is of type y, then δ y-balls and λ z-balls are added; and if the sampled ball is of type z, then θ z-balls are added. These numbers are given in the reinforcement matrix RM below.

In order to model the positive dependence between the risky and the default states, we use a particular sampling rule, expressed by the following balanced triangular reinforcement matrix, which is constant over time:

$$RM = \begin{matrix} & \begin{matrix} x & y & w \end{matrix} \\ \begin{matrix} x \\ y \\ w \end{matrix} & \begin{bmatrix} \theta & 0 & 0 \\ 0 & \delta & \lambda \\ 0 & 0 & \theta \end{bmatrix} \end{matrix}, \quad \text{where } \lambda = \theta - \delta. \tag{10}$$

In particular, we will also assume two simple conditions. Let

$$\theta\delta \neq 0 \tag{11}$$

to avoid degenerate cases, and let

$$\lambda = \theta - \delta \geq 0 \tag{12}$$

to have a non-negative dependence between y and w balls.

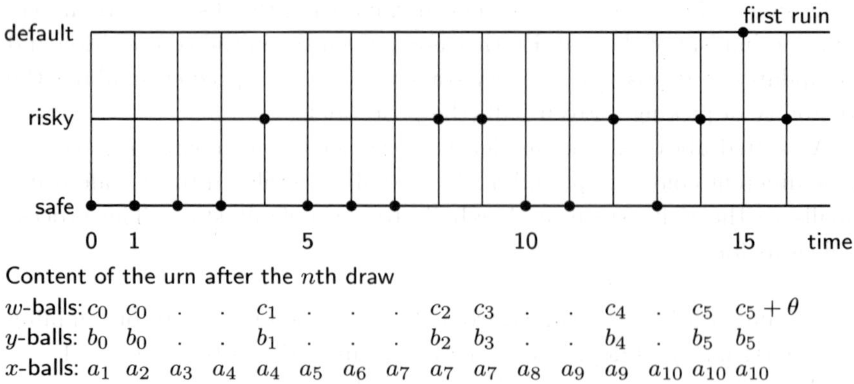

Fig. 2. Realization of the urn model process (UbGESM) with the configuration of the urn.

The distribution and the main properties of the urn process can be described analytically through the analysis of its generating function*; see the details in Cirillo and Hüsler (2009). In particular, we can quickly state the following theorems.

Theorem 4.1. *(Cirillo and Hüsler, 2009) Consider an urn process with a triangular reinforcement matrix RM that satisfies all the conditions introduced above and with an initial composition* (a_0, b_0, c_0) *of balls. The 4-variable generating function of urn histories is:*

$$H(z;x,y,w) = x^{a_0} y^{b_0} w^{c_0} (1-\theta x^{\theta} z)^{-\frac{a_0}{\theta}} (1-\theta w^{\theta} z)^{-\frac{c_0}{\theta}} \times \left(1 - y^{\delta} w^{-\delta}\left(1-(1-\theta w^{\theta} z)^{\frac{\delta}{\theta}}\right)\right)^{-\frac{b_0}{\delta}}.$$

Theorem 4.2. *(Cirillo and Hüsler, 2009) Let* X_n, Y_n *and* W_n *represent the number of* x-, y-, *and* w-*balls in the urn at time* n. *Their moments may be shown to be hypergeometric functions, that is, finite linear combinations of products and quotients of Euler gamma functions. In particular,*

*The generating function of urn histories (GFUH) is an exponential generating function of the form $H(z;x,y,w) = \sum_{n=0}^{\infty} f_n(x,y,w)\frac{z^n}{n!}$, where $f_n(x,y,w)$ is a counting function that counts the number of x-, y-, and w-balls in sampling sequences of length n. In other words, the GFUH is that generating function which enumerates all the possible compositions of the urn at time n, given its reinforcement matrix and initial composition.

the moments of order l are given by

$$E\left[(X_n)_l\right] = \theta^l \frac{\left(\frac{a_0}{\theta}\right)^{(l)}}{\left(\frac{t_0}{\theta}\right)^{(l)}} n^l + O(n^{l-1}),$$

$$E\left[(Y_n)_l\right] = \delta^l \frac{\left(\frac{b_0}{\delta}\right)^{(l)}}{\left(\frac{t_0}{\theta}\right)^{(l\frac{\delta}{\theta})}} n^{l\frac{\delta}{\theta}} + O(n^{(l-1)\frac{\delta}{\theta}}),$$

$$E\left[(W_n)_l\right] = \lambda^l \frac{\left(\frac{t_0-a_0}{\theta}\right)^{(l)}}{\left(\frac{t_0}{\theta}\right)^{(l\frac{\lambda}{\theta})}} n^{l\frac{\delta}{\theta}} + O(n^{(l-1)\frac{\delta}{\theta}}),$$

where $t_0 = a_0 + b_0 + c_0$, $\lambda = \theta - \delta$, $(x)^{(n)} = \frac{\Gamma(x+n)}{\Gamma(x)}$ represents the standard Pochhammer formula, and $(x)_l = \frac{\Gamma(x)}{\Gamma(x-l)}$.

Other results about the limit law of X_n, Y_n, and W_n can be found in Cirillo and Hüsler (2009), together with results about the asymptotic exchangeability of the triangular urn process.

For the application in the next section, we need the first and second moments of the number of balls in the urn, where the urn process starts with some initial configuration. Of interest is the first time when a w-ball is selected, which indicates a breakdown or default. For prediction of a default, we use the configuration of the urn at time n to predict the possible outcome at the next time point, $n+1$. The details are given in the next section. This model is called an urn-based generalized extreme shock model (UbGESM).

So for applications, we only have to fix and estimate the initial number of x-, y-, and w-balls and the parameters of the reinforcement matrix RM, which means θ, δ, and λ. Different estimation procedures for these parameters could be applied. Maximum likelihood estimation is quite involved because the joint distribution function of the numbers of colored balls is rather involved, and the algorithm for solving the ML-equations is slow. We apply simpler estimation methods.

5. Using the UbGESM to study firms' defaults

We now present an application of the urn-based generalized extreme shock model to study firms' defaults. We also show a first comparison of our model with a standard scoring model used by practitioners, the so-called Z-score approach, introduced in Altman (1968) and further developed in many subsequent studies. Here we use the Z^*-score specification of Altman (2005) as a benchmark.

Despite its simplicity, the UbGESM seems to predict failures and bankruptcies quite well. All of this represents a good incentive for future research.

5.1. *The Data*

For our analysis, we have used data from the CEBI database. CEBI is a comprehensive database first developed by the Bank of Italy and now maintained by Centrale dei Bilanci Srl (http://www.centraledeibilanci.it). It represents the biggest Italian industrial dataset and it contains firm-level observations and balance sheets of thousands of firms.

Our experiment has been performed with a simple random sample of 380 manufacturing firms, from a population of about 6000 companies, that respect the following conditions:

(1) All firms' data are active in the period 1982-2000.
(2) For every firm, we have reliable information about capital and financial ratios, and each company has more than 100 employees.
(3) They have been under bank control for possible insolvency at least once.

For every firm, we have considered the following standard balance ratios: working capital/total assets (r_1), retained earnings/total assets (r_2), operating income/total assets (r_3), market value of equity/book value of total liabilities (r_4), sales/total assets (r_5), equity ratio (r_6) and debt ratio (r_7). The use of these quantities to study firms' behavior and defaults is quite common in the industrial economic literature (e.g., Philosophov and Philosophov, 2002).

5.2. *Initialization of the process*

In order to use our urn model in applications, it is fundamental to decide the way the process is initialized and reinforced.

As regards the initial composition of firms' urns, we have developed the following heuristic method, based on well-known stylized facts of industrial economics. Suitable references can be found in Fujiwara (2004), delli Gatti *et al.* (2007), and Phlips (1998).

We first consider the equity ratio as a proxy for the proportion of $x-$balls. In fact, it is known that a firm with an equity ratio greater than 0.5 can be considered financially robust (see Fujiwara, 2004). Hence, for every firm, we trivially have $a_0 = \text{round}(r_6 * 100)$.

Table 1. Example of initial urn composition for some firms of the dataset.

firm code	year	equity ratio r_6	debt ratio r_7	a_0	b_0	c_0	default in $t+1$
$IM223A$	1982	0.42	0.71	42	41	17	0
$IM298A$	1982	0.62	0.52	62	20	18	0
$IM567B$	1982	0.68	0.37	68	12	20	1
$IM1031B$	1982	0.39	0.66	39	40	21	1
$IM1988A$	1982	0.72	0.57	72	16	12	0

Regarding the y- and w-balls, expressing the risky and the default states, we combine the debt ratio with the complement of the equity ratio. Unlike the equity ratio, a higher debt ratio can be read as a signal of danger for firms' reliability because of the excessive financial exposure (Fujiwara, 2004; Phlips, 1998). In particular, we have set

$$b_0 = \text{round}\left[r_7 * (1 - r_6) * 100\right] \text{ and } c_0 = \text{round}\left[(1 - r_6)(1 - r_7) * 100\right].$$

It is easy to verify that, for every firm, the initial number of balls in the urn is equal to 100.

Let us analyze some firms in order to better understand the data and the procedure. Table 1 summarizes the fundamental data for five different firms. Using the equity ratio and the debt ratio, it is simple to derive a_0, b_0, and c_0 by simply applying the formula we have proposed. The last column represents a dummy variable we have created to control for defaults (0 = no default, 1 = default).

As regards the reinforcement matrix, we have used the method of moments to estimate the values of its parameters, that is, θ and δ.

First of all, using the equity ratio as a control index, we have divided the firms into three groups of riskiness. The firms with $r_6 \geq 0.50$ can be considered the most reliable ones and form group A. Group B, on the other hand, is made up of firms with $0.35 \leq r_6 < 0.50$. All other firms, with $r_6 < 0.35$, are in group C, the worst one.

In total, we have 82 firms in A, 165 in B, and 53 in C. The majority of firms are in group B because, on average, around 0.4 is a standard equity ratio value for Italian firms (e.g., Philosophov and Philosophov, 2002, and see also the industrial reports of the Italian Institute of Statistics at www.istat.it). For every group, we have estimated θ and δ.

If we consider the variable Y_n, we know from Theorem 4.2 that

$$E\left[Y_n\right] = b_0 n^{\frac{\delta}{\theta}} \frac{\Gamma\left(\frac{t_0}{\theta}\right)}{\Gamma\left(\frac{t_0+\delta}{\theta}\right)} + O\left(n^{\frac{\delta}{\theta}-1}\right).$$

Table 2. Average initial composition and estimated θ and δ for the three groups of firms for $n = 1, 2$.

Group	$\mathbf{a}_0$	$\mathbf{b}_0$	$\mathbf{c}_0$	$\widehat{\theta}_{n=1}$	$\widehat{\delta}_{n=1}$	$\widehat{\theta}_{n=2}$	$\widehat{\delta}_{n=2}$
A	52	41	9	2.06	0.98	2.36	1.17
B	43	44	13	4.22	2.45	4.16	2.18
C	34	48	18	5.17	3.12	4.98	3.34

Furthermore, with tedious but simple calculations, it is possible to compute the variance as

$$Var\left[Y_n\right] = b_0 \left[(b_0 + \delta) \frac{\Gamma\left(\frac{t_0}{\theta}\right)}{\Gamma\left(\frac{t_0+2\delta}{\theta}\right)} - b_0 \left(\frac{\Gamma\left(\frac{t_0}{\theta}\right)}{\Gamma\left(\frac{t_0+\delta}{\theta}\right)} \right)^2 \right] + O\left(n^{\frac{\delta}{\theta}}\right). \quad (13)$$

Unfortunately, given the hypergeometric form of the urn moments, it is not possible to find explicit formulas for the reinforcement matrix parameters θ and δ, so it is necessary to use numerical solutions for the resulting equations.

In particular, given the average values of a_0, b_0 and c_0 in every group, we have estimated θ and δ for every group for $n = 1, 2$. Table 2 contains all the values of interest: it is interesting to notice that, after rounding, there is no difference among the estimates for $n = 1, 2$ (and even 3), indicating that perhaps in this short period, the constancy of the reinforcement matrix is not a ludicrous assumption.

Another simulation-based method, which could represent an efficient and robust alternative to the method of moments, could be indirect inference, a simulation-based estimation technique introduced in Gouriéroux and Monfort (1996) and widely used in the computational literature. In any case, viewing this as a first application, we prefer not to complicate our first computations.

Assuming that the reinforcement matrix is constant over time (as in our theoretical model), we have studied firms' defaults, computing for every firm the average urn composition at time $t + 1$, $t + 2$, and so on, for all the available periods of observation.

5.3. *The benchmark: the Z^*-score*

For every firm, in every period, we have compared our predictions about failures with actual data and Altman's estimates. Altman's Z-score (Altman, 1968) is a popular measure, based on discriminant analysis, to classify firms' riskiness. In particular, Altman's original formulation is

$$Z = 0.012r_1 + 0.014r_2 + 0.033r_3 + 0.006r_4 + 0.999r_5.$$

According to this score, a firm is likely to default if $Z < 1.8$, it is safe if $Z > 3$, while it is in a "gray" condition otherwise.

In this paper, we use a more recent version of the Z-score, which has been introduced in Altman (2005) and which we call the Z^*-score. The original specification is

$$Z^* = 6.56r_1 + 3.26r_2 + 6.72r_3 + 1.05r_4 + 3.25.$$

Obviously, as required by this kind of model, we have directly estimated the Z^*-score on the CEBI data to understand if its general formulation can be used on Italian data. The result we have obtained using standard regression techniques is

$$Z^*_{CEBI} = 4.87r_1 + 2.98r_2 + 6.01r_3 + 1.16r_4 + 1.42. \tag{14}$$

It is clear that our estimates are different from the original Altman's Z^*-score, since we only preserve the signs of the parameters. This variability is indeed one of the main weaknesses of the scoring models based on discriminant analysis. Generally, every time one changes the data, the specification of the Z-score-like models changes too. In the literature (e.g., Altman and Narayanan, 1997), different specifications are proposed and applied.

In our first comparison, we use the specification of Equation (14) as a benchmark. In particular, we have that a firm is safe if $Z^* >= 4.75$, is unsafe for $Z^* < 3.25$, and is in the gray area otherwise.

5.4. *Results*

Given the initial composition of every firm and the reinforcement matrix in its different specifications, for every firm we are able to compute all the probabilities we are interested in.

It is possible to derive the probability of having a certain number of x-balls or the probability of picking a w-ball (that represents default) at time n and we can thus verify if a firm is in a safe, risky, or default state. In particular, in this experiment, we assume that a firm fails at time $n + 1$ if the probability of extracting a w-ball is equal to or greater than 0.20 at time n, which represents a well-informed threshold value in studying firms' dynamics (delli Gatti *et al.*, 2007; Philosophov and Philosophov, 2002).

For every firm, in every period, we have compared the predictions of the two different approaches—Z^*-score and UbGesm—with actual data, finding interesting results.

For 263 firms out of 380, the two approaches seem to behave in the same way. In particular, both methods correctly predict firms' defaults. An example is given in Figure 3, in which both the UbGesm and the Z^*-score are able to predict the failure of a firm in 1989-90. For the UbGesm, we report the probability of picking a w-ball.

As regards the remaining 137 firms, the UbGesm seems to behave slightly better. In fact, it is able to correctly predict 42 more firms, while the Z^*-score is able to forecast the behavior of 17 extra firms. This means that the proportion of correct predictions is about 0.80 for the UbGesm and 0.73 for the Z^*-score. It is interesting to notice that, for several firms, the two methods give contrasting results, indicating the extreme complexity of forecasting business failures with simple models[†]. Figure 4, for example, shows the divergent prediction of the two models for a firm that failed in 1995. While the Z^*-score only shows a general worsening of the health conditions of the firm, the UbGesm clearly predicts failure.

Concerning the firms whose fates the two models are not able to accurately foresee (75 for the UbGesm and 100 for the Z-score), we notice an interesting characteristic of the urn-based generalized extreme shock model. In fact, while the Z^*-score generally underestimated risk, the UbGesm does the opposite. In other words, our model tends to anticipate defaults by about 2 periods. Even if it is not correct, such behavior can be considered as a positive result if we take the point of view of a bank or of another financial institution. In fact, more prudent behavior is required for this kind of company (consider Basel II regulations, for example).

Table 3. Analysis of the not correctly predicted firms for the two models (75 vs 100). In parentheses is the average number of periods of difference between actual defaults and predictions.

	UbGesm (75)	Z^*-score (100)
overestimated	71% (2.1)	36% (1.6)
underestimated	29% (−1.4)	64% (−2.6)

Table 3 shows the number of overestimated and underestimated defaults for the two models. In particular, we show how many defaults are wrongly

[†] A good idea for future research could be to find a combination of the two models for classifying firms' riskiness and to predict their failures.

expected before actual failure and how many after, also giving information about the average period of error.

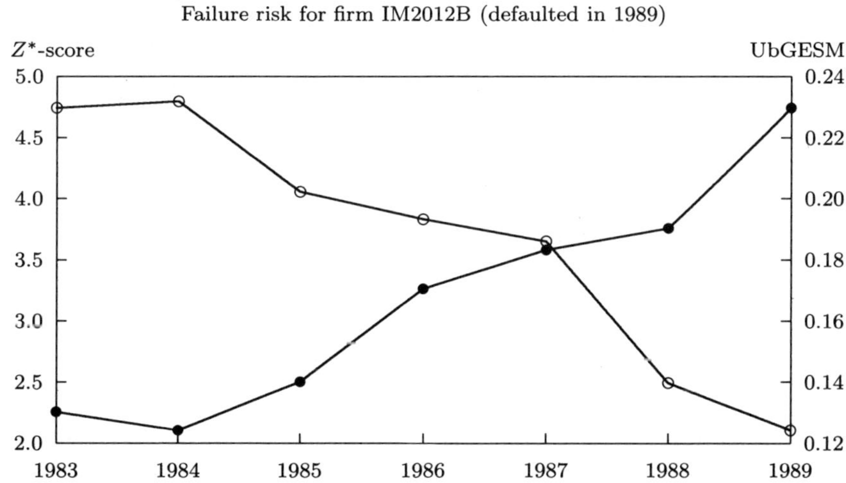

Fig. 3. Comparison of the Z^*-score (open circles) and the UbGesm (solid dots) for a firm correctly predicted by both methods.

Hence, this simple application has shown the good capabilities of the urn-based generalized extreme shock model in studying shock and default phenomena. In particular, the results obtained are promising, considering that our model does perform better than (or at least as well as) one of the most widely-used scoring methods on the market.

Finally, it must be stressed that it would be interesting to extend the application to more general firms. In fact, considering firms that have been under bank control at least once surely simplifies our analysis, since their probability of defaults is surely higher than the average, given their fragile microstructure.

Acknowledgments

This paper was started when the second author was visiting the University of Uppsala. So the second author would like to thank Allan Gut for his stimulating discussion and his kind hospitality. We are also grateful to the referees for several comments to improve the review paper.

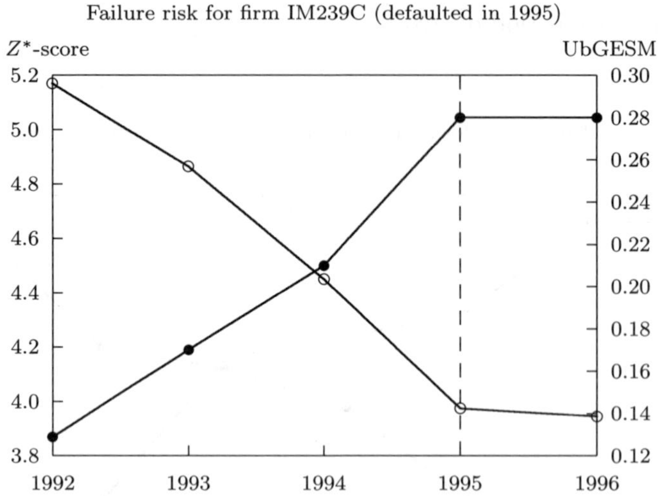

Fig. 4. Comparison of the Z^*-score (open circles) and the UBGesm (solid dots) for a firm in the Z^*-score gray area. Here the UbGesm behaves better, forecasting default in 1995.

References

Altman, E. I. (1968). Financial ratio, discriminant analysis and prediction of corporate bankruptcy, *Journal of Finance* **4**, pp. 589–610.

Altman, E. I. (2005). An emerging market credit scoring system for corporate bonds, *Emerging Markets Review* **6**, pp. 311–323.

Altman, E. I. and Narayanan, P. (1997). An international survey of business failure classification models, *Financial Markets, Institutions and Instruments* **6**, pp. 1–57.

Cirillo, P. and Hüsler, J. (2009). An urn approach to generalized extreme shock models, *Statistics and Probability Letters* **79**, 7, pp. 969–976.

delli Gatti, D., Gaffeo, E., Gallegati, M., Giulioni, G., Kirman, A., Palestrini, A. and Russo, A. (2007). Complex dynamics and empirical evidence, *Information Sciences* **177**, pp. 1204–1221.

Fujiwara, Y. (2004). Zipf law in firms bankruptcy, *Physica A* **337**, pp. 219–230.

Gouriéroux, C. and Monfort, A. (1996). *Simulation-Based Econometric Methods* (Oxford University Press, Oxford).

Gut, A. (1990). Cumulative shock models, *Advances in Applied Probability* **22**, pp. 504–507.

Gut, A. (2001). Mixed shock models, *Bernoulli* **7**, pp. 541–555.

Gut, A. and Hüsler, J. (1999). Extreme shock models, *Extremes* **2**, pp. 293–305.

Gut, A. and Hüsler, J. (2005). Realistic variation of shock models, *Statistics and Probability Letters* **74**, pp. 187–204.

Philosophov, L. V. and Philosophov, V. L. (2002). Corporate bankruptcy prognosis: an attempt at a combined prediction of the bankruptcy event and time interval of its occurence, *International Review of Financial Analysis* **11**, pp. 375–406.

Phlips, L. (1998). *Applied Industrial Economics* (Cambridge University Press, Cambridge).

Kernel Density Estimation with Missing Data: Misspecifying the Missing Data Mechanism

Suzanne R. Dubnicka

Department of Statistics, Kansas State University
Manhattan, KS 66506, USA
E-mail: dubnicka@ksu.edu

This paper explores additional properties of an inverse propensity score weighted kernel density estimator for estimating the density of incomplete data. This estimator is based on the Horvitz-Thompson estimator and requires estimating the propensity score assuming the response variable is missing at random. Nonparametric methods are used to estimate the propensity scores. Implications of misspecifying the missing data mechanism on the performance of the density estimator are discussed and evaluated. In addition, an augmented inverse propensity score weighted kernel density estimator, which is not influenced by this misspecification, is proposed and evaluated.

Keywords: Horvitz-Thompson estimator; Missing at random; Robustness

1. Introduction

Kernel density estimators are important as both descriptive and inferential tools. As a descriptive tool, kernel density estimators are essentially smooth versions of histograms and approximate the distribution of the variable of interest using a random sample collected on that variable. However, one complication with large studies is incomplete data. That is, the variable of interest may not be observed for all subjects. In such cases, a kernel density estimator constructed from the data at hand, i.e., the complete cases, simply estimates the distribution of the observed data. This distribution may be quite different from the true underlying distribution because the observed data may not be a random sample on the variable of interest. Therefore, one needs to address the relationship between the missingness of the variable of interest and the study variables, which is known as the missing data mechanism.

A semiparametric approach to handling missing data was introduced by Robins *et al.* (1994) in their seminal paper on weighted estimating equations. Rather than simply carrying out an estimation procedure on the complete cases, i.e., the completely observed subjects, Robins *et al.* (1994) proposed weighting each term of the complete-case estimating equation with the inverse of the probability that the subject is completely observed, known as the propensity score. This approach was derived from the idea of the Horvitz-Thompson estimator used in finite population sampling. Complete-case estimates are known to be biased unless the data are missing completely at random (MCAR) (Little and Rubin, 2002). The MCAR assumption is a missing data mechanism that states that the probability a subject is completely observed does not depend on the observed or unobserved data so that the complete cases are simply a random sample from intended population. This inverse probability weighting approach of Robins *et al.* (1994) is shown to effectively correct for this bias and produce consistent parameter estimates when the data are missing at random (MAR). The MAR assumption is a weaker, and more realistic, assumption than MCAR and states that the probability a subject is completely observed depends on the observed data but not the unobserved data.

Using this inverse propensity score weighting, Dubnicka (2009) developed kernel density estimators that are also based on the Horvitz-Thompson estimator and that are appropriate when the variable of interest is MAR. To fix notation, let $y_1, \ldots, y_n$ be a random sample from a distribution with smooth density g. Also, let $r_i = 1$ or 0 if y_i is observed or missing, respectively, and let π_i be the probability that y_i is observed, i.e., the propensity score. The Horvitz-Thompson kernel density estimator is given by

$$\hat{g}_h^{HT}(y) = \frac{1}{n}\sum_{i=1}^{n}\frac{r_i}{\hat{\pi}_i}K_h(y - y_i), \tag{1}$$

and the modified Horvitz-Thompson kernel density estimator is

$$\hat{g}_h^{HTM}(y) = \frac{1}{n^*}\sum_{i=1}^{n}\frac{r_i}{\hat{\pi}_i}K_h(y - y_i), \tag{2}$$

where $K_h(\cdot) = (1/h)K(\cdot/h)$ is a kernel with bandwidth h, $\hat{\pi}_i$ is an estimate of the true propensity score π_i, and $n^* = \sum_{i=1}^{n} r_i/\hat{\pi}_i$. Note that (1) is technically not a density because it does not integrate to one. In empirical studies, Dubnicka (2009) showed that (1) and (2) perform similarly in terms of mean integrated squared error (MISE).

The propensity score π_i is the probability that the response for subject i is observed. These propensity scores, and especially their estimates, are important components of the Horvitz-Thompson-type kernel density estimators. The inverse propensity score weighting in (1) and (2) adjusts for the additional bias that is present in the complete-case kernel density estimator

$$\hat{g}_h^{CC}(y) = \frac{1}{r}\sum_{i=1}^{n} r_i K_h(y - y_i), \tag{3}$$

where $r = \sum_{i=1}^{n} r_i$ is the number of observed responses. Dubnicka (2009) assumed that data on additional variables, called auxiliary variables, are available and that the responses are MAR so that the probability that the response for subject i is missing depends only on the observed data, which consists of data collected on the auxiliary variables in the study. Therefore, under the MAR assumption, the propensity score for subject i is denoted by

$$\pi_i = \pi(\boldsymbol{x}_i) = P(r_i = 1|\boldsymbol{x}_i), \tag{4}$$

where $\boldsymbol{x}_i$ denotes the vector of auxiliary variables for subject i. Dubnicka (2009) used both local mean estimation (a Nadaraya-Watson estimator) and local logistic regression to estimate the unknown propensity scores. The method of estimating the propensity scores did not prove to be important provided that consistent estimates were produced.

However, in many studies, the MAR assumption is not met. It is not uncommon for the probability of observing a variable to depend on both the observed and unobserved data. In this case, the data are said to be not missing at random (NMAR), or nonignorably missing, and the propensity scores are given in this setting by

$$\pi_i = \pi(y_i, \boldsymbol{x}_i) = P(r_i = 1|y_i, \boldsymbol{x}_i). \tag{5}$$

However, since the response y is not observed for all subjects, it is not possible to estimate these propensity scores directly. Misspecifying the missing data mechanism, that is, assuming that the response is MAR when it is NMAR, will likely produce a density estimate that does not accurately reflect the true distribution of the response. This paper explores the effect of misspecifying the missing data mechanism on the Horvitz-Thompson-type kernel density estimator developed by Dubnicka (2009). In addition, an augmented Horvitz-Thompson kernel density estimator, which is based on the idea of double robustness (Bang and Robins, 2005), is discussed.

In Section 2, we will discuss propensity score estimation and the motivation behind the current paper. Section 3 includes a more detailed description of the NMAR assumption and the potential effect on the Horvitz-Thompson-type kernel density estimators. In Section 4, the augmented

Horvitz-Thompson-type kernel density estimator is discussed. The results of a simulation study are reported in Section 5, and concluding remarks are in Section 6.

2. Propensity Score Estimation

Inverse propensity score weighting is important in the construction of the Horvitz-Thompson kernel density estimator as it produces a density estimator with the same expectation, and thus bias, as the kernel density estimator for the full data, i.e., the data that would have been observed if none of the values were missing. Suppose that the propensity scores are known and correctly specified so that the Horvitz-Thompson kernel density estimator is given by

$$\hat{g}_h^{HTt}(y) = \frac{1}{n}\sum_{i=1}^{n} \frac{r_i}{\pi_i} K_h(y - y_i). \tag{6}$$

The expectation of this density estimator is

$$\begin{aligned} \mathrm{E}\{\hat{g}_h^{HTt}(y)\} &= \frac{1}{n}\sum_{i=1}^{n} \mathrm{E}_{y_i,\boldsymbol{x}_i}\left[\mathrm{E}_{r_i|y_i,\boldsymbol{x}_i}\left\{\frac{r_i}{\pi_i}K_h(y - y_i)\right\}\right] \\ &= \frac{1}{n}\sum_{i=1}^{n} \mathrm{E}_{y_i,\boldsymbol{x}_i}\left[K_h(y - y_i)\mathrm{E}_{r_i|\boldsymbol{x}_i}\left\{\frac{r_i}{\pi_i}\right\}\right] \qquad (7) \\ &= \mathrm{E}\left\{\frac{1}{n}\sum_{i=1}^{n} K_h(y - y_i)\right\} \;=\; \mathrm{E}\{\hat{g}_h^F(y)\}. \qquad (8) \end{aligned}$$

Expression (7) follows from a simple conditioning argument and the missing at random assumption, and expression (8) follows because $\pi_i = P(r_i = 1|\boldsymbol{x}_i) = \mathrm{E}(r_i|\boldsymbol{x}_i)$ under the MAR assumption. Note that (8) is the expectation of the full data kernel density estimator. If the propensity scores were misspecified, then $\mathrm{E}(r_i|\boldsymbol{x}_i)$ would no longer equal π_i, and the expectation of the Horvitz-Thompson kernel density estimator, using true propensity scores, would no longer be the same as that of the full data kernel density estimator. As the full data kernel density estimator is already biased, additional bias would be introduced.

Typically, the propensity scores are unknown and must be estimated. If the propensity scores are consistently estimated, then the expectation of the corresponding Horvitz-Thompson kernel density estimator will be approximately the same as that of the full data kernel density estimator in large samples. Dubnicka (2009) considered two possible estimators of the propensity scores under the MAR assumption: the Nadaraya-Watson estimator

$$\hat{\pi}_{NWi} = \hat{\pi}_{NW}(\boldsymbol{x}_i) = \frac{\sum_{j=1}^{n} r_j K^*_{h_1}(\boldsymbol{x}_i - \boldsymbol{x}_j)}{\sum_{j=1}^{n} K^*_{h_1}(\boldsymbol{x}_i - \boldsymbol{x}_j)} \tag{9}$$

and the local logistic regression predicted values $\hat{\pi}_{LLi} = \hat{\pi}_{LL}(\boldsymbol{x}_i)$. See Dubnicka (2009) for more details regarding the local logistic regression estimator in this setting, or see Tibshirani and Hastie (1987) for details regarding local likelihood estimation in general.

These estimators can produce somewhat different estimates of the propensity scores. Consider a random sample of 300 pairs of responses and auxiliary variables for which 35% of the responses are MAR with missingness depending on the value of the auxiliary variable. The relationship between the true probability that the response is observed (the propensity score) and the auxiliary variable is given in panel (a) of Figure 1. Nadaraya-Watson and local logistic regression estimates of the propensity scores both provide consistent estimates of the true propensity scores. However, from panel (b) of Figure 1, it is clear that the estimates are not the same, as the points do not fall on a line through the origin with slope one. Despite the difference in propensity score estimates provided by these two methods, Dubnicka (2009) showed in a simulation study that the distributions of integrated squared errors for the two corresponding Horvitz-Thompson-type kernel density estimators were approximately the same. The integrated squared error, $\int(\hat{g} - g)^2$, specifically the mean integrated squared error (MISE), $\mathrm{E}\{\int(\hat{g} - g)^2\}$, combines information on bias and variability of the density estimator.

Dubnicka (2009) also showed that, asymptotically, the variance of the Horvitz-Thompson kernel density estimator with Nadaraya-Watson estimates of the propensity scores (1) is smaller than that of the Horvitz-Thompson kernel density estimator with true propensity scores (6). Heuristically, this is true because only the responses from the complete cases are used when estimating the density with true propensity scores; however, through the estimation of the propensity scores, all of the available data is used when estimating the density with estimated propensity scores. Together, the asymptotic and empirical results seem to imply that the method of estimating the propensity scores is not important, as long as they are consistently estimated.

One issue with inverse propensity score weighting occurs when one or more observed subjects have very small propensity scores. This will result in very large weights being placed on such individuals in the construction

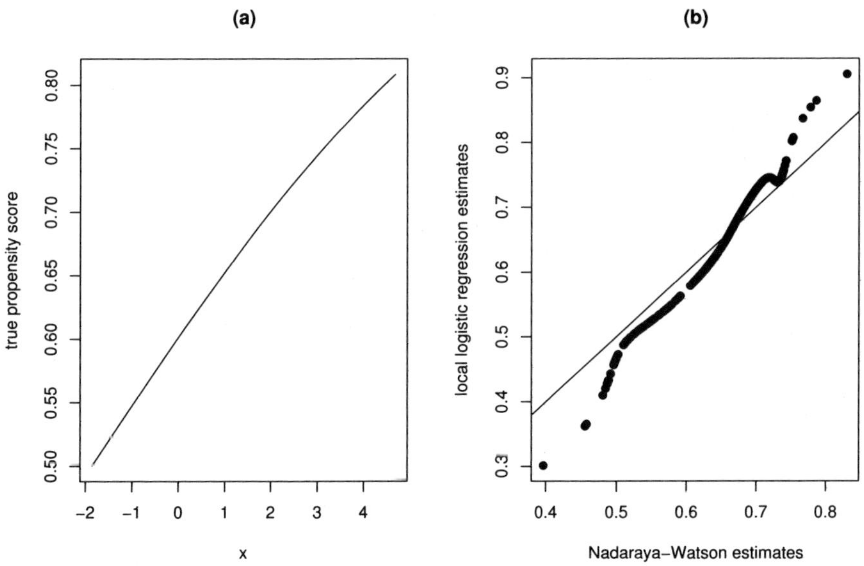

Fig. 1. (a) True propensity scores vs. auxiliary variable x; (b) Nadaraya-Watson estimates vs. local logistic regression estimates.

of the Horvitz-Thompson kernel density estimate. If the true propensity score is indeed small and estimated correctly, this results in an estimator with an inflated variance. However, if the propensity score was mistakenly estimated to be very small, then excess weight is being placed on that subject in the kernel density estimator, inflating the bias. Therefore, the Horvitz-Thompson kernel density estimator, as well as any method based on this inverse propensity score weighting, can be unstable. This has been well documented by Kang and Schafer (2007). A possible correction for this is to use the stabilized weights proposed by Robins *et al.* (2000). The stabilized weight for a subject is the ratio of an unbiased estimate of the unconditional probability that the subject was observed to an estimate of the propensity score. This has the effect of decreasing the weight of that subject.

One question that remains is how the Horvitz-Thompson-type kernel density estimators (1) and (2) perform when the propensity scores are misspecified. That is, we would like to determine how an incorrect MAR assumption affects the integrated squared errors of the Horvtiz-Thompson-type kernel density estimators. In the following section, we further discuss the NMAR assumption and its potential impact on the Horvitz-Thompson kernel density estimator.

3. Misspecifying the Missing Data Mechanism

We now turn our attention to the effects of misspecifying the missing data mechanism on the Horvitz-Thompson-type kernel density estimators. To simplify calculations, we focus for the moment on the Horvitz-Thompson estimator with known propensity scores (6), even though it is not technically a density. To distinguish the case in which the missing data mechanism is correctly specified from the case in which it is not, let π_i^* denote the true (correct) propensity score for subject i, and let π_i be the assumed (incorrect) propensity score. Since, in reality, the responses are NMAR, the true propensity scores are

$$\pi_i^* = \pi^*(y_i, \boldsymbol{x}_i) = P(r_i = 1 | y_i, \boldsymbol{x}_i), \ i = 1, \ldots, n. \tag{10}$$

However, we assume that the responses are MAR so that the assumed propensity scores are

$$\pi_i = \pi(\boldsymbol{x}_i) = P(r_i = 1 | \boldsymbol{x}_i), \ i = 1, \ldots, n. \tag{11}$$

Therefore, the correct Horvitz-Thompson kernel density estimator is given by

$$\hat{g}_h^{HTc}(y) = \frac{1}{n} \sum_{i=1}^{n} \frac{r_i}{\pi_i^*} K_h(y - y_i), \tag{12}$$

but we mistakenly use

$$\hat{g}_h^{HTi}(y) = \frac{1}{n} \sum_{i=1}^{n} \frac{r_i}{\pi_i} K_h(y - y_i). \tag{13}$$

(The superscript c means correct missing data mechanism and propensity scores, while i means incorrect.)

If the responses are NMAR, then the correct Horvitz-Thompson kernel density estimator (12) has the same expectation as the full data kernel density estimator. This follows from the same arguments that result in expressions (7) and (8), with π_i^* replacing π_i. However, the incorrect

Horvitz-Thompson kernel density estimator has expectation

$$\begin{aligned}
\mathrm{E}\{\hat{g}_h^{HTi}(y)\} &= \mathrm{E}\left\{\frac{1}{n}\sum_{i=1}^{n}\frac{r_i}{\pi_i}K_h(y-y_i)\right\} \\
&= \frac{1}{n}\sum_{i=1}^{n}\mathrm{E}_{y_i,\boldsymbol{x}_i}\left[K_h(y-y_i)\mathrm{E}_{r_i|y_i,\boldsymbol{x}_i}\left\{\frac{r_i}{\pi_i}\right\}\right] \\
&= \mathrm{E}_{y_i,\boldsymbol{x}_i}\left\{\frac{1}{n}\sum_{i=1}^{n}K_h(y-y_i)\frac{\pi^*(y_i,\boldsymbol{x}_i)}{\pi(\boldsymbol{x}_i)}\right\}. \qquad (14)
\end{aligned}$$

This expectation cannot be evaluated further without additional information. However, it is enough to note that (14) is not the expectation of the full data kernel density estimator. Furthermore, the additional bias introduced by misspecifying the missing data mechanism will depend heavily upon the relationship between the true and assumed propensity scores.

The situation is complicated further because the true propensity scores under the NMAR assumption cannot be estimated. In the regression of r on y and $\boldsymbol{x}$, the "covariate" y is not completely observed, so none of the standard parametric (e.g., logistic regression) and nonparametric (e.g., Nadaraya-Watson estimator) methods can be used directly to estimate the propensity scores. However, it is possible that, with many auxiliary variables that are important in predicting missingness of the response variable and other variables that are highly correlated with the incomplete response, we may still be able to obtain reasonable estimates of the true propensity scores even though the response variable has been omitted from the estimation procedure. These other variables, which are highly correlated with the response, will be referred to as surrogate variables, although with incomplete response information it is difficult to assess the strength of the correlation in practice. Furthermore, the inclusion of the interaction terms among the auxiliary variables should be considered when selecting "covariates" for estimating the propensity scores. Variable selection techniques can be used to choose "covariates" from among the auxiliary variables, their interactions, and the potential surrogate variables. In the end, we hope to show that the performance of the corresponding Horvitz-Thompson-type kernel density estimator will not be unduly influenced by the misspecification of the propensity scores with the appropriate choice of auxiliary and surrogate variables.

Consider the estimation of the density of CD4 counts at final follow-up in the AIDS Clinical Trial Group (ACTG) Study 175 (Hammer *et al.*, 1996). The ACTG Study was a large, randomized clinical trial designed to compare four antiretroviral therapies for improving the immune systems of AIDS patients. CD4 counts are one indication of the strength of a person's immune system, with low CD4 counts indicating impairment of the immune system. The final follow-up of the study occurred at 96 ± 5 weeks, and 37% of the patients had dropped out of the study by this time. Dropout could be directly related to a patient's health status, as measured by the CD4 counts at final follow-up. Therefore, it would not be unreasonable to assume that the CD4 counts at final follow-up were missing for reasons related directly to the CD4 count that would have been observed at final follow-up had the patient stayed in the study.

In a previous analysis of the ACTG data, Davidian *et al.* (2005) assumed that the CD4 counts at final follow-up were MAR and indicated that a host of auxiliary variables, such as weight, indicators of HIV symptoms and prior antiretroviral therapy, Karnofsky score (a measure of a patient's ability to carry out day-to-day activities), intermediate measures of CD4 and CD8 counts, and off-treatment status, were useful in predicting the missingness of CD4 counts at final follow-up. Dubnicka (2009) also made these assumptions in her estimation of the CD4 counts at final follow-up using a Horvitz-Thompson-type kernel density estimator. However, it is likely that several of these variables are closely related to CD4 counts at final follow-up as well. For example, patients with high CD4 counts are healthier and more likely to be able to carry out ordinary daily activities, so they would probably have high Karnofsky scores. Hence, although the MAR assumption may be questionable in this case, the addition of auxiliary variables that are highly correlated with the response of interest may allow for a reasonable analysis of the data.

Before constructing the kernel density estimator, we need to determine whether the propensity score estimates are within a reasonable range so that a few subjects do not dominate the density estimate through the inverse propensity score weights. The boxplots in Figure 3 show (a) the distribution of all propensity score estimates using the Nadaraya-Watson estimator and (b) the distribution of propensity score estimates for the observed subjects. In terms of inverse propensity score weighted methods such as the Horvitz-Thompson kernel density estimator, the propensity score estimates for the observed subjects are most relevant. However, for these data, the smallest propensity score estimate for observed subjects is 0.213, which corresponds to a weight of 4.7, which is not excessively large. Therefore, the use of stabilized weights is not necessary. As an aside, the smallest propensity score among all subjects is 0.127, which corresponds to a weight of 7.8.

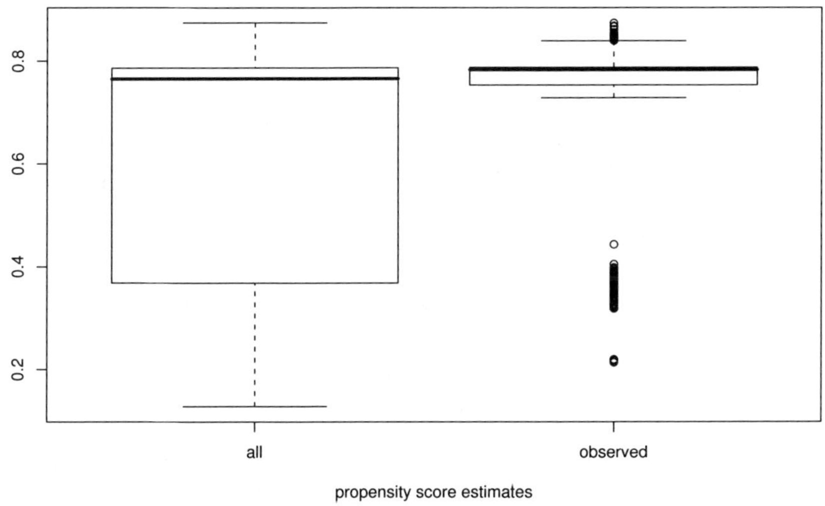

Fig. 2. Propensity score estimates for all subjects and observed subjects in the AIDS Clinical Trial Group study.

Figure 3 shows two estimates of the density of CD4 counts at final follow-up for the ACTG Study: the complete-case kernel density estimator (3) and the modified Horvitz-Thompson kernel density estimator (2) using Nadaraya-Watson estimates (9) of the propensity scores under the MAR assumption based on the auxiliary variables listed above. Boundary bias was addressed by using a reflection technique. The basic shape of each density estimator is the same, with the exception of the "bump" in the modified Horvitz-Thompson kernel density estimator near zero. Recall that the complete-case kernel density estimator simply estimates the density of the observed CD4 counts and is not likely to be a reflection of the density of the CD4 counts for all patients. The modified Horvitz-Thompson density estimator adjusts for the additional bias of the complete-case estimator and indicates that the unobserved CD4 counts were likely to be low values. This seems to reinforce the idea that CD4 counts may not be missing at random in this study.

Other variables that are not thought to directly affect the missingness of the response but that are highly correlated with the response may prove useful in obtaining reasonable propensity score estimates under the NMAR assumption. These surrogate variables attempt to fill the role that the response variable would in estimating the propensity scores. Therefore, it is

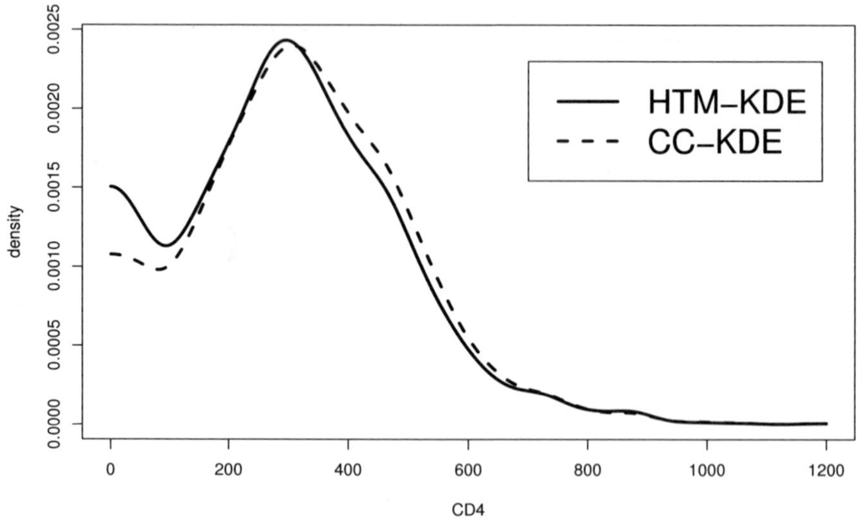

Fig. 3. Comparison of complete-case and modified Horvtiz-Thompson kernel density estimators for AIDS Clinical Trial Group data.

useful to explore (a) the effect of auxiliary variables already believed to effect missingness as a function of the strength of their association with the missing response and (b) the effect of including a surrogate variable on the estimation of the propensity scores when the response is not missing at random on the corresponding kernel density estimate. A simulation study making these assessments is given in Section 5.

In practice, it may be difficult to distinguish between an auxiliary variable and a surrogate variable. The term auxiliary variable is being used to describe those additional variables upon which missingness directly depends but that are not necessarily strongly associated with the response variable. On the other hand, surrogate variables are strongly associated with the response variable that, under the NMAR assumption, influences its own missingness. Certainly, a researcher might believe that response variable missingness depends on these surrograte variables as well, but their real utility is in serving as placeholders, so to speak, for the response variable that is not completely observed and that cannot be used to estimate the propensity scores.

4. Augmenting the Kernel Density Estimator

Before continuing with our evaluation of the Horvitz-Thompson-type kernel density estimators, we consider one possible improvement to these estimators. In the previous section, we showed that the expectation of the Horvitz-Thompson kernel density estimator with true propensity scores (6) only has the same expectation as the full data kernel density estimator when the missing data mechanism is correctly specified. In this section, we discuss an augmented Horvtitz-Thompson kernel density estimator that draws on the idea of double robustness in an attempt to construct a kernel density estimator for which the expectation is equal to that of the full-data kernel density estimator even when the missing data mechanism is misspecified.

Noting that the ordinary weighted estimating equations are inherently inefficient because they only use information from the complete cases, Robins *et al.* (1994) also included more efficient weighted estimating equations, which we will refer to as augmented weighted estimating equations and which attempt to use all of the available data. Applying the same approach to kernel density estimation with missing data, we define the augmented Horvitz-Thompson kernel density estimator to be

$$\hat{g}_h^{HTA}(y) = \frac{1}{n}\sum_{i=1}^{n}\left[\frac{r_i}{\hat{\pi}_i}K_h(y-y_i) + \left(1-\frac{r_i}{\hat{\pi}_i}\right)\hat{q}(\boldsymbol{x}_i)\right], \tag{15}$$

where, again, $\hat{\pi}_i$ represents an estimate of the true propensity score π_i under the MAR assumption and $\hat{q}(\boldsymbol{x}_i)$ is an estimate of the conditional expectation $\mathrm{E}\{K_h(y-y_i)|\boldsymbol{x}_i\}$.

Estimators of the same form as (15) are called doubly robust estimators (Bang and Robins, 2005). The name follows from the fact that such estimators are unbiased if either the missing data mechanism or the data distribution is modeled correctly. For the augmented Horvitz-Thompson estimator to have the same expectation as the full data kernel density estimator, we need to correctly specify the propensity scores or the conditional expectation $\mathrm{E}\{K_h(y-y_i)|\boldsymbol{x}_i\}$. This is possible under the MAR assumption. Kang and Schafer (2007) showed, however, that even the doubly robust estimator cannot recover from the inclusion of extremely small propensity score estimates; additional bias will still be introduced.

To partially evaluate this proposed kernel density estimator, consider the augmented Horvitz-Thompson kernel density estimator with known but misspecified propensity scores and known conditional expectation $q(\boldsymbol{x}_i) =$

$\mathrm{E}\{K_h(y - y_i)|\boldsymbol{x}_i\}$:

$$\hat{g}_h^{HTAi}(y) = \frac{1}{n}\sum_{i=1}^{n}\left[\frac{r_i}{\pi_i}K_h(y - y_i) + \left(1 - \frac{r_i}{\pi_i}\right)\mathrm{E}\{K_h(y - y_i)|\boldsymbol{x}_i\}\right], \quad (16)$$

where $\pi_i = \pi(\boldsymbol{x}_i) = P(r_i = 1|\boldsymbol{x}_i)$ under the incorrect MAR assumption. However, if the responses are actually NMAR, so that the correct propensity scores are given by $\pi_i^* = \pi^*(y_i, \boldsymbol{x}_i) = P(r_i = 1|y_i, \boldsymbol{x}_i)$, then the expectation of the augmented Horvitz-Thompson kernel density estimator (16) is

$$\begin{aligned}
\mathrm{E}\{\hat{g}_h^{HTAi}(y)\} &= \frac{1}{n}\sum_{i=1}^{n}\mathrm{E}_{y_i,\boldsymbol{x}_i}\left\{K_h(y - y_i)\mathrm{E}_{r_i|y_i,\boldsymbol{x}_i}\left(\frac{r_i}{\pi_i}\right)\right\} \\
&\quad + \frac{1}{n}\sum_{i=1}^{n}\mathrm{E}_{y_i,\boldsymbol{x}_i}\left[\mathrm{E}\{K_h(y - y_i)|\boldsymbol{x}_i\}\mathrm{E}_{r_i|y_i,\boldsymbol{x}_i}\left(1 - \frac{r_i}{\pi_i}\right)\right] \\
&= \frac{1}{n}\sum_{i=1}^{n}\mathrm{E}_{\boldsymbol{x}_i}\left[\mathrm{E}_{y_i|\boldsymbol{x}_i}\left\{K_h(y - y_i)\frac{\pi_i^*}{\pi_i}\right\}\right] \\
&\quad + \frac{1}{n}\sum_{i=1}^{n}\mathrm{E}_{\boldsymbol{x}_i}\left[\mathrm{E}_{y_i|\boldsymbol{x}_i}\left\{K_h(y - y_i)\left(1 - \frac{\pi_i^*}{\pi_i}\right)\right\}\right] \\
&= \frac{1}{n}\sum_{i=1}^{n}\mathrm{E}\{K_h(y - y_i)\} = \mathrm{E}\{\hat{g}_h^F(y)\}.
\end{aligned}$$

Therefore, the augmented Horvitz-Thompson kernel density estimator with known but misspecified propensity scores (16) has the same expectation as the full data kernel density estimator. That is, even when the missing data mechanism is misspecified, (16) does not introduce additional bias as the ordinary Horvitz-Thompson kernel density estimator in (13) does, provided that the conditional expectation $\mathrm{E}\{K_h(y - y_i)|\boldsymbol{x}_i\}$ is correctly specified.

Practically, the primary obstacle in using the augmented Horvitz-Thompson kernel density estimator and achieving this robustness against misspecification of the missing data mechanism lies in the estimation of the conditional expectation $q(\boldsymbol{x}_i) = \mathrm{E}\{K_h(y - y_i)|\boldsymbol{x}_i\}$. Robustness to misspecifications of the missing data mechanism technically depends on this conditional expectation being correctly modeled. Furthermore, estimating this conditional expectation adds computational complexity to the problem.

Drawing on the work of Robins *et al.* (1994), Kennedy (2007) evaluated the use of inverse propensity score weighted local polynomial regression when either the response or a covariate is MAR. This approach can also be used to estimate the conditional expectation $\mathrm{E}\{K_h(y-y_i)|\boldsymbol{x}_i\}$, where y is a fixed evaluation point for the kernel density estimation. Let $z_i(y) = K_h(y-$

$y_i)$, $i = 1, \ldots, n$, and note that not all of $z_1(y), \ldots, z_n(y)$ are observed. When there is only one auxiliary variable, the weighted local polynomial regression estimate of $\mathrm{E}\{K_h(y - y_i)|x_i\}$ is the height of the fit, $\hat{\beta}_0$, where $\hat{\boldsymbol{\beta}} = (\hat{\beta}_0, \ldots, \hat{\beta}_p)^T$ minimizes

$$\sum_{j=1}^{n} \frac{r_j}{\hat{\pi}_j} \{z_j(y) - \beta_0 - \beta_1(x_j - x_i) - \cdots - \beta_p(x_j - x_i)^p\}^2 K_{h*}(x_j - x_i)$$

and $K_{h^*}(\cdot) = (1/h^*)K(\cdot/h^*)$ is another kernel function with bandwidth h^*. The extension to more than one auxiliary variable is natural.

Kennedy (2007) showed that the inverse propensity score weighted local linear regression approach ($p = 1$) was effective in estimating the true relationship between the response and the covariates under the MAR assumption. Of course, if the MAR assumption does not hold, the performance of this technique for estimating the conditional expectation $\mathrm{E}\{K_h(y - y_i)|\boldsymbol{x}_i\}$ may be substantially compromised. With the inclusion of appropriate surrogate variables, it is again possible that we may obtain a reasonable estimate of this conditional expectation. However, if we know that the missing data mechanism has been misspecified, both the propensity scores and this conditional expectation will be estimated incorrectly using the methods outlined in this paper, even if both estimates are deemed to be "reasonable." Certainly, it is important to evaluate Kennedy's method under the NMAR assumption. Our focus here will be the effect that his estimation procedure has on the augmented Horvitz-Thompson kernel density estimator (15).

Even under MAR, the use of this approach is complicated by the fact that the local polynomial regression must be performed for each evaluation point y. This is computationally time-consuming but, in addition, there is potentially a different bandwidth $h*$ to be chosen for each evaluation point, which can further increase the computational burden. However, in comparing the use of a common bandwidth versus different bandwidths for each evaluation point, we made two observations. First, some bandwidth selection procedures, such as the approximate degrees-of-freedom method (Anscombe, 1967; Tukey, 1967), will lead to the same bandwidth for different evaluation points. Second, for bandwidth selection procedures that lead to different bandwidths for each evaluation point, such as the corrected AIC method (McQuarrie and Tsai, 1998), most of the bandwidths are quite similar except at the boundaries. Therefore, to ease the computational burden, we suggest choosing a common bandwidth for all evaluation points by computing the bandwidth from the completely observed data $(\boldsymbol{x}_i, z_i(y^*))$, where y^* is some sample quantile, such as the median, of the observed responses.

In the simulation study, we use weighted local linear regression with common bandwidth h^*, based on the corrected AIC method, as well as the same estimates of the propensity scores that are already used in estimating the density, to estimate $\mathrm{E}\{K_h(y-y_i)|\boldsymbol{x}_i\}$. In the subsequent section, the resulting kernel density estimator is compared to those introduced by Dubnicka (2009).

5. Simulation Study

One of the primary purposes of this work is to determine how the Horvitz-Thompson-type kernel density estimators (KDEs) perform when the response is not missing at random but it is assumed that the response is missing at random. The proposed augmented Horvitz-Thompson KDE should, in theory, provide a better estimate of the density when the missing data mechanism is misspecified. Therefore, a simulation study was conducted to evaluate the performance of the ordinary (1) and (2) and augmented (15) Horvitz-Thompson-type KDEs under misspecification of the missing data mechanism and to compare those estimators to each other and to the full data KDE. The performance was measured in terms of the integrated squared error (ISE) between the density estimator $\hat{g}$ and the true density g, $\int(\hat{g}-g)^2$, and in particular, the observed median ISE. The median ISE was used rather than the more common mean ISE because the distribution of ISEs is right skewed.

Throughout the simulation study, the response variable y is assumed to be not missing at random, and missingness depends on both y and an auxiliary variable x_1. Additionally, in most studies, data on other variables are often collected. Therefore, we assume that we have access to data on another variable x_2, called a surrogate variable, that does not directly affect the missingness of y but may be correlated with y. Since, in practice, the propensity scores need to be estimated, the Nadaraya-Watson estimator of the propensity scores was used in all cases. However, two different estimators were computed: one that depends on the auxiliary variable x_1 only and another that depends on both x_1 and the surrogate x_2. In the first case, we have naively assumed that the response is missing at random. In the second case, we recognize that the response is not missing at random and we have included another variable, x_2, which we believe to be a good surrogate for y.

We generated the triples $\{(x_{1i}, x_{2i}, y_i) : i = 1, \ldots, n\}$ by first generating y_i distributed as F_Y, x_{1i}^* distributed as $F_{X_1^*}$, and x_{2i}^* distributed as $F_{X_2^*}$, independently, $i = 1, \ldots, n$. We then let $x_{1i} = \rho_1 y_i + \sqrt{1-\rho_1^2} x_{1i}^*$ and $x_{2i} = \rho_2 y_i + \sqrt{1-\rho_2^2} x_{2i}^*$, where $-1 < \rho_1, \rho_2 < 1$, so that the variables within each pair (y, x_1) and (y, x_2) are correlated. We will refer to ρ_1 and ρ_2 as correlations even though this would only be true if (y, x_1, x_2) was multivariate normal. To create an incomplete set of responses that are not missing at random, we then generated $r_i \sim \text{Bernoulli}(\pi_i)$, $i = 1, \ldots, n$, where $\text{logit}(\pi_i) = \gamma_0 + \gamma_1 x_{1i} + \gamma_2 y_i$, $-\infty < \gamma_0, \gamma_1 < \infty$.

A moderate sample size of $n = 300$ was used. The set of responses, $\{y_1, \ldots, y_n\}$, was generated from one of three distributions: symmetric unimodal, symmetric bimodal, or left-skewed unimodal. Mixtures of normal distributions were used because of their flexibility and the ease of computing the ISE when using a normal kernel (Marron and Wand, 1992). The symmetric unimodal distribution was $f_1(y) = \phi(y; 3/2, 3/2)$, the symmetric bimodal distribution was $f_2(y) = 0.5\phi(y; 0, 1) + 0.5\phi(y; 4, 1)$, and the skewed distribution was $f_3(y) = 0.4\phi(y; 2, 2) + 0.6\phi(y; 37/12, 10/9)$, where $\phi(y; \theta, \sigma)$ represents the normal density with mean θ and standard deviation σ. We generated $\{x_{11}^*, \ldots, x_{1n}^*\}$ and $\{x_{21}^*, \ldots, x_{2n}^*\}$ independently from standard normal distributions, and we set ρ_1 and ρ_2 equal to one of 0, 0.5, and 0.8. In generating the indicators $\{r_1, \ldots, r_n\}$, we selected values of $\{\gamma_0, \gamma_1, \gamma_2\}$ to achieve approximately 25% missing data; the exact values of these parameters depended on the distribution of the responses and the correlation between the responses and auxiliary variables. It should be noted that the probability of observing the response variable increased as y and x_1 increased.

The different combinations of response variable distribution and correlations resulted in 27 different simulation settings. For each setting, the ISE of each density estimate was computed for each of the 501 data sets generated. The following density estimators were compared: full data (Full); modified Horvitz-Thompson with true propensity scores (HTt), with Nadaraya-Watson propensity score estimates using only x_1 (HTMO, O=one auxiliary variable), and with Nadaraya-Watson estimates using both x_1 and x_2 (HTME, E=extra surrogate variable); and augmented Horvitz-Thompson with Nadaraya-Watson propensity score estimates using x_1 only (HTAO) and with Nadaraya-Watson estimates using both x_1 and x_2 (HTAE). The complete-case estimator was also included in the simulation study; the results are omitted here since the observed median ISE was always much larger than any of the Horvitz-Thompson-type KDEs. Also, note that each of the aforementioned KDEs was computed using the normal kernel and using a bandwidth determined by the Sheather-Jones method (Sheather and Jones, 1991) based on the complete cases; the bandwidth for the full-data KDE was based on the full data. Furthermore, the expectation in the second term of the augmented Horvitz-Thompson KDEs was computed using weighted local linear regression and a common bandwidth, using the corrected AIC method (McQuarrie and Tsai, 1998), for each evaluation point.

Dubnicka (2009) evaluated four bandwidth selection procedures under the MAR assumption: normal optimal smoothing, least squares cross validation, the Sheather-Jones plug-in method, and a weighted least squares cross validation method that was introduced in that paper. In a simulation study that compared the ISEs of each method, the Sheather-Jones method was found to be at least as good as the other methods for all KDEs considered under a variety of conditions. A smaller-scale study was also conducted under the NMAR assumption, with similar results. Certainly, in practice, one has the luxury of evaluating the effect of several different bandwidths on the chosen KDE, whether using a particular bandwidth selection procedure or not. In a simulation study, the bandwidth selection process must be automated.

Figure 4 shows a plot of the observed median ISEs for each of the above methods under the various combinations of response distribution, correlation between y and x_1 (ρ_1), and correlation between y and x_2 (ρ_2). The different rows represent the different response distributions, while the different columns represent the different values of correlation between y and x_1 considered. Within a graph, the horizontal axis represents the correlation between y and x_2 and the vertical axis is the observed median ISE.

One goal of this study was to determine how the (ordinary) modified Horvitz-Thompson KDEs perform when we assume the response is missing at random and it is actually not missing at random. Therefore, we first focus on the ordinary Horvitz-Thompson kernel density estimators, HTMO and HTME, which use one auxiliary variable and both the auxiliary variable and the surrogate variable, respectively. From Figure 4, it appears that the performance, in terms of observed median ISEs, depends somewhat on the true response distribution. For the symmetric response distributions, the median ISE for HTMO is greater than that of HTt unless y and x_1 are highly correlated ($\rho_1 = 0.8$). Also, the median ISE for HTME is greater than that of HTt unless (y, x_2) are highly correlated, i.e., $\rho_2 = 0.8$. In that case, the ISEs are approximately the same or that of HTt is slightly larger. For the left-skewed distribution, however, the median ISE of HTMO is greater than that of HTt unless y and x_1 are at least moderately corre-

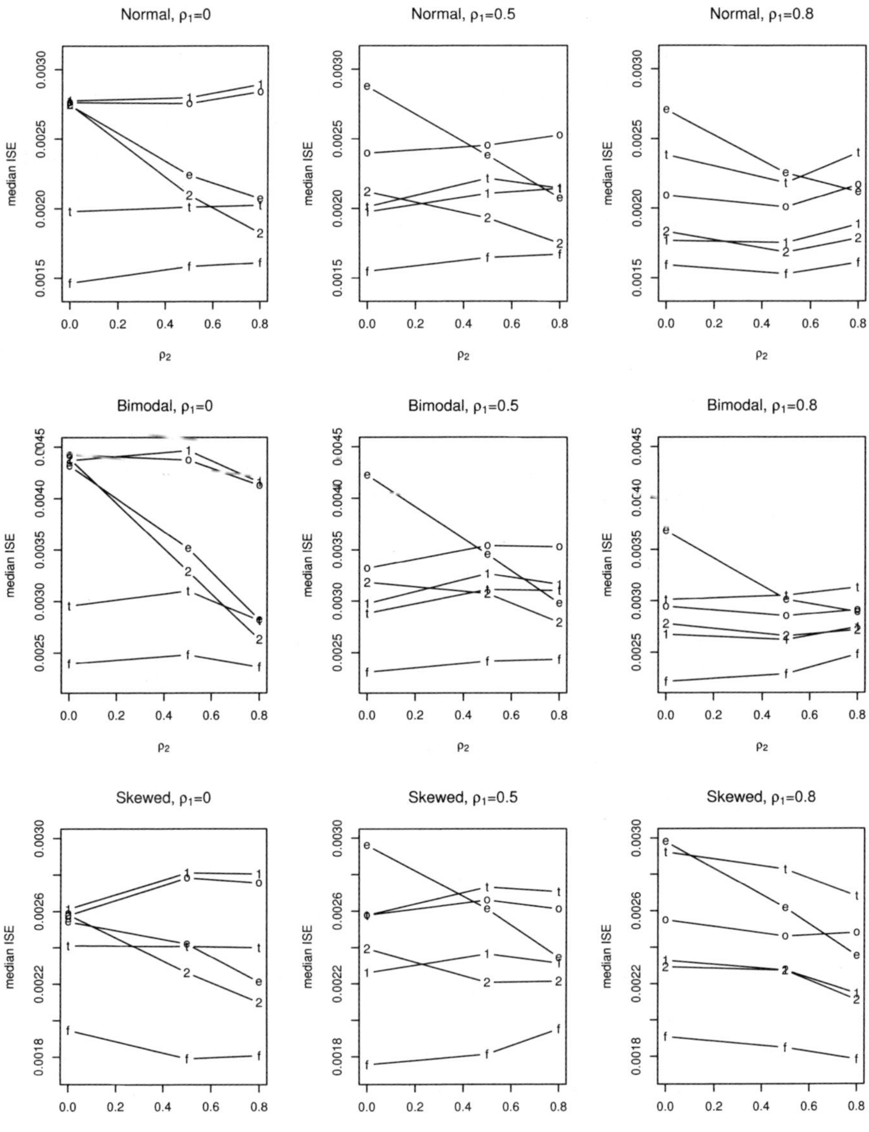

Fig. 4. Observed median integrated squared errors for various kernel density estimators: f=Full, t=HTt, o=HTMO, e=HTME, 1=HTAO, and 2=HTAE.

lated ($\rho_1 \geq 0.5$), and the median ISE of HTME is greater than that of HTt unless y is at least moderately correlated with either x_1 or x_2 ($\rho_1 \geq 0.5$ or

$\rho_2 \geq 0.5$). Therefore, unless the response is highly correlated with the auxiliary variables, the modified Horvitz-Thompson kernel density estimators with estimated propensity scores did not perform as well as the modified Horvitz-Thompson kernel density estimator with true propensity scores. Furthermore, regardless of the response variable distribution, the modified Horvitz-Thompson KDE with estimated propensity scores using the surrogate variable actually performed worse than if only the one (correct) auxiliary variable was used if that surrogate variable was uncorrelated with the response. However, for larger correlations between the response and the surrogate variable, the reverse was true.

The empirical results regarding the modified Horvitz-Thompson KDE highlight the need for an improved kernel density estimator under the not missing at random assumption. Thus, we next focus on comparisons among all the KDEs considered. When the response is uncorrelated with either the auxiliary variable or the surrogate variable, all of the Horvitz-Thompon-type KDEs with estimated propensity scores had approximately the same median ISE, which was greater than the median ISEs of both the modified Horvitz-Thompson KDE with true propensity scores (HTMt) and the full data KDE (Full). When y and x_1 were uncorrelated, regardless of the correlation between y and the surrogate x_2, the median ISEs of HTMO and HTAO were approximately the same and remained constant. Under the same conditions, the median ISEs of HTAE is less than or equal to that of HTME, and both decreased as the correlation between y and x_2 increased. Furthermore, the median ISEs of HTME and HTAE were less than or equal to that of HTt when y and x_2 were highly correlated. It is reasonable to conclude that when the response y and the auxiliary variable x_1 are uncorrelated, it is worthwhile to include the surrogate variable x_2 in estimating the propensity scores. Also, there is a slight improvement in performance when using the augmented Horvitz-Thompson KDE over the modified Horvitz-Thompson KDE.

As the correlation between the response y and the auxiliary variable x_1 increases, however, somewhat different patterns emerge. First, consider the case when y and x_1 are moderately correlated, $\rho_1 = 0.5$. As one would expect, as the correlation between y and x_2 increases, the median ISEs of HTME and HTAE decrease while the median ISEs of HTMO and HTAO remain relatively constant. Whether or not the extra auxiliary is used in estimating the propensity scores, the median ISE of the augmented Horvitz-Thompson KDE (HTAO and HTAE) is smaller that that of the corresponding modified Horvitz-Thompson KDE (HTMO and HTME). Furthermore, when the correlation between y and x_2 is moderate to large, the median ISEs of the augmented Horvitz-Thompson KDEs, using either one or both auxiliary variables in estimating the propensity scores, are as least as small as that of the Horvitz-Thompson KDE with true propensity scores, and using both the auxiliary variable and the surrogate variable provides the greatest improvement.

Finally, consider the case when the response y and the auxiliary variable x_1 are highly correlated, $\rho_1 = 0.8$. Except for HTME, all Horvitz-Thompson-type KDEs with estimated propensity scores have smaller median ISEs than that of the Horvitz-Thompson KDE with true propensity scores, regardless of the correlation between the response y and the surrogate variable x_2. The median ISE of HTME is only less than that of HTt when the correlation between y and x_2 is at least moderately large, so that adding a surrogate variable that is not correlated with the response actually decreases the performance of HTME. Also, because the response y and the auxiliary variable x_1 are highly correlated, the median ISEs are approximately the same for the two augmented Horvitz-Thompson KDEs; adding the surrogate variable x_2 does not improve performance.

6. Summary

In this paper, we evaluated the Horvitz-Thompson kernel density estimators proposed by Dubnicka (2009) for the estimation of the density of an incomplete response variable when the missing data mechanism was misspecified. In particular, we considered the situation in which the responses were incorrectly assumed to be missing at random. We also discussed an augmented Horvitz-Thompson kernel density estimator that was designed to better address the issue of missing data mechanism misspecification. The modified Horvitz-Thompson kernel density estimator with Nadaraya-Watson estimates performed well under some circumstances when the missing data mechanism was misspecified, depending on the underlying distribution and the availability of additional variables to aid in estimating the propensity scores. However, under other circumstances, this estimator was the worst among all those considered. It is possible that additional auxiliary or surrogate variables would improve these estimators if they are available, but further evaluation is needed.

The proposed augmented Horvitz-Thompson kernel density estimator performed well under misspecification of the missing data mechanism, in some cases outperforming the modified Horvitz-Thompson kernel density estimator with true propensity scores, despite the theoretical reasons for it to fail. The performance of the proposed density estimator was especially good when a surrogate variable was available to aid in estimating the propensity scores or when the auxiliary variable was moderately correlated with the response variable. Despite the additional computational burden, the proposed augmented Horvitz-Thompson kernel density estimator, with propensity scores estimated via the Nadaraya-Watson estimator using all useful auxiliary and surrogate variables, is recommended over the modified Horvitz-Thompson kernel density estimator with the same propensity score estimates when the response variable could possibly be not missing at random. A small simulation study, not presented here, also indicated that the augmented estimator performed similarly to the modified estimator when the response was correctly assumed to be missing at random.

References

Anscombe, F. J. (1967). Topics in the investigation of linear relations fitted by the method of least squares, *Journal of the Royal Statistical Society, Series B* **29**, pp. 1–52.

Bang, H. and Robins, J. M. (2005). Doubly robust estimation in missing data and causal inference models, *Biometrics* **61**, pp. 962–972.

Davidian, M., Tsiatis, A. A. and Leon, S. (2005). Semiparametric estimation of treatment effect in a pretest-posttest study with missing data, *Statistical Science* **20**, pp. 261–301.

Dubnicka, S. R. (2009). Kernel density estimation with missing data and auxiliary variables, *Australian and New Zealand Journal of Statistics* **51**, 3, pp. 247–270.

Hammer, S. M., Katzenstein, D. A., Hughes, M. D., Gundaker, H., Schooley, R. T., Haubrich, R. H., Henry, W. K., Lederman, M. M., Phair, J. P., Niu, M., Hirsch, M. S., Merigan, T. C. and for The AIDS Clinical Trials Group Study 175 Study Team (1996). A trial comparing nucleoside monotherapy with combination therapy in HIV-infected adults with CD4 cell counts from 200 to 500 per cubic millimeter, *New England Journal of Medicine* **335**, pp. 1081–1090.

Kang, J. D. Y. and Schafer, J. L. (2007). Demystifying double robustness: a comparison of alternative strategies for estimating a population mean form incomplete data (with discussion), *Statistical Science* **22**, pp. 523–580.

Kennedy, K. F. (2007). *A weighted estimating equation approach to local linear regression with missing covariate data*, Master's thesis, Kansas State University.

Little, R. J. A. and Rubin, D. B. (2002). *Statistical Analysis with Missing Data, second ed.* (John Wiley and Sons, New Jersey).

Marron, J. S. and Wand, M. P. (1992). Exact mean integrated squared error, *Annals of Statistics* **20**, pp. 712–736.

McQuarrie, A. D. R. and Tsai, C. L. (1998). *Regression and Time Series Model*

Selection. (World Scientific Publishing Co. Pte. Ltd., New Jersey).

Robins, J. M., Hernan, M. A. and Brumback, B. (2000). Marginal structural models and causal inference in epidemiology, *Epidemiology* **11**, pp. 550–560.

Robins, J. M., Rotnitzky, A. and Zhao, L. P. (1994). Estimation of regression coefficients when some regressors are not always observed, *Journal of the American Statistical Association* **89**, pp. 846–866.

Sheather, S. J. and Jones, M. C. (1991). A reliable data-based bandwidth selection method for kernel density estimation, *Journal of the Royal Statistical Society, Series B* **53**, pp. 683–690.

Tibshirani, R. and Hastie, T. (1987). Local likelihood estimation, *Journal of the American Statistical Association* **82**, pp. 559–567.

Tukey, J. W. (1967). Discussion of: Topics in the investigation of linear relations fitted by the method of least squares, *Journal of the Royal Statistical Society, Series B* **29**, pp. 47–48.

On the Non-Gaussian Asymptotics of the Likelihood Ratio Test Statistic for Homogeneity of Covariance

Marc Hallin*

ECARES, Institut de Recherche en Statistique, and Département de Mathématique
Université Libre de Bruxelles, B-1050 Bruxelles, Belgium
E-mail: mhallin@ulb.ac.be

The likelihood ratio test for m-sample homogeneity of covariance is notoriously sensitive to violations of the Gaussian assumptions. Its asymptotic behavior under non-Gaussian densities has been the subject of an abundant literature. In a recent paper, Yanagihara *et al.* (2005) show that the asymptotic distribution of the likelihood ratio test statistic, under arbitrary elliptical densities with finite fourth-order moments, is that of a linear combination of two mutually independent chi-square variables. Their proof is based on characteristic function methods, and only allows for convergence in distribution conclusions. Moreover, they require homokurticity among the m populations. Exploiting the findings of Hallin and Paindaveine (2009), we reinforce that convergence-in-distribution result into a convergence-in-probability one—that is, we explicitly decompose the likelihood ratio test statistic into a linear combination of two variables that are asymptotically independent chi-square—and moreover extend it to the heterokurtic case.

Keywords: Elliptical densities; Homogeneity of covariances; Local asymptotic normality; Locally asymptotically most stringent tests; Multivariate analysis of variance

1. Introduction

Likelihood ratio tests (LRTs) for covariance matrices are notoriously sensitive to violations of strict Gaussian assumptions, and the problem of extending their validity to more general classes of distributions has generated a huge amount of literature. In a classical reference, Muirhead and Waternaux (1980) provide a general study of the problem of turning such tests into pseudo-Gaussian ones remaining valid under elliptical densities with adequate moment assumptions (typically, finite fourth-order moments). They

*Also member of the Académie Royale de Belgique and extra-muros Fellow, CenTER, Tilburg University

clearly distinguish some "easy" cases—tests of sphericity, tests of equality of a subset of the characteristic roots of the covariance matrix (i.e., *subspace sphericity*), tests of block-diagonality—and some "harder" ones, among which are the (apparently simpler) one-sample test of the hypothesis that the covariance matrix $\mathbf{\Sigma}$ takes some given value $\mathbf{\Sigma}_0$, the two-sample test of equality of covariance matrices, and the corresponding m-sample tests. The "easy cases" only require multiplying the traditional LRT statistic by some factor involving a consistent estimator of kurtosis; these cases have been fully characterized and solved by Shapiro and Browne (1987). As for the "hard" ones, Muirhead and Waternaux's helpless conclusion is that *"it is not possible in the more general elliptical case to adjust the (Gaussian likelihood ratio) test so that its limiting distribution agrees with that obtained under the normality assumption".*

Muirhead and Waternaux, as we shall see, were too pessimistic, and solutions since then have been obtained for some of the "hard" cases too—see Hallin and Paindaveine (2008a) for a general method. Nevertheless, it took another twenty years of uninterrupted efforts for such results (Schott, 2001; Hallin and Paindaveine, 2008b, 2009, for homogeneity of covariances) to appear, with contributions by Gupta and Tang (1984), Browne (1984), Hayakawa (1986), Shapiro and Browne (1987), Wakaki *et al.* (1990), Nagao and Srivastava (1992), Zhang and Boos (1992, 1993), Yuan and Bentler (1999), Tonda and Wakaki (2003), Yanagihara *et al.* (2005), Gupta and Xu (2006), Hallin and Paindaveine (2006), and Paindaveine (2008), to name only a few.

Due to its fundamental role in multivariate analysis of variance and covariance, the problem of testing the null hypothesis $\mathcal{H}_0$ of homogeneity of covariances, on which we concentrate here, is of special interest for applications. Denoting by $(\mathbf{X}_{i1}, \ldots, \mathbf{X}_{in_i})$, $i = 1, \ldots, m$, a collection of m mutually independent samples of i.i.d. random k-dimensional vectors with location parameters $\boldsymbol{\theta}_i$ and covariance matrices $\mathbf{\Sigma}_i$, this null hypothesis is written $\mathcal{H}_0 : \mathbf{\Sigma}_1 = \ldots = \mathbf{\Sigma}_m$. The Gaussian likelihood ratio test $\phi^{(n)}_{\mathrm{LRT}}$ for this problem was first obtained by Wilks (1932) and rejects $\mathcal{H}_0$ for large values of $Q^{(n)}_{\mathrm{LRT}} := -2\log\Lambda^{(n)}$, with

$$\Lambda^{(n)} := \frac{\prod_{i=1}^{m} |\mathbf{W}_i/n_i|^{n_i/2}}{|\mathbf{W}/n|^{n/2}} =: \frac{\prod_{i=1}^{m} |\mathbf{S}_i|^{n_i/2}}{|\mathbf{S}|^{n/2}}, \tag{1}$$

where $\bar{\mathbf{X}}_i := n_i^{-1}\sum_{j=1}^{n_i}\mathbf{X}_{ij}$, $\mathbf{W}_i := \sum_{j=1}^{n_i}(\mathbf{X}_{ij} - \bar{\mathbf{X}}_i)(\mathbf{X}_{ij} - \bar{\mathbf{X}}_i)' =: n_i\mathbf{S}_i$, and $\mathbf{W} := \sum_{i=1}^{m}\mathbf{W}_i =: n\mathbf{S}$. Even under Gaussian assumptions, this LRT is actually biased, and one therefore usually relies on the Bartlett (1937) modified LRT $\phi^{(n)}_{\text{MLRT}}$, based on $Q^{(n)}_{\text{MLRT}} := -2\log\dot{\Lambda}^{(n)}$, where

$$\dot{\Lambda}^{(n)} := \frac{\prod_{i=1}^{m} |\mathbf{W}_i/\dot{n}_i|^{\dot{n}_i/2}}{|\mathbf{W}/\dot{n}|^{\dot{n}/2}} =: \frac{\prod_{i=1}^{m} |\dot{\mathbf{S}}_i|^{\dot{n}_i/2}}{|\dot{\mathbf{S}}|^{\dot{n}/2}}, \tag{2}$$

with $\dot{n}_i := n_i - 1$ and $\dot{n} := \sum_{i=1}^{m} \dot{n}_i = n - m$; the statistics $Q^{(n)}_{\text{MLRT}}$ and $Q^{(n)}_{\text{LRT}}$ being asymptotically equivalent, however, we do not distinguish between the modified and the "unmodified" LRT anymore in the sequel, with a unique notation $Q^{(n)}_{\text{Wilks}}$.

2. A generalization of a result by Yanagihara *et al.* (2005)

In this context of testing covariance homogeneity, Yanagihara *et al.* (2005) show that, under *homokurtic* elliptical densities (when referring to *homo-* or *heterokurticity*, we of course tacitly assume the existence of finite fourth-order moments; the Yanagihara *et al.* result moreover holds under a slightly more general family of *generalized elliptical densities*), the asymptotic null distribution of $Q^{(n)}_{\text{Wilks}}$ is provided by

$$Q^{(n)}_{\text{Wilks}} \xrightarrow{\mathcal{L}} (1+\kappa)\left\{\left[1 + \frac{k\kappa}{2(1+\kappa)}\right] Y_1 + Y_2\right\}, \tag{3}$$

where Y_1 and Y_2 are independent chi-square random variables with $(m-1)$ and $(m-1)(k-1)(k+2)/2$ degrees of freedom, respectively, κ stands for the common radial kurtosis of the m underlying elliptical distributions, and $\xrightarrow{\mathcal{L}}$ for convergence in distribution under the null. In the multinormal case, $\kappa = 0$, and (3) yields the well-known Gaussian result of Wilks (1932) that $Q^{(n)}_{\text{Wilks}}$ under the null hypothesis is asymptotically chi-square with $(m-1)k(k+1)/2$ degrees of freedom. But for $\kappa \neq 0$, (3) is no longer asymptotically chi-square (see also Gupta and Xu, 2006).

The $(1+\kappa)$ factor sitting in front of (3) is not uncommon in the context of likelihood ratio testing for covariance matrices (see Theorem 1 of Shapiro and Browne, 1987, for a general result about this) and is very easily dealt with by dividing $Q^{(n)}_{\text{Wilks}}$ by some consistent estimator $(1+\hat{\kappa})$. The presence of κ in the coefficient of Y_1, however, is more problematic. Indeed, a corrected version of $Q^{(n)}_{\text{Wilks}}$ would require replacing Y_1 and Y_2 with $[1+(k+2)\hat{\kappa}/2]^{-1}Y_1$ and $(1+\hat{\kappa})^{-1}Y_2$, respectively. The result in Yanagihara *et al.* (2005) being of a purely distributional nature, with proofs based on characteristic function methods, it does not provide any definition of Y_1 and Y_2 in terms of the observations, and such a correction therefore cannot be implemented.

Instead of (3), which is a convergence-in-distribution result, we provide here a convergence-in-probability result, with an explicit decomposition of $Q^{(n)}_{\text{Wilks}}$ into a linear combination of explicitly defined and well-interpretable component variables. Our decomposition moreover allows for heterokurtic populations.

As mentioned before, we throughout assume that all populations are elliptically symmetric and possess finite moments of order four. More precisely, defining, for $q \geq 2$,

$$\mathcal{F}^q := \left\{ h : \mathbb{R}_0^+ \to \mathbb{R}^+ : \mu_{k+q-1;h} < \infty \right\} \quad \text{and}$$
$$\mathcal{F}_1^q := \left\{ h \in \mathcal{F}^q \ : \frac{\mu_{k+1;h}}{\mu_{k-1;h}} = k \right\},$$

respectively, where $\mu_{\ell;h} := \int_0^\infty r^\ell h(r)\, dr$, we require the following.
ASSUMPTION (A). The observations $\mathbf{X}_{ij}$, $j = 1, \ldots, n_i$, are mutually independent, with probability density function

$$\mathbf{x} \mapsto c_{k,f_i} |\mathbf{\Sigma}_i|^{-1/2} f_i \left(\left((\mathbf{x} - \boldsymbol{\theta}_i)' \mathbf{\Sigma}_i^{-1} (\mathbf{x} - \boldsymbol{\theta}_i) \right)^{1/2} \right), \quad i = 1, \ldots, m, \qquad (4)$$

for some k-dimensional vector $\boldsymbol{\theta}_i$ (*location*), some positive definite $(k \times k)$ covariance matrix $\mathbf{\Sigma}_i$, and some f_i in the class $\mathcal{F}_1^4$ of *standardized radial densities* with finite fourth-order moments.

Define (throughout, $\mathbf{\Sigma}^{1/2}$ stands for the symmetric root of $\mathbf{\Sigma}$) the *elliptical coordinates*

$$\mathbf{U}_{ij}(\boldsymbol{\theta}_i, \mathbf{\Sigma}_i) := \frac{\mathbf{\Sigma}_i^{-1/2}(\mathbf{X}_{ij} - \boldsymbol{\theta}_i)}{\|\mathbf{\Sigma}_i^{-1/2}(\mathbf{X}_{ij} - \boldsymbol{\theta}_i)\|} \quad \text{and} \quad d_{ij}(\boldsymbol{\theta}_i, \mathbf{\Sigma}_i) := \|\mathbf{\Sigma}_i^{-1/2}(\mathbf{X}_{ij} - \boldsymbol{\theta}_i)\|. \qquad (5)$$

Under Assumption (A), the unit vectors $\mathbf{U}_{ij}$, $j = 1, \ldots, n_i$ and $i = 1, \ldots, m$, are i.i.d. uniform over the unit sphere in $\mathbb{R}^k$, and the *standardized elliptical distances* d_{ij} are independent of the $\mathbf{U}_{ij}$, with density $\tilde{f}_{ik}(r) := (\mu_{k-1;f_i})^{-1} r^{k-1} f_i(r)$ (justifying the terminology *standardized radial density* for f_i) and distribution function $\tilde{F}_{ik}$. The condition that $f_i \in \mathcal{F}^4$ is equivalent to the finiteness of d_{ij}'s fourth-order moments, while $\mathcal{F}_1^4 \subset \mathcal{F}_1^2$ implies that f_i is standardized in such a way that $\mathrm{E}[d_{ij}^2(\boldsymbol{\theta}_i, \mathbf{\Sigma}_i)] = k$, hence that $\mathbf{\Sigma}_i = \mathrm{Var}[\mathbf{X}_{ij}]$ is the covariance matrix in population i.

Although, for the sake of notational simplicity, we do not mention it explicitly, we actually consider sequences of statistical experiments, with triangular arrays of observations of the form $(\mathbf{X}_{11}^{(n)}, \ldots, \mathbf{X}_{1n_1^{(n)}}^{(n)}, \mathbf{X}_{21}^{(n)}, \ldots, \mathbf{X}_{2n_2^{(n)}}^{(n)}, \ldots, \mathbf{X}_{m1}^{(n)}, \ldots, \mathbf{X}_{mn_m^{(n)}}^{(n)})$ indexed by the total sample size n, where the m sequences $n_i^{(n)}$, $n \in \mathbb{N}$ of sample sizes satisfy the following assumption.

ASSUMPTION (B). For all $i = 1, \ldots, m$, $n_i = n_i^{(n)} \to \infty$ as $n \to \infty$.

Denoting by

$$\kappa_i := [k(k+2)]^{-1} \mathrm{E}_{\boldsymbol{\vartheta}}[d_{ij}^4(\boldsymbol{\theta}_i, \boldsymbol{\Sigma}_i)] - 1$$
$$= [k(k+2)]^{-1} \int_0^1 (\tilde{F}_{ik}^{-1}(u))^4 du - 1$$

the kurtosis coefficient in population i (see, e.g., page 54 of Anderson, 2003), let, for all $i \neq i'$,

$$\kappa_{i,i'} := \kappa_{i,i'}^{(n)} := \frac{n_i}{n_i + n_{i'}} \kappa_{i'} + \frac{n_{i'}}{n_i + n_{i'}} \kappa_i. \tag{6}$$

Proposition 2.1. *Let Assumptions (A) and (B) hold. Then, under the null hypothesis $\mathcal{H}_0^{(n)}$ of covariance homogeneity,*

$$Q_{\mathrm{Wilks}}^{(n)} = \frac{1}{2kn} \sum_{1 \leq i < i' \leq m} n_i n_{i'} \mathrm{tr}^2[\mathbf{S}^{-1}(\mathbf{S}_i - \mathbf{S}_{i'})] \tag{7}$$
$$+ \frac{1}{2n} \sum_{1 \leq i < i' \leq m} n_i n_{i'} \Big\{ \mathrm{tr}[(\mathbf{S}^{-1}(\mathbf{S}_i - \mathbf{S}_{i'}))^2] - \frac{1}{k} \mathrm{tr}^2[\mathbf{S}^{-1}(\mathbf{S}_i - \mathbf{S}_{i'})] \Big\}$$
$$+ o_{\mathrm{P}}(1)$$
$$= \frac{1}{n} \sum_{1 \leq i < i' \leq m} (n_i + n_{i'}) \Big(1 + (k+2)\kappa_{i,i'}/2 \Big) Q_{\dagger;i,i'}^{(n)\ \mathrm{scale}}$$
$$+ \frac{1}{n} \sum_{1 \leq i < i' \leq m} (n_i + n_{i'})(1 + \kappa_{i,i'}) Q_{\dagger;i,i'}^{(n)\ \mathrm{shape}} + o_{\mathrm{P}}(1), \tag{8}$$

where the $m(m-1)/2$ statistics

$$Q_{\dagger;i,i'}^{(n)\ \mathrm{scale}} := \frac{n_i n_{i'}}{2k(n_i + n_{i'})} \frac{1}{1 + (k+2)\kappa_{i,i'}/2} \mathrm{tr}^2[\mathbf{S}^{-1}(\mathbf{S}_i - \mathbf{S}_{i'})] \tag{9}$$

are asymptotically chi-square, each with one degree of freedom (not mutually independent, though), in such a way that

$$Q_{\dagger}^{(n)\ \mathrm{scale}} := \frac{1}{n} \sum_{1 \leq i < i' \leq m} (n_i + n_{i'}) Q_{\dagger;i,i'}^{(n)\ \mathrm{scale}} \tag{10}$$

is asymptotically chi-square with $(m-1)$ degrees of freedom and the $m(m-1)/2$ statistics

$$Q_{\dagger;i,i'}^{(n)\ \mathrm{shape}} := \frac{n_i n_{i'}}{2(n_i + n_{i'})} \frac{1}{1 + \kappa_{i,i'}} \Big\{ \mathrm{tr}[(\mathbf{S}^{-1}(\mathbf{S}_i - \mathbf{S}_{i'}))^2] - \frac{1}{k} \mathrm{tr}^2[\mathbf{S}^{-1}(\mathbf{S}_i - \mathbf{S}_{i'})] \Big\} \tag{11}$$

are asymptotically chi-square, with $(k-1)(k+2)/2$ degrees of freedom each (but not mutually independent), in such a way that

$$Q_{\dagger}^{(n)\ \mathrm{shape}} := \frac{1}{n} \sum_{1 \leq i < i' \leq m} (n_i + n_{i'}) Q_{\dagger;i,i'}^{(n)\ \mathrm{shape}} \tag{12}$$

is asymptotically chi-square with $(m-1)(k-1)(k+2)/2$ degrees of freedom. Moreover, the $Q_{\dagger;i,i'}^{(n)\ \mathrm{scale}}$ and $Q_{j,j'}^{(n)\ \mathrm{shape}}$ statistics are mutually independent for all i, i', j, j'; hence $Q_{\dagger}^{(n)\ \mathrm{scale}}$ and $Q_{\dagger}^{(n)\ \mathrm{shape}}$ are also mutually independent.

The proof of this proposition is provided in the Appendix.

An immediate consequence of this proposition is that, unless the $\kappa_{i,i'}$ all vanish (which happens under Gaussian conditions), the asymptotic distribution of $Q_{\mathrm{Wilks}}^{(n)}$ fails to be chi-square: Wilks' test thus loses its validity under non-Gaussian densities—which includes, in particular, all heterokurtic situations.

The statistics $Q_{\dagger;i,i'}^{(n)\ \mathrm{scale}}$ and $Q_{\dagger;i,i'}^{(n)\ \mathrm{shape}}$ are actually the test statistics of the optimal pseudo-Gaussian tests obtained by Hallin and Paindaveine (2009) for testing the subhypotheses $|\mathbf{\Sigma}_i| = |\mathbf{\Sigma}_{i'}|$ (equality of the generalized variances in populations i and i') and $\mathbf{\Sigma}_i/|\mathbf{\Sigma}_i|^{1/k} = \mathbf{\Sigma}_{i'}/|\mathbf{\Sigma}_{i'}|^{1/k}$ (equality of the *shape* matrices in populations i and i'), respectively. The null hypothesis of covariance homogeneity clearly holds if and only if all those subhypotheses, each of which is associated with an "elementary contrast", hold. The asymptotic equivalence (8) thus also provides an interesting insight into the structure of Wilks' statistic by decomposing $Q_{\mathrm{Wilks}}^{(n)}$ into a weighted sum of quadratic forms addressing those elementary contrasts. The way these contrasts are weighted, however, depends on the (unknown) kurtoses of the underlying populations; in particular, the relative importance given to scale and shape depends on the κ_i, which does not correspond to any sound decision-theoretical principle. The same reproach can be addressed to the bootstrapped versions of $Q_{\mathrm{Wilks}}^{(n)}$ developed in Zhang and Boos (1992, 1993).

It follows that the adequate correction, turning $Q_{\mathrm{Wilks}}^{(n)}$ into a pseudo-Gaussian test statistic with asymptotically chi-square distribution with $(m-1)k(k+1)/2$ degrees of freedom under arbitrary elliptical densities with finite fourth-order moments, is

$$\hat{Q}_{\dagger}^{(n)} := \frac{1}{2kn} \sum_{1 \le i < i' \le m} \frac{n_i n_{i'}}{1 + (k+2)\hat{\kappa}_{i,i'}/2} \mathrm{tr}^2[\mathbf{S}^{-1}(\mathbf{S}_i - \mathbf{S}_{i'})]$$
$$+ \frac{1}{2n} \sum_{1 \le i < i' \le m} \frac{n_i n_{i'}}{1 + \hat{\kappa}_{i,i'}} \Big\{ \mathrm{tr}\left[(\mathbf{S}^{-1}(\mathbf{S}_i - \mathbf{S}_{i'}))^2\right]$$
$$- \frac{1}{k} \mathrm{tr}^2\left[\mathbf{S}^{-1}(\mathbf{S}_i - \mathbf{S}_{i'})\right] \Big\} \qquad (13)$$
$$= \frac{1}{n} \sum_{1 \le i < i' \le m} (n_i + n_{i'}) \left[\hat{Q}_{\dagger;i,i'}^{(n)\,\mathrm{scale}} + \hat{Q}_{\dagger;i,i'}^{(n)\,\mathrm{shape}}\right],$$

where, denoting by $\hat{\kappa}_i$, $i = 1, \ldots, m$, consistent estimators of the kurtosis coefficients κ_i,

$$\hat{\kappa}_{i,i'} := \hat{\kappa}_{i,i'}^{(n)} := \frac{n_i}{n_i + n_{i'}} \hat{\kappa}_{i'} + \frac{n_{i'}}{n_i + n_{i'}} \hat{\kappa}_i, \quad i = 1, \ldots, m,$$

have been substituted for the $\kappa_{i,i'}$ in $Q_{\dagger;i,i'}^{(n)\,\mathrm{scale}}$ and $Q_{\dagger;i,i'}^{(n)\,\mathrm{shape}}$ defined in (9) and (11), respectively. Contrary to the bootstrapping approach, this correction preserves $Q_{\mathrm{Wilks}}^{(n)}$'s nature of an unweighted sum of chi-squared variables and does not affect its decision-theoretic properties.

More precisely, $Q_{\mathrm{Wilks}}^{(n)}$, although formally defined [see Eq. (7)] as an unweighted sum over $i < i'$, actually behaves asymptotically, under non-Gaussian densities, as a kurtosis-weighted sum of independent chi-square terms, each of which is associated with a particular direction in the space of local alternatives [see Eq. (8)]. This not only affects (still, under non-Gaussian densities) its asymptotic chi-square distribution; it also ruins the maximin features associated with the "equal treatment" aspects of unweighted sums of chi-squares, by privileging, in a totally uncontrolled way, some of the terms, hence the corresponding local alternatives. The weights we are introducing (based on empirical kurtosis) in the definition (13) of $\hat{Q}_{\dagger}^{(n)}$ are carefully designed in order to neutralize that undesirable imbalance and restore "equal treatment" among all terms, hence among all local alternatives.

Note that $\hat{Q}_{\dagger}^{(n)}$ is precisely the test statistic resulting from the general result of Hallin and Paindaveine (2008a) and proposed in Section 5.2 of Hallin and Paindaveine (2009).

3. The homokurtic case

In the homokurtic case—that is, when all kurtosis parameters κ_i have some common value κ—definition (6) yields $\kappa_{i,i'}^{(n)} = \kappa$ for all n, and (8) reduces to

$$
\begin{aligned}
Q^{(n)}_{\text{Wilks}} &= \Big(1+(k+2)\kappa/2\Big)\frac{1}{n}\sum_{1\le i<i'\le m}(n_i+n_{i'})Q^{(n)\,\text{scale}}_{\dagger;i,i'} \\
&\quad +(1+\kappa)\frac{1}{n}\sum_{1\le i<i'\le m}(n_i+n_{i'})Q^{(n)\,\text{shape}}_{\dagger;i,i'}+o_{\mathrm{P}}(1) \\
&= (1+(k+2)\kappa/2)Q^{(n)\,\text{scale}}_{\dagger}+(1+\kappa)Q^{(n)\,\text{shape}}_{\dagger} \\
&= (1+\kappa)\Big\{\Big[1+\frac{k\kappa}{2(1+\kappa)}\Big]Q^{(n)\,\text{scale}}_{\dagger}+Q^{(n)\,\text{shape}}_{\dagger}\Big\}, \qquad (14)
\end{aligned}
$$

to be compared with (3). Denoting by $\hat\kappa = \hat\kappa^{(n)}$ a consistent estimator of the common kurtosis κ, the corrected version of $Q^{(n)}_{\text{Wilks}}$ is thus

$$
\begin{aligned}
\hat Q^{(n)}_{*} &= \frac{1}{2kn}(1+(k+2)\hat\kappa/2)^{-1}\sum_{1\le i<i'\le m} n_i n_{i'}\,\mathrm{tr}^2[\mathbf{S}^{-1}(\mathbf{S}_i-\mathbf{S}_{i'})] \\
&\quad +\frac{1}{2n}(1+\hat\kappa)^{-1}\sum_{1\le i<i'\le m} n_i n_{i'}\Big\{\mathrm{tr}\left[(\mathbf{S}^{-1}(\mathbf{S}_i-\mathbf{S}_{i'}))^2\right] \\
&\qquad\qquad -\frac{1}{k}\mathrm{tr}^2\left[\mathbf{S}^{-1}(\mathbf{S}_i-\mathbf{S}_{i'})\right]\Big\} \qquad (15) \\
&= \frac{1}{n}\sum_{1\le i<i'\le m}(n_i+n_{i'})\Big[\hat Q^{(n)\,\text{scale}}_{*;i,i'}+\hat Q^{(n)\,\text{shape}}_{*;i,i'}\Big],
\end{aligned}
$$

which coincides with the homokurtic pseudo-Gaussian test statistic of Section 5.1 of Hallin and Paindaveine (2009).

Appendix A. Proof of Proposition 1

This appendix is devoted to the proof of Proposition 1. We first show that, under $\mathcal{H}^{(n)}_0$,

$$
Q^{(n)}_{\text{Wilks}} = \frac{1}{2n}\sum_{1\le i<i'\le m} n_i n_{i'}\ \mathrm{tr}\Big[(\mathbf{S}^{-1}(\mathbf{S}_i-\mathbf{S}_{i'}))^2\Big]+o_{\mathrm{P}}(1) \qquad \text{(A.1)}
$$

as $n\to\infty$. For notational convenience, we do not work with the Bartlett corrected version, but the result clearly remains valid for this version. Letting

$$
\mathbf{\Sigma}^{1/2}\mathbf{Z}_i\mathbf{\Sigma}^{1/2} := n_i^{1/2}(\mathbf{S}_i-\mathbf{\Sigma}) =: n_i^{1/2}\mathbf{\Sigma}\mathbf{Y}_i,
$$

where $\mathbf{\Sigma}$ stands for the common null value of the $\mathbf{\Sigma}_i$, and taking into account the fact that, as $\|\mathbf{A}\|\to 0$, $\log|\mathbf{I}_k+\mathbf{A}| = \mathrm{tr}\,\mathbf{A}-\frac{1}{2}\mathrm{tr}\,(\mathbf{A}^2)+o(\|\mathbf{A}\|^2)$, we have that

$$
\begin{aligned}
Q_{\mathrm{Wilks}}^{(n)} := -2\log\Lambda^{(n)} &= -\sum_{i=1}^{m} n_i \log|\mathbf{S}_i| + n\log|\mathbf{S}| \\
&= -\sum_{i=1}^{m} n_i \log|\mathbf{\Sigma}(\mathbf{I}_k + \mathbf{Y}_i)| + n\,\log\Big|\mathbf{\Sigma}\Big(\mathbf{I}_k + \Big(\frac{1}{n}\sum_{i=1}^{m} n_i\mathbf{Y}_i\Big)\Big)\Big| \\
&= \frac{1}{2}\left\{\sum_{i=1}^{m} n_i \,\mathrm{tr}\left[\mathbf{Y}_i^2\right] - \frac{1}{n}\,\mathrm{tr}\left[\Big(\sum_{i=1}^{m} n_i\mathbf{Y}_i\Big)^2\right]\right\} + o_{\mathrm{P}}(1) \\
&= \frac{1}{2}\sum_{i=1}^{m} n_i \,\mathrm{tr}\left[\Big(\mathbf{Y}_i - \Big(\frac{1}{n}\sum_{i'=1}^{m} n_{i'}\mathbf{Y}_{i'}\Big)\Big)^2\right] + o_{\mathrm{P}}(1) \\
&= \frac{1}{2}\sum_{i=1}^{m} n_i \,\mathrm{tr}\left[\left(\mathbf{\Sigma}^{-1}\left(\mathbf{S}_i - \mathbf{S}\right)\right)^2\right] + o_{\mathrm{P}}(1) \\
&= \frac{1}{2}\sum_{i=1}^{m} n_i \,\mathrm{tr}\left[\left(\mathbf{S}^{-1}\left(\mathbf{S}_i - \mathbf{S}\right)\right)^2\right] + o_{\mathrm{P}}(1) \qquad \text{(A.2)}
\end{aligned}
$$

as $n \to \infty$, under the null hypothesis of covariance homogeneity and any elliptical density with finite fourth-order moments. Now,

$$
\begin{aligned}
\frac{1}{2}\sum_{i=1}^{m} n_i \,\mathrm{tr}\left[(\mathbf{S}^{-1}(\mathbf{S}_i - \mathbf{S}))^2\right] &= \frac{1}{2}\sum_{i=1}^{m} n_i \,\mathrm{tr}\left[(\mathbf{S}^{-1}\sum_{i'=1}^{m}\frac{n_{i'}}{n}(\mathbf{S}_i - \mathbf{S}_{i'}))^2\right] \\
&= \frac{1}{2n^2}\sum_{i,i',i''=1}^{m} n_i n_{i'} n_{i''} \\
&\quad \times \mathrm{tr}\left[\mathbf{S}^{-1}(\mathbf{S}_i - \mathbf{S}_{i'})\mathbf{S}^{-1}(\mathbf{S}_i - \mathbf{S}_{i''})\right].
\end{aligned}
$$

Splitting $\mathbf{S}_i - \mathbf{S}_{i''}$ into $(\mathbf{S}_i - \mathbf{S}_{i'}) + (\mathbf{S}_{i'} - \mathbf{S}_{i''})$ then yields

$$
\begin{aligned}
\frac{1}{2}\sum_{i=1}^{m} n_i \,\mathrm{tr}\left[(\mathbf{S}^{-1}(\mathbf{S}_i - \mathbf{S}))^2\right] &= \frac{1}{2n}\sum_{i,i'=1}^{m} n_i n_{i'} \,\mathrm{tr}\left[(\mathbf{S}^{-1}(\mathbf{S}_i - \mathbf{S}_{i'}))^2\right] \\
&\quad -\frac{1}{2}\sum_{i=1}^{m} n_i \,\mathrm{tr}\left[(\mathbf{S}^{-1}(\mathbf{S}_i - \mathbf{S}))^2\right].
\end{aligned}
$$

Plugging this into (A.2), we obtain

$$
\begin{aligned}
Q_{\mathrm{Wilks}}^{(n)} &= \frac{1}{4n}\sum_{i,i'=1}^{m} n_i n_{i'} \,\mathrm{tr}\left[(\mathbf{S}^{-1}(\mathbf{S}_i - \mathbf{S}_{i'}))^2\right] + o_{\mathrm{P}}(1) \\
&= \frac{1}{2n}\sum_{1\le i<i'\le m} n_i n_{i'} \,\mathrm{tr}\left[(\mathbf{S}^{-1}(\mathbf{S}_i - \mathbf{S}_{i'}))^2\right] + o_{\mathrm{P}}(1),
\end{aligned}
$$

which establishes (A.1), hence also (7). Inserting the statistics defined in (9) and (11), (7) takes the form (8). The asymptotic distributions of $Q_{\dagger}^{(n)\ \mathrm{scale}}$ and $Q_{\dagger}^{(n)\ \mathrm{shape}}$ in (10) and (12), and their relation to the subhypotheses of scale and shape homogeneity, respectively, have been derived in Section 6.1 of Hallin and Paindaveine (2009). As for the asymptotic distributions of $Q_{\dagger;i,i'}^{(n)\ \mathrm{scale}}$ and $Q_{\dagger;i,i'}^{(n)\ \mathrm{shape}}$, they follow from those of $Q_{\dagger}^{(n)\ \mathrm{scale}}$ and $Q_{\dagger}^{(n)\ \mathrm{shape}}$ by letting $m = 2$.

Acknowledgments

The author thanks Davy Paindaveine for his careful and pertinent comments on the original manuscript. Supported by the Sonderforschungsbereich "Statistical modelling of non-linear dynamic processes" (SFB 823) of the Deutsche Forschungsgemeinschaft, and by a Discovery Grant of the Australian Research Council.

References

Anderson, T. W. (2003). *An Introduction to Multivariate Statistical Analysis, 3rd edition* (Wiley-Interscience, New York).

Bartlett, M. S. (1937). Properties of sufficiency and statistical tests, *Journal of the Royal Statistical Society, Series A* **160**, pp. 268–282.

Browne, M. W. (1984). Asymptotivally distribution-free methods or the analysis of covariance structures, *British Journal of Mathematical and Statistical Psychology* **37**, pp. 62–83.

Gupta, A. K. and Tang, J. (1984). Distribution of likelihood ratio statistic for testing equality of covariance matrices of multivariate gaussian models, *Annals of Statistics* **71**, pp. 555–559.

Gupta, A. K. and Xu, J. (2006). On some tests of the covariance matrix under general conditions, *Annals of the Institute of Statistical Mathematics* **58**, pp. 101–114.

Hallin, M. and Paindaveine, D. (2006). Parametric and semiparametric inference for shape: the role of the scale functional, *Statistics and Decisions* **24**, pp. 1001–1023.

Hallin, M. and Paindaveine, D. (2008a). A general method for constructing pseudo-gaussian tests, *Journal of the Japan Statistical Society* **38 (Celebration Volume for Hirotugu Akaike)**, pp. 27–40.

Hallin, M. and Paindaveine, D. (2008b). Optimal rank-based tests for homogeneity of scatter, *Annals of Statistics* **36**, pp. 1261–1298.

Hallin, M. and Paindaveine, D. (2009). Optimal tests for homogeneity of covariance, scale, and shape, *Journal of Multivariate Analysis* **100**, pp. 422–444.

Hayakawa, T. (1986). On testing hypotheses of covariance matrices under an elliptical population, *Journal of Statistical Planning and Inference* **13**, pp. 193–202.

Muirhead, R. J. and Waternaux, C. M. (1980). Asymptotic distributions in canonical correlation analysis and other multivariate procedures for nonnormal populations, *Biometrika* **67**, pp. 31–43.

Nagao, H. and Srivastava, M. S. (1992). On the distributions of some test criteria for a covariance matrix under local alternatives and bootstrap approximations, *Journal of Multivariate Analysis* **43**, pp. 331–350.

Paindaveine, D. (2008). A canonical definition of shape, *Statistics and Probability Letters* **78**, pp. 2240–2247.

Schott, J. R. (2001). Some tests for the equality of covariance matrices, *Journal of Statistical Planning and Inference* **94**, pp. 25–36.

Shapiro, A. and Browne, M. W. (1987). Analysis of covariance structures under elliptical distributions, *Journal of the American Statistical Association* **82**, pp. 1092–1097.

Tonda, T. and Wakaki, H. (2003). Asymptotic expansion of the null distribution of the likelihood ratio statistic for testing the equality of variances in a nonormal one-way ANOVA model, *Hiroshima Mathematical Journal* **33**, pp. 113–126.

Wakaki, H., Eguchi, S. and Fujikoshi, Y. (1990). A class of tests for a general covariance structure, *Journal of Multivariate Analysis* **32**, pp. 313–325.

Wilks, S. S. (1932). Certain generalizations in the analysis of variance, *Biometrika* **24**, pp. 471–494.

Yanagihara, H., Tonda, T. and Matsumoto, C. (2005). The effects of nonnormality on asymptotic distributions of some likelihood ratio criteria for testing covariance structures under normal assumption, *Journal of Multivariate Analysis* **96**, pp. 237–264.

Yuan, K. H. and Bentler, P. M. (1999). On normal theory and associated test statistics in covariance structure analysis under two classes of nonnormal distributions, *Statistica Sinica* **9**, pp. 831–853.

Zhang, J. and Boos, D. D. (1992). Bootstrap critical values for testing homogeneity of covariance matrices, *Journal of the American Statistical Association* **87**, pp. 425–429.

Zhang, J. and Boos, D. D. (1993). Testing hypotheses about covariance matrices using bootstrap methods, *Communications in Statistics—Theory and Methods* **22**, pp. 723–739.

Deconvolution Density Estimation on the Space of Positive Definite Symmetric Matrices

Peter T. Kim

Department of Mathematics and Statistics, University of Guelph
Guelph, Ontario N1G 2W1, Canada
E-mail: pkim@uoguelph.ca

Donald St. P. Richards

Department of Statistics, Pennsylvania State University
University Park, PA 16802, USA
E-mail: richards@stat.psu.edu

Motivated by applications in microwave engineering and diffusion tensor imaging, we study the problem of deconvolution density estimation on the space of positive definite symmetric matrices. We develop a nonparametric estimator for the density function of a random sample of positive definite matrices. Our estimator is based on the Helgason-Fourier transform and its inversion, the natural tools for analysis of compositions of random positive definite matrices. Under several smoothness conditions on the density of the intrinsic error in the random sample, we derive upper bounds on the rates of convergence of our nonparametric estimator to the true density.

Keywords: Diffusion tensor imaging; Deconvolution; Harish-Chandra c-function; Helgason-Fourier transform; inverse problems; Laplace-Beltrami operator; magnetic resonance imaging; Sobolev spaces; Wishart distribution.

1. Introduction

In this paper, we are motivated by applications in microwave engineering and medical imaging to study the problem of deconvolution density estimation on the symmetric space of positive definite (symmetric) matrices.

In microwave engineering, researchers have studied problems in which random observations arise as positive definite matrices. Terras (1985, p. 156 *ff.*) reviews the problem of transmitting electrical signals over a long, lossless line containing random inhomogeneities and shows how the problem may be modeled using mathematical analysis on the space of 2×2 positive definite matrices. All in all, Terras (1985, sections 3.1–3.2) contains a fascinating account of the appearance of sequences of random positive definite matrices in a real-world engineering problem.

The appearance in medical imaging of sequences of random positive definite matrices has become commonplace due to developments in *diffusion tensor imaging* (DTI), a method of imaging based upon the observation that water molecules in biological tissue are always in motion. For the purposes of mathematical modeling, it is generally assumed that the diffusion of water molecules at any given location in biological tissue follows a Brownian motion. A diffusion tensor image then is represented by the 3×3 positive-definite covariance matrix of the local diffusion process at the given location. DTI seeks to measure the diffusion of water protons between and within distinct tissue cells, and to derive estimates of the dominant orientation and direction of the Brownian motion (Le Bihan, 1995; Hasan *et al.*, 2001).

In DTI brain imaging, the diffusion of water molecules within and between *voxels*, the three-dimensional volume elements that constitute an image, reveals both the orientation of fibers comprising white-matter tracts in the brain and the coherence of fibers, the extent to which fibers are aligned together. DTI may be the only non-invasive, *in vivo* procedure that enables the study of deep brain white-matter fibers. Consequently, DTI has been found to be highly promising for comparing the human brain in normal states with abnormal states caused by strokes, epileptic seizures, tumors, white-matter abnormalities, multiple sclerosis lesions, HIV infection, traumatic brain injuries, aging, Alzheimer's disease, alcoholism, and developmental disorders; and there are potential applications to psychiatric conditions including schizophrenia, autism, and cognitive and learning disabilities (Neumann-Haefelin *et al.*, 2000; Rosenbloom *et al.*, 2003; Pomara *et al.*, 2001; Matthews and Arnold, 2001). In computational anatomy (Grenander and Miller, 1998), it has been observed that DTI-based methods can provide directional information about changes in shape; these methods have been found recently (Leporé *et al.*, 2008) to be predictive in MRI-based studies of HIV/AIDS, and DTI methods have been applied in research on the pathology of organ and tissue types such as the human breast, kidney, lingual, cardiac, and skeletal muscles, and spinal cord (Damon *et al.*, 2002).

It is well-known that magnetic resonance imaging, from which diffusion tensors are derived, is endowed inherently with random noise. Potential sources of noise include measurement noise within individual subjects, variability between subjects, and co-registration algorithm errors. Consequently, DTI data also contain noise (Basu *et al.*, 2006), and it is natural that statistical inferential issues arise in the analysis of DTI data (Koltchinskii *et al.*, 2007; Schwartzman *et al.*, 2008a,b; Zhu *et al.*, 2007).

In this paper, we study the problem of estimating the probability density function of a population of positive definite matrices based on a random sample from that population. An instance in which this problem arises may be obtained from Schwartzman *et al.* (2008a), who studied the two-sample comparison of twelve children divided into two groups according to reading ability, where the issue is to compare physical characteristics of brain tissue of the two groups on the basis of DTI images. In addition to comparing the population parameters, it is natural to seek estimators of the underlying density functions, and then it will be important to estimate the rates of convergence of the density estimators.

The deconvolution density estimation problem has been widely studied on Euclidean spaces (Diggle and Hall, 1993; Fan, 1992; Koo, 1993; Mair and Ruymgaart, 1996). In this classical setting, the commutative nature of the underlying mathematical operations renders the problem amenable to classical mathematical methods. The deconvolution problem has also been studied on certain compact manifolds (Healy *et al.*, 1998; Hendriks, 1990; Kim, 1998; Kim and Koo, 2000, 2002; Kim and Richards, 2001); in that setting, the problem is solvable using well-known generalizations of classical Fourier analysis.

By contrast, $\mathcal{P}_m$, the set of $m \times m$ positive definite matrices, is a symmetric space and a noncompact Riemannian manifold, and has an intrinsically non-commutative nature. Therefore, it is natural to expect that the deconvolution problem will be more difficult in this setting. To the best of our knowledge, Pesenson (2006) is the only author who has studied the deconvolution problem on $\mathcal{P}_m$, albeit under band-limiting restrictions on the underlying density functions. Thus, no results were available before now for the general deconvolution problem on $\mathcal{P}_m$, and in this article we provide proofs of the results that were announced in Kim and Richards (2008).

To date, the primary statistical emphases regarding mathematical methods on $\mathcal{P}_m$ are motivated by properties of the Wishart distribution (Muirhead, 1982; Letac and Massam, 2004; Richards, 1985, 1989), estimation problems associated with the Wishart distribution (Haff, 1979, 1980, 1991), and high-dimensional random matrices and their eigenvalue distributions (Johnstone, 2001; Takemura and Sheena, 2005). To solve the general deconvolution problem on $\mathcal{P}_m$, more advanced mathematical methods are required. While much is known in the mathematical literature about the necessary methods (Helgason, 1984; Terras, 1988), virtually nothing about those methods has appeared in the statistical literature. Using these new methods, we develop nonparametric estimation methods for solving the deconvolution problem on $\mathcal{P}_m$.

In summary, Section 2 provides notation and introduces the *Helgason-Fourier transform* and its inversion formula, involving the *Harish-Chandra c-function*; we have provided the necessary details of these concepts so as to make the paper fully accessible to readers who are new to this area. We formalize in Section 3 the statistical procedure in terms of measurement errors on $\mathcal{P}_m$ and present the main results on general deconvolution density estimation on $\mathcal{P}_m$. In Section 4, we provide the explicit details in the case of the Wishart distribution. Finally, all proofs are provided in Section 5.

2. Preliminaries

Throughout the paper, we denote by G the general linear group $\mathrm{GL}(m,\mathbb{R})$ of $m \times m$ nonsingular real matrices and by K the group $\mathrm{O}(m)$ of $m \times m$ orthogonal matrices. The group G acts transitively on $\mathcal{P}_m$, the space of $m \times m$ positive definite matrices, by the action

$$G \times \mathcal{P}_m \to \mathcal{P}_m, \quad (g,x) \mapsto g'xg, \tag{1}$$

$g \in G$, $x \in \mathcal{P}_m$, where g' denotes the transpose of g. Under this action, the isotropy group of the identity in G is K, hence the homogeneous space $K\backslash G$ can be identified with $\mathcal{P}_m$ by the "natural" mapping

$$K\backslash G \to \mathcal{P}_m, \quad Kg \mapsto g'g. \tag{2}$$

In distinguishing between left and right cosets, we place the quotient operation on the left and right of the group, respectively.

A random matrix $X \in \mathcal{P}_m$ is said to be *K-invariant* if $X \stackrel{\mathcal{L}}{=} k'Xk$ for all $k \in K$, where "$\stackrel{\mathcal{L}}{=}$" denotes equality in distribution. A function f on $\mathcal{P}_m$ is called K-invariant if $f(k'xk) = f(x)$ for all $k \in K$, $x \in \mathcal{P}_m$; we will indicate that f is K-invariant by writing its domain as $\mathcal{P}_m/K$, with a similar notation for K-invariant positive definite random matrices.

By means of the relationship (2) between $K\backslash G$ and $\mathcal{P}_m$, we identify K-invariant functions on $\mathcal{P}_m$ with *K-biinvariant* functions on G, i.e., functions $\widetilde{f} : G \to \mathbb{C}$ that satisfy $\widetilde{f}(g) = \widetilde{f}(k_1 g k_2)$ for all $k_1, k_2 \in K$ and $g \in G$. In particular, $\mathcal{P}_m/K \simeq K\backslash G/K$, where "$\simeq$" denotes diffeomorphic equivalence.

Consider random matrices $X, \varepsilon \in \mathcal{P}_m$ with corresponding group elements $\widetilde{X}, \widetilde{\varepsilon} \in G$, respectively. By the natural map (2), $K\widetilde{X} \mapsto X$, equivalently $\widetilde{X}'\widetilde{X} = X$ and, similarly, $K\widetilde{\varepsilon} \mapsto \varepsilon$. Then $\widetilde{X}\widetilde{\varepsilon} \in G$ is mapped via (2) to

$$\widetilde{X}\widetilde{\varepsilon} \mapsto (\widetilde{X}\widetilde{\varepsilon})'(\widetilde{X}\widetilde{\varepsilon}) \equiv \widetilde{\varepsilon}'\widetilde{X}'\widetilde{X}\widetilde{\varepsilon} = \widetilde{\varepsilon}'X\widetilde{\varepsilon} \in \mathcal{P}_m. \tag{3}$$

If $\widetilde{\varepsilon}$ is K-biinvariant, i.e., $\widetilde{\varepsilon} \stackrel{\mathcal{L}}{=} k_1\widetilde{\varepsilon}k_2$ for all $k_1, k_2 \in K$, then $\widetilde{\varepsilon}'\widetilde{\varepsilon} = \varepsilon = k'\Lambda k$, where $k \in K$ and $\Lambda \in \mathcal{P}_m$ is the diagonal matrix of eigenvalues of ε. Consequently, we define $\varepsilon^{1/2} = k'\Lambda^{1/2}k$ and note that, by K-biinvariance of $\widetilde{\varepsilon}$, the relationship (3) on the group G corresponds in distribution to $\varepsilon^{1/2}X\varepsilon^{1/2} \in \mathcal{P}_m$, where $X \in \mathcal{P}_m$ and $\varepsilon \in \mathcal{P}_m/K$. Bearing this in mind, we formally make the following definition.

Definition 2.1. Suppose that $X \in \mathcal{P}_m$ and $\varepsilon \in \mathcal{P}_m/K$ are random matrices. Then the *composition of* X *and* ε is

$$X \circ \varepsilon = \varepsilon^{1/2}X\varepsilon^{1/2}, \tag{4}$$

where $\varepsilon^{1/2}$ is the positive definite square root of ε.

We remark that the composition operation has arisen in the context of central limit theorems on $\mathcal{P}_m$; see Terras (1985, 1988), Richards (1989), and Graczyk (1992).

There is an alternative approach leading to composition (4). We begin by noting that $\mathcal{P}_m$ is a Riemannian manifold, hence each pair $X, Y \in \mathcal{P}_m$ defines a unique geodesic path. Since $\mathcal{P}_m$ is also a homogeneous space, *viz.*, $\mathcal{P}_m = K\backslash G$, there exists $V \in G$ such that

$$Y = VXV'. \tag{5}$$

This model can be viewed as a multiplicative analog of classical regression analysis, with Y serving as the dependent variable, X as the independent variable, and V as the error variable. We can also study this relationship between X and Y using the logarithm map on $\mathcal{P}_m$. To that end, each observation on Y may be transformed into an observation on y where $\exp(y) = Y$ with $y \in T(\mathcal{P}_m)$, the tangent space of $\mathcal{P}_m$. Similarly, each measurement on X may be transformed into a measurement on $x \in T(\mathcal{P}_m)$ where $\exp(x) = X$. By postulating the existence of a "small" error $v \in T(\mathcal{P}_m)$ such that $y = x + v$, and by exponentiating this linear "regression" relationship between y and x, we are led naturally to (5); see Terras (1988), section 4.1–4.2 for details on the exponentiation of such linear relationships on $T(\mathcal{P}_m)$.

If we assume that the measurement error V is *isotropic* or has no preferred orientation, i.e., V has a K-invariant distribution, then we obtain the model (5) where $V \in G/K$. We refer to Schwartzman *et al.* (2008b, 2009) for justification for the assumption of an isotropic distribution for V. As noted by Rosenbloom *et al.* (2003, p. 148, Figure 1), in the context of DTI, the assumption that water molecules diffuse isotropically is appropriate for regions such as the ventricles, which are large fluid-filled spaces deep in the brain. On the other hand, water molecules located in white-matter fiber are constrained by the axon sheath; this forces greater movement along the longitudinal axes of fibers than across the axes, and then diffusion may be isotropic only at sufficiently small scales.

Returning to (5), we apply polar coordinates on G, *viz.*, $V = k\varepsilon^{1/2}$ where $k \in K$ and $\varepsilon \in \mathcal{P}_m$. Since V is K-invariant, $V \stackrel{\mathcal{L}}{=} k'V = k'k\varepsilon^{1/2} = \varepsilon^{1/2}$; hence (5) reduces to $Y \stackrel{\mathcal{L}}{=} \varepsilon^{1/2}X\varepsilon^{1/2} \equiv X \circ \varepsilon$, which agrees with Definition 2.1 because $\varepsilon = V'V$ is K-invariant. Thus, the problem is to estimate nonparametrically the density of X based on a random sample from $Y = X \circ \varepsilon$ where the error matrix ε is isotropic.

2.1. *The Helgason-Fourier Transform*

Let $C_c^\infty(\mathcal{P}_m)$ denote the collection of complex-valued, infinitely differentiable, compactly supported functions f on $\mathcal{P}_m$. For $w \in \mathcal{P}_m$ and $j = 1, \ldots, m$, denote by $|w_j|$ the principal minor of order j of w. For $s = (s_1, \ldots, s_m) \in \mathbb{C}^m$, the *power function* $p_s : \mathcal{P}_m \to \mathbb{C}$ is defined by

$$p_s(w) = \prod_{j=1}^{m} |w_j|^{s_j}, \tag{6}$$

where $w \in \mathcal{P}_m$.

Let $\mathrm{d}k$ denote the Haar measure on K, normalized to have total volume equal to one. Then

$$h_s(w) = \int_K p_s(k'wk)\,\mathrm{d}k, \tag{7}$$

$w \in \mathcal{P}_m$, $s \in \mathbb{C}^m$, is a *zonal spherical function* on $\mathcal{P}_m$. It is well-known that the spherical functions play a fundamental role in harmonic analysis on symmetric spaces; see Helgason (1984); in particular, (7) is a special case of Harish-Chandra's formula for the general spherical functions. If $s_1, \ldots, s_m$ are nonnegative integers then, except for a constant factor, (7) is an integral formula for the zonal polynomials that arise often in aspects of multivariate distribution theory; see Muirhead (1982, pp. 231–232).

Let $w = (w_{ij}) \in \mathcal{P}_m$. Then, up to a constant factor, the unique G-

invariant measure on $\mathcal{P}_m$ is

$$\mathrm{d}_* w = |w|^{-(m+1)/2} \prod_{1 \le i \le j \le m} \mathrm{d}w_{ij}, \tag{8}$$

where $|w|$ is the determinant of w.

Definition 2.2. For $s = (s_1, \ldots, s_m) \in \mathbb{C}^m$ and $k \in K$, the *Helgason-Fourier transform* (Terras, 1988, p. 87) of a function $f \in C_c^\infty(\mathcal{P}_m)$ is

$$\mathcal{H}f(s,k) = \int_{\mathcal{P}_m} f(w)\, \overline{p_s(k'wk)}\, \mathrm{d}_* w, \tag{9}$$

where $\overline{p_s(k'wk)}$ denotes complex conjugation and $\mathrm{d}_* w$ is the G-invariant measure (8).

For the case in which $f \in C_c^\infty(\mathcal{P}_m/K)$, we make the change of variables $w \to k_1' w k_1$ in (9), $k_1 \in K$, and integrate with respect to the Haar measure $\mathrm{d}k_1$. Applying the invariance of f and the formula (7) for the zonal spherical function, we deduce that $\mathcal{H}f(s,k)$ does not depend on k; specifically, $\mathcal{H}f(s,k) = \widehat{f}(s)$, where

$$\widehat{f}(s) = \int_{\mathcal{P}_m} f(w)\, \overline{h_s(w)}\, \mathrm{d}_* w \tag{10}$$

is the *zonal spherical transform* of f.

2.2. *The Inversion Formula for the Helgason-Fourier Transform*

Let

$$A = \{\mathrm{diag}(a_1, \ldots, a_m) : a_j > 0, j = 1, \ldots, m\}$$

denote the group of diagonal positive definite matrices in G and let

$$N = \{n = (n_{ij}) \in G : n_{ij} = 0, 1 \le j < i \le m; n_{jj} = 1, j = 1, \ldots, m\}$$

be the subgroup of G consisting of upper-triangular matrices with all diagonal entries equal to 1. It is well-known (see Terras, 1988, p. 20) that each $g \in G$ can be decomposed uniquely as $g = kan$ where $k \in K$, $a \in A$, and $n \in N$. This result is the *Iwasawa decomposition*, and (k, a, n) are called the *Iwasawa coordinates* of g.

For $a, b \in \mathbb{C}$ with $\mathrm{Re}(a), \mathrm{Re}(b) > 0$, let

$$B(a,b) = \frac{\Gamma(a)\,\Gamma(b)}{\Gamma(a+b)}$$

denote the well-known beta function, where $\Gamma(\cdot)$ is the classical gamma function. For $s = (s_1, \ldots, s_m) \in \mathbb{C}^m$, the *Harish-Chandra c-function* is

$$c_m(s) = \prod_{1 \le i < j \le m-1} \frac{B\left(\frac{1}{2}, s_i + \cdots + s_j + \frac{1}{2}(j-i+1)\right)}{B\left(\frac{1}{2}, \frac{1}{2}(j-i+1)\right)} . \tag{11}$$

Let

$$\rho \equiv \left(\tfrac{1}{2}, \ldots, \tfrac{1}{2}, \tfrac{1}{4}(1-m)\right), \tag{12}$$

and set

$$\omega_m = \frac{\prod_{j=1}^m \Gamma(j/2)}{(2\pi i)^m \, \pi^{m(m+1)/4} \, m!} . \tag{13}$$

We shall use the notation

$$\mathbb{C}^m(\rho) = \{s \in \mathbb{C}^m : \mathrm{Re}(s) = -\rho\} \tag{14}$$

because this subset of $\mathbb{C}^m$ arises frequently in the sequel, and we also define

$$\mathrm{d}_* s = \omega_m \, |c_m(s)|^{-2} \, \mathrm{d}s_1 \cdots \mathrm{d}s_m \tag{15}$$

since this measure is ubiquitous in our development.

Let $M = \{\mathrm{diag}(\pm 1, \ldots, \pm 1)\}$ be the collection of $m \times m$ diagonal matrices with entries ± 1 on the diagonal; then M is a subgroup of K and is of order 2^m. By factorizing the Haar measure $\mathrm{d}k$ on K, it may be shown (Terras, 1988, p. 88) that there exists an invariant measure $\mathrm{d}\bar{k}$ on the coset space K/M such that

$$\int_{\bar{k} \in K/M} \mathrm{d}\bar{k} = 1.$$

In stating the inversion formula for the Helgason-Fourier transform, we make particular use of the notation (11)–(15). The inversion formula, then, is the following result.

Theorem 2.1. *(Helgason, 1984) For $f \in C_c^\infty(\mathcal{P}_m)$ and $w \in \mathcal{P}_m$,*

$$f(w) = \int_{\mathbb{C}^m(\rho)} \int_{\bar{k} \in K/M} \mathcal{H}f(s,k) \, p_s(k'wk) \, \mathrm{d}\bar{k} \, \mathrm{d}_* s. \tag{16}$$

In particular, if $f \in C_c^\infty(\mathcal{P}_m/K)$ then

$$f(w) = \int_{\mathbb{C}^m(\rho)} \widehat{f}(s) \, h_s(w) \, \mathrm{d}_* s. \tag{17}$$

We refer to Terras (1988, p. 87 *ff.*) for a detailed treatment of this inversion formula and many references to the literature.

For the case in which $m = 1$, the Helgason-Fourier transform is the classical Mellin transform and (16) reduces to the corresponding classical inverse Mellin transform, *viz.*, if $f : \mathbb{R}_+ \to \mathbb{C}$ and

$$\widehat{f}(s) = \int_0^\infty t^{s-1}\, f(t)\, \mathrm{d}t$$

then, formally,

$$f(t) = (2\pi i)^{-1} \int_{\mathrm{Re}(s)=\alpha} t^{-s}\, \widehat{f}(s)\, \mathrm{d}s,$$

where α is a constant.

2.3. *The Convolution Property of the Helgason-Fourier Transform*

For $f \in L^1(\mathcal{P}_m)$ and $h \in L^1(\mathcal{P}_m/K)$, we define $f * h$, the convolution of f and h, by

$$(f * h)(w) = \int_{\mathcal{P}_m} f(z)\, h(z^{-1/2} w z^{-1/2})\, \mathrm{d}_* z, \tag{18}$$

$w \in \mathcal{P}_m$. Thus, if f and h are the density functions of independent random matrices $X \in \mathcal{P}_m$ and $\varepsilon \in \mathcal{P}_m/K$, respectively, then $f * h$ is the density function of the composition $X \circ \varepsilon$.

For $f \in C_c^\infty(\mathcal{P}_m)$ and $h \in C_c^\infty(\mathcal{P}_m/K)$, the *convolution property of the Helgason-Fourier transform* is that

$$\mathcal{H}(f * h)(s, k) = \mathcal{H}f(s, k)\, \widehat{h}(s), \tag{19}$$

$s \in \mathbb{C}^m$, $k \in K$; see Terras (1988, p. 88, Theorem 1).

2.4. *Eigenvalues, the Laplacian, and Sobolev Spaces*

For $w = (w_{ij}) \in \mathcal{P}_m$, let $\partial/\partial w$ be the $m \times m$ matrix

$$\frac{\partial}{\partial w} = \Big(\tfrac{1}{2}(1 + \delta_{ij}) \frac{\partial}{\partial w_{ij}}\Big),$$

where δ_{ij} denotes Kronecker's delta. The Laplacian, Δ, on $\mathcal{P}_m$ can be written in terms of the local coordinates w_{ij} as

$$\Delta = -\mathrm{tr}\Big(\big(w \frac{\partial}{\partial w}\big)^2\Big);$$

see Terras (1988, p. 106). The power function p_s is an eigenfunction of Δ: Let

$$r_j = s_j + s_{j+1} + \cdots + s_m + \tfrac{1}{4}(m - 2j + 1),$$

$j = 1, \ldots, m$; then

$$\Delta p_s(w) = \lambda_s p_s(w),$$

where

$$\lambda_s = -\big(r_1^2 + \cdots + r_m^2 - \tfrac{1}{48} m(m^2 - 1)\big).$$

Since $\mathrm{Re}(s) = -\rho$, each r_j, $j = 1, \ldots, m$, is purely imaginary, so $\lambda_s > 0$. See Terras (1988, p. 49) for the explicit calculation of λ_s; Muirhead (1982, p. 229) and Richards (1985, p. 283) also provide alternative approaches to that calculation.

The Helgason-Fourier transform has the property of changing the effect of invariant differential operators on functions to simple pointwise multiplication. That is, for $f \in C_c^\infty(\mathcal{P}_m)$,

$$\mathcal{H}(\Delta f)(s,k) = \lambda_s \mathcal{H} f(s,k), \tag{20}$$

$s \in \mathbb{C}^m$, $k \in K$ (see Terras, 1988, p. 88). For $\sigma > 0$, we therefore define $\Delta^{\sigma/2}$, the $\sigma/2$-fractional power of Δ, as the operator that satisfies the identity

$$\mathcal{H}(\Delta^{\sigma/2} f)(s,k) = \lambda_s^{\sigma/2} \mathcal{H} f(s,k) \tag{21}$$

for all $f \in C_c^\infty$, $s \in \mathbb{C}^m$, $k \in K$. In turn, we define the *Sobolev class* of functions,

$$H_\sigma(\mathcal{P}_m) = \big\{ f \in C^\infty(\mathcal{P}_m) : \|\Delta^{\sigma/2} f\|^2 < \infty \big\}, \tag{22}$$

where $2\sigma > \dim \mathcal{P}_m = m(m+1)/2$ and, for $f \in C^\infty(\mathcal{P}_m)$,

$$\|f\| = \Big(\int_{\mathcal{P}_m} |f(w)|^2 \, \mathrm{d}_* w \Big)^{1/2}$$

denotes the $L^2(\mathcal{P}_m)$-norm with respect to the invariant measure $\mathrm{d}_* w$. For $Q > 0$, we also define the bounded Sobolev class,

$$H_\sigma(\mathcal{P}_m, Q) = \big\{ f \in C^\infty(\mathcal{P}_m) : \|\Delta^{\sigma/2} f\|^2 < Q \big\}, \tag{23}$$

where $2\sigma > \dim \mathcal{P}_m = m(m+1)/2$.

2.5. *The Plancherel Formula for the Helgason-Fourier Transform*

The Plancherel formula (Terras, 1988, p. 88, Theorem 1) states that for $f \in C_c^\infty(\mathcal{P}_m)$,

$$\int_{\mathcal{P}_m} |f(w)|^2 \, \mathrm{d}_* w = \int_{\mathbb{C}^m(\rho)} \int_{\bar{k} \in K/M} |\mathcal{H}f(s, \bar{k})|^2 \, \mathrm{d}\bar{k} \, \mathrm{d}_* s. \tag{24}$$

For the case in which $f \in C_c^\infty(\mathcal{P}_m/K)$, (24) simplifies via (10) to

$$\int_{\mathcal{P}_m} |f(w)|^2 \, \mathrm{d}_* w = \int_{\mathbb{C}^m(\rho)} |\widehat{f}(s)|^2 \, \mathrm{d}_* s. \tag{25}$$

3. Deconvolution Density Estimation on $\mathcal{P}_m$

On the group G, the deconvolution problem arises from the statistical model

$$\widetilde{Y} \stackrel{\mathcal{L}}{=} \widetilde{X} \widetilde{\varepsilon}, \tag{26}$$

where $\widetilde{X}$ is a random unobservable, $\widetilde{\varepsilon}$ is an independent random error, and $\widetilde{Y}$ is the observed random measurement. We assume that $f_{\widetilde{\varepsilon}}$, the density of $\widetilde{\varepsilon}$, is known and K-biinvariant and that the unknown densities $f_{\widetilde{X}}$ and $f_{\widetilde{Y}}$ of $\widetilde{X}$ and $\widetilde{Y}$, respectively, are K-invariant. Under the equivalence $K \backslash G \simeq \mathcal{P}_m$, we have $\widetilde{\varepsilon} \mapsto \varepsilon$, $\widetilde{X} \mapsto X$, and $\widetilde{Y} \mapsto Y$ together with the identification $f_\varepsilon(\widetilde{g}) = f_\varepsilon(\widetilde{g}'\widetilde{g})$, $f_X(\widetilde{g}) = f_X(\widetilde{g}'\widetilde{g})$, and $f_Y(\widetilde{g}) = f_Y(\widetilde{g}'\widetilde{g})$, $g \in G$. Since $\widetilde{X}$ and $\widetilde{\varepsilon}$ are independent, X and ε also are independent, and (26) implies that

$$Y \stackrel{\mathcal{L}}{=} \varepsilon^{1/2} X \varepsilon^{1/2}; \tag{27}$$

hence $f_Y = f_X * f_\varepsilon$.

Applying to (27) the convolution property (19) of the Helgason-Fourier transform, we obtain

$$\mathcal{H}f_Y(s, k) = \mathcal{H}f_X(s, k) \, \widehat{f_\varepsilon}(s), \tag{28}$$

$s \in \mathbb{C}^m$, $k \in K$. Given a random sample $Y_1, \ldots, Y_n$ from Y, we estimate the density function f_X as follows. We form $\mathcal{H}_n f_Y$, the *empirical Helgason-Fourier transform*,

$$\mathcal{H}_n f_Y(s, k) = \frac{1}{n} \sum_{\ell=1}^{n} \overline{p_s(k' Y_\ell k)}. \tag{29}$$

Substituting (29) in (28), together with the assumption that $\widehat{f_\varepsilon}(s) \neq 0$, $s \in \mathbb{C}^m$, we obtain

$$\mathcal{H}_n f_X(s, k) = \frac{\mathcal{H}_n f_Y(s, k)}{\widehat{f_\varepsilon}(s)}, \tag{30}$$

$s \in \mathbb{C}^m$, $k \in K$.

Two elementary facts about $\mathcal{H}_n f_Y$ are:

$$\mathbb{E}\mathcal{H}_n f_Y(s,k) = \mathcal{H} f_Y(s,k), \tag{31}$$

i.e., $\mathcal{H}_n f_Y$ is an unbiased estimator of $\mathcal{H} f_Y$; and, with ρ as defined in (12), we have

$$\mathbb{V}ar\,(\mathcal{H}_n f_Y(s,k)) = \frac{1}{n}\left(\mathcal{H} f_Y(-2\rho,k) - |\mathcal{H} f_Y(s,k)|^2\right) \tag{32}$$

when $\mathrm{Re}(s) = -\rho$.

In analogy with classical Euclidean deconvolution, we introduce a smoothing parameter $T = T(n)$ where $T(n) \to \infty$ as $n \to \infty$, and then we apply the Helgason-Fourier inversion formula (16) using a spectral cut-off.

We introduce the notation

$$\mathbb{C}^m(\rho, T) = \{s \in \mathbb{C}^m(\rho) : \lambda_s < T\}, \tag{33}$$

where $C^m(\rho)$ is defined in (14). As an estimator of the population density f_X, we define the density estimator f_X^n given by

$$f_X^n(w) = \int_{\mathbb{C}^m(\rho,T)} \int_{\bar{k}\in K/M} \frac{\mathcal{H}_n f_Y(s,\bar{k})}{\widehat{f_\varepsilon}(s)}\, p_s(\bar{k}' w \bar{k})\, \mathrm{d}\bar{k}\, \mathrm{d}_* s, \tag{34}$$

$w \in \mathcal{P}_m$. The estimator f_X^n will serve as our nonparametric deconvolution estimator of the density f_X.

We note that

$$\begin{aligned}\mathbb{E} f_X^n(w) &= \int_{\mathbb{C}^m(\rho,T)} \int_{\bar{k}\in K/M} \frac{\mathbb{E}\mathcal{H}_n f_Y(s,\bar{k})}{\widehat{f_\varepsilon}(s)}\, p_s(\bar{k}' w \bar{k})\, \mathrm{d}\bar{k}\, \mathrm{d}_* s \\ &= \int_{\mathbb{C}^m(\rho,T)} \int_{\bar{k}\in K/M} \frac{\mathcal{H} f_Y(s,\bar{k})}{\widehat{f_\varepsilon}(s)}\, p_s(\bar{k}' w \bar{k})\, \mathrm{d}\bar{k}\, \mathrm{d}_* s \\ &= \int_{\mathbb{C}^m(\rho,T)} \int_{\bar{k}\in K/M} \mathcal{H} f_X(s,\bar{k})\, p_s(\bar{k}' w \bar{k})\, \mathrm{d}\bar{k}\, \mathrm{d}_* s. \end{aligned} \tag{35}$$

Under certain smoothness assumptions on f_ε, (35) converges to $f_X(w)$ as $T \to \infty$.

Let $\{a_n\}$ and $\{b_n\}$ be sequences of real numbers. We write $a_n \ll b_n$ to mean $a_n \leq C b_n$ for some constant $C > 0$ as $n \to \infty$ (the Vinogradov notation). We use the notation $a_n = o(b_n)$ to mean $a_n/b_n \to 0$ as $n \to \infty$. We also write $a_n \asymp b_n$ if both $a_n \ll b_n$ and $b_n \ll a_n$; and we write $a_n \sim b_n$ if $a_n/b_n \to 1$ as $n \to \infty$.

For technical reasons, we will also assume the moment condition

$$\int_{\mathcal{P}_m} |w_1|^{-1} \cdots |w_{m-1}|^{-1} \, |w|^{(m-1)/2} \, f_Y(w) \, \mathrm{d}_* w < \infty \tag{36}$$

on the principal minors $|y_1|, \ldots, |y_m|$ of $y \in \mathcal{P}_m$. This assumption will be maintained throughout the rest of the paper.

Theorem 3.1. *Suppose there exists $\beta \geq 0$ such that*

$$|\widehat{f}_\varepsilon(s)|^{-2} \ll T^\beta \tag{37}$$

as $T \to \infty$ for all $s \in \mathbb{C}^m(\rho, T)$. If $f_X \in H_\sigma(\mathcal{P}_m, Q)$ and $\sigma > \frac{1}{2} \dim \mathcal{P}_m \equiv m(m+1)/4$ then, as $n \to \infty$,

$$\mathbb{E} \, \| f_X^n - f_X \|^2 \ll n^{-2\sigma/(2\sigma + 2\beta + \dim \mathcal{P}_m)}.$$

As an example, consider the special case in which

$$\widehat{f}_\varepsilon(s) = (1 + \gamma \lambda_s)^{-\beta}, \tag{38}$$

$s \in \mathbb{C}^m$, where $\gamma > 0$ is a scale parameter. Since $\widehat{f}_\varepsilon$ does not depend on $k \in K$, the underlying density function f_ε is K-invariant. By the Helgason-Fourier inversion formula (17),

$$f_\varepsilon(w) = \int_{\mathbb{C}^m(\rho)} (1 + \gamma \lambda_s)^{-\beta} \, h_s(w) \, \mathrm{d}_* s, \tag{39}$$

$w \in \mathcal{P}_m$. Letting $\beta \to 0$, we obtain $\widehat{f}_\varepsilon(s) \to 1$, so the underlying probability distribution approaches the Dirac measure concentrated at $\mathbf{I}_m$, the identity matrix in $\mathcal{P}_m$, and the interpretation here is that observations are made without error. This result corresponds to the special case of Theorem 3.1 in which $\beta = 0$, and hence we obtain the following result.

Corollary 3.1. *Suppose the distribution of ε is concentrated at $\mathbf{I}_m$. If $f_X \in H_\sigma(\mathcal{P}_m, Q)$, where $\sigma > \frac{1}{2} \dim \mathcal{P}_m$, then as $n \to \infty$,*

$$\mathbb{E} \, \| f_X^n - f_X \|^2 \ll n^{-2\sigma/(2\sigma + \dim \mathcal{P}_m)}.$$

We shall also obtain a result for the situation in which the hypothesis (37) in Theorem 3.1 is replaced by an exponential bound. In such a situation, we have the following result.

Theorem 3.2. *Suppose there exist $\beta, \gamma > 0$ such that*

$$|\widehat{f}_\varepsilon(s)|^{-2} \ll \exp(T^\beta / \gamma), \tag{40}$$

as $T \to \infty$, *for all* $s \in \mathbb{C}^m(\rho, T)$. *If* $f_X \in H_\sigma(\mathcal{P}_m, Q)$ *with* $\sigma > \frac{1}{2}\dim \mathcal{P}_m$ *then, as* $n \to \infty$,

$$\mathbb{E}\,\|f_X^n - f_X\|^2 \ll (\log n)^{-\sigma/\beta}.$$

In this situation, we consider the special case in which

$$\widehat{f_\varepsilon}(s) = \exp(-\gamma^{-1}\,\lambda_s^\beta), \tag{41}$$

$s \in \mathbb{C}^m$, where $\gamma > 0$ is a scale parameter. Again by the inversion formula (17), the underlying density function is

$$f_\varepsilon(w) = \int_{\mathbb{C}^m(\rho)} \exp(-\gamma^{-1}\,\lambda_s^\beta)\, h_s(w)\,\mathrm{d}_* s, \tag{42}$$

$w \in \mathcal{P}_m$. The case in which $\beta = 1$ is particularly important and is called the *heat* or *Gaussian kernel*, since the latter is the fundamental solution to the heat equation on $\mathcal{P}_m$: $(\Delta - (\partial/\partial t))\phi = 0$, $t > 0$; see Terras (1988, pp. 106–107). As a consequence, we obtain the following result.

Corollary 3.2. *Suppose that* f_ε *is Gaussian. If* $f_X \in H_\sigma(\mathcal{P}_m, Q)$, *where* $\sigma > \frac{1}{2}\dim \mathcal{P}_m$, *then, as* $n \to \infty$,

$$\mathbb{E}\,\|f_X^n - f_X\|^2 \ll (\log n)^{-\sigma}.$$

4. The Wishart Distribution

A case that is familiar in multivariate statistics is that of the Wishart distribution, $W_m(N, \Sigma)$, where $\Sigma \in \mathcal{P}_m$ and $N > m-1$. For $s = (s_1, \ldots, s_m) \in \mathbb{C}^m$, define the multivariate gamma function,

$$\Gamma_m(s_1, \ldots, s_m) = \pi^{m(m-1)/4} \prod_{j=1}^{m} \Gamma\big(s_j + \cdots + s_m - \tfrac{1}{2}(j-1)\big), \tag{43}$$

where $\mathrm{Re}(s_j + \cdots + s_m) > (j-1)/2$, $j = 1, \ldots, m$. Relative to the invariant measure $\mathrm{d}_* w$ in (8), the probability density function of the standard Wishart distribution $W_m(N, \mathbf{I}_m)$ is

$$f_\varepsilon(w) = \frac{1}{\Gamma_m(0, \ldots, 0, N/2)}\, |\tfrac{1}{2}w|^{N/2} \exp\big(-\tfrac{1}{2}\mathrm{tr}\, w\big), \tag{44}$$

$w \in \mathcal{P}_m$. We note that (44) is K-invariant and its Helgason-Fourier transform

$$\widehat{f_\varepsilon}(s) = \frac{\Gamma_m((0, \ldots, 0, N/2) + s^*)}{\Gamma_m(0, \ldots, 0, N/2)}\, h_s(\tfrac{1}{2}\mathbf{I}_m), \tag{45}$$

where $s^* = (s_{m-1}, s_{m-2}, \ldots, s_2, s_1, -(s_1 + \cdots + s_m))$, is well-known; see Muirhead (1982, p. 248) or Terras (1988, pp. 85–86).

We have the following result.

Lemma 4.1. *For $N > m-1$, the Wishart distribution $W_m(N, \mathbf{I}_m)$ satisfies*

$$|\widehat{f}_\varepsilon(s)|^{-2} \ll \exp\left(\pi T^{1/2}\right),$$

as $T \to \infty$, where $s \in \mathbb{C}^m(\rho, T)$.

Consequently, we deduce the following result.

Theorem 4.1. *Suppose that ε follows the Wishart distribution (44) with $N > m-1$. If $f_X \in H_\sigma(\mathcal{P}_m, Q)$ with $\sigma > \dim \mathcal{P}_m/2$ then, as $n \to \infty$,*

$$\mathbb{E}\|f_X^n - f_X\|^2 \ll (\log n)^{-2\sigma}.$$

From this result, one can see that the Wishart distribution has faster convergence in its Helgason-Fourier transform as compared to the Gaussian distribution. This of course results in a slower recovery in the corresponding deconvolution problem.

5. Proofs

The strategy of the proofs is, first, to decompose the integrated mean-squared error into its variance and bias components, *viz.*,

$$\begin{aligned} \mathbb{E}\big\|f_X^n - f_X\big\|^2 &= \mathbb{E}\big\|(f_X^n - \mathbb{E}f_X^n) + (\mathbb{E}f_X^n - f_X)\big\|^2 \\ &= \mathbb{E}\big\|f_X^n - \mathbb{E}f_X^n\big\|^2 + \big\|\mathbb{E}f_X^n - f_X\big\|^2, \end{aligned} \tag{46}$$

and, last, to estimate each component separately using estimates based on the Plancherel formula and the inversion formula for the Helgason-Fourier transform.

5.1. *The Integrated Bias*

Lemma 5.1. *Suppose that $f_X \in H_\sigma(\mathcal{P}_m, Q)$ and $\sigma > \dim \mathcal{P}_m/2$. Then*

$$\|\mathbb{E}f_X^n - f_X\|^2 \ll T^{-\sigma}.$$

Proof. By (35), we have for $w \in \mathcal{P}_m$

$$\begin{aligned} \mathbb{E} f_X^n(w) - f_X(w) &= \int_{\mathbb{C}^m(\rho,T)} \int_{\bar{k}\in K/M} \mathcal{H} f_X(s,\bar{k})\, p_s(\bar{k}' w \bar{k})\, \mathrm{d}\bar{k}\, \mathrm{d}_* s \\ &\quad - \int_{\mathbb{C}^m(\rho)} \int_{\bar{k}\in K/M} \mathcal{H} f_X(s,\bar{k})\, p_s(\bar{k}' w \bar{k})\, \mathrm{d}\bar{k}\, \mathrm{d}_* s \\ &= - \int_{\lambda_s > T, \mathrm{Re}(s) = -\rho} \int_{\bar{k}\in K/M} \mathcal{H} f_X(s,\bar{k})\, p_s(\bar{k}' w \bar{k})\, \mathrm{d}\bar{k}\, \mathrm{d}_* s. \end{aligned} \tag{47}$$

Applying the Plancherel formula, (24), to (47), we obtain

$$\|\mathbb{E} f_X^n - f_X\|^2 = \int_{\lambda_s > T, \mathrm{Re}(s) = -\rho} \int_{K/M} |\mathcal{H}(f_X)(s,\bar{k})|^2\, \mathrm{d}\bar{k}\, \mathrm{d}_* s.$$

Consequently,

$$\begin{aligned} \|\mathbb{E} f_X^n - f_X\|^2 &= \int_{\lambda_s \geq T, \mathrm{Re}(s) = -\rho} \int_{K/M} |\mathcal{H} f_X(s,\bar{k})|^2\, \mathrm{d}\bar{k}\, \mathrm{d}_* s \\ &\leq T^{-\sigma} \int_{\lambda_s \geq T, \mathrm{Re}(s) = -\rho} \int_{K/M} \lambda_s^\sigma\, |\mathcal{H} f_X(s,\bar{k})|^2\, \mathrm{d}\bar{k}\, \mathrm{d}_* s \\ &\leq T^{-\sigma} \int_{\mathbb{C}^m(\rho)} \int_{K/M} \lambda_s^\sigma\, |\mathcal{H} f_X(s,\bar{k})|^2\, \mathrm{d}\bar{k}\, \mathrm{d}_* s. \end{aligned} \tag{48}$$

By (21),

$$\lambda_s^\sigma\, |\mathcal{H} f_X(s,k)|^2 \equiv |\lambda_s^{\sigma/2}\, \mathcal{H} f_X(s,k)|^2 = |\mathcal{H}(\Delta^{\sigma/2} f_X)(s,k)|^2,$$

therefore

$$\begin{aligned} \|\mathbb{E} f_X^n - f_X\|^2 &\leq T^{-\sigma} \int_{\mathbb{C}^m(\rho)} \int_{\bar{k}\in K/M} |\mathcal{H}(\Delta^{\sigma/2} f_X)(s,k)|^2\, \mathrm{d}\bar{k}\, \mathrm{d}_* s \\ &= T^{-\sigma} \int_{\mathcal{P}_m} |\Delta^{\sigma/2} f_X(w)|^2\, \mathrm{d}_* w, \end{aligned}$$

where the equality follows from the Plancherel formula. By assumption, $f_X \in H_\sigma(\mathcal{P}_m, Q)$ and $\sigma > \frac{1}{2} \dim \mathcal{P}_m$; therefore the latter integral is bounded above by Q, so we obtain

$$\|\mathbb{E} f_X^n - f_X\|^2 \leq Q\, T^{-\sigma},$$

and the proof is complete. □

5.2. *The Integrated Variance*

To obtain bounds for the integrated variance, several preliminary calculations are needed. In particular, we begin with the variance calculation of the empirical Helgason-Fourier transform, which has similarities to the usual empirical characteristic function; see Feuerverger and Mureika (1977).

Lemma 5.2. *For $s \in \mathbb{C}^m(\rho)$ and $k \in K/M$,*

$$\mathbb{E}\big|\mathcal{H}_n f_Y(s,k) - \mathbb{E}\mathcal{H}_n f_Y(s,k)\big|^2 = \frac{1}{n}\left(|\mathcal{H}f_Y(-2\rho,k)|^2 - |\mathcal{H}f_Y(s,k)|^2\right).$$

Proof. By (29),

$$\begin{aligned}
|\mathcal{H}_n f_Y(s,k)|^2 &= \overline{\mathcal{H}_n f_Y(s,k)}\,\mathcal{H}_n f_Y(s,k) \\
&= \frac{1}{n^2}\sum_{j,\ell=1}^{n} \overline{p_s(k'Y_jk)}\,p_s(kY_\ell k) \\
&= \frac{1}{n^2}\Big\{\sum_{j=1}^{n}|p_s(k'Y_jk)|^2 + \sum_{j\neq\ell}\overline{p_s(k'Y_jk)}p_s(k'Y_\ell k)\Big\}. \qquad (49)
\end{aligned}$$

Observe also that

$$\begin{aligned}
p_s(w)\overline{p_s(w)} &= |w_1|^{s_1}\cdots|w_m|^{s_m}\,\overline{|w_1|^{s_1}\cdots|w_m|^{s_m}} \\
&= |w_1|^{2\,\mathrm{Re}(s_1)}\cdots|w_m|^{2\,\mathrm{Re}(s_m)} \\
&= p_{-2\rho}(w),
\end{aligned}$$

since $\mathrm{Re}(s) = -\rho$. Applying this result to (49) and taking expectations, we obtain

$$\begin{aligned}
\mathbb{E}|\mathcal{H}_n f_Y(s,k)|^2 &= \frac{1}{n^2}\mathbb{E}\Big\{\sum_{j=1}^{n}|p_s(k'Y_jk)|^2 + \sum_{j\neq\ell}\overline{p_s(k'Y_jk)}p_s(k'Y_\ell k)\Big\} \\
&= \frac{1}{n^2}\Big\{\sum_{j=1}^{n}\mathbb{E}p_{-2\rho}(k'Y_jk) + \sum_{j\neq\ell}\overline{\mathbb{E}\,p_s(k'Y_jk)}\,\mathbb{E}\,p_s(k'Y_\ell k)\Big\} \\
&= \frac{1}{n}\mathcal{H}f_Y(-2\rho,k) + \frac{n-1}{n}|\mathcal{H}f_Y(s,k)|^2,
\end{aligned}$$

where the last equality follows from the fact that $Y_1,\ldots,Y_n$ are independent and identically distributed as Y, and because $\mathbb{E}\,p_s(k'Yk) = \mathcal{H}f_Y(s,k)$ by (31). □

This allows us to establish the following bound.

Lemma 5.3. *As* $T \to \infty$,

$$\mathbb{E}\left\|f_X^n - \mathbb{E}f_X^n\right\|^2 \ll \sup_{s\in\mathbb{C}^m(\rho)} |\widehat{f_\varepsilon}(s)|^{-2}\, \frac{T^{\dim \mathcal{P}_m/2}}{n}.$$

Proof. By the Plancherel formula, (24),

$$\begin{aligned}
&\mathbb{E}\left\|f_X^n - \mathbb{E}f_X^n\right\|^2 \\
&= \int_{\mathbb{C}^m(\rho,T)} \int_{K/M} \mathbb{E}\left|\mathcal{H}_n f_X(s,k) - \mathbb{E}\mathcal{H}_n f_X(s,k)\right|^2 \mathrm{d}\bar{k}\, \mathrm{d}_* s \\
&= \int_{\mathbb{C}^m(\rho,T)} \int_{K/M} \mathbb{E}\left|\mathcal{H}_n f_Y(s,k) - \mathbb{E}\mathcal{H}_n f_Y(s,k)\right|^2 \mathrm{d}\bar{k}\, |\widehat{f_\varepsilon}(s)|^{-2}\, \mathrm{d}_* s \\
&\le \frac{1}{n} \sup_{\lambda_s<T,\mathrm{Re}(s)=-\rho} |\widehat{f_\varepsilon}(s)|^{-2} \int_{K/M} |\mathcal{H} f_Y(-2\rho,\bar{k})|^2 d\bar{k} \int_{\mathbb{C}^m(\rho,T)} \mathrm{d}_* s \\
&\ll \sup_{\lambda_s<T,\mathrm{Re}(s)=-\rho} |\widehat{f_\varepsilon}(s)|^{-2}\, \frac{T^{\dim \mathcal{P}_m/2}}{n}
\end{aligned}$$

as $T \to \infty$.

In the above calculation, we used the moment bound (36) together with the fact that $|\mathcal{H} f_Y(-2\rho,\bar{k})|^2$ varies continuously with respect to $\bar{k} \in K/M$, with the latter being compact. In addition, we used the fact that, as $T \to \infty$,

$$\sup_{\mathbb{C}^m(\rho,T)} |c_m(s)|^{-2} \ll T^{m(m-1)/4},$$

a result which follows from Helgason (1984, p. 450, Proposition 7.2). □

The proofs of all the theorems now are obtained by applying Lemma 5.1 and Lemma 5.3 to (46).

5.3. *Proof of Theorem 3.1*

By the condition of Theorem 3.1, (46) reduces to

$$\mathbb{E}\left\|f_X^n - f_X\right\|^2 \ll n^{-1}\, T^{\beta+\dim \mathcal{P}_m/2} + T^{-\sigma}$$

as $T \to \infty$. Thus, we choose $T \asymp T^{2/(2\beta+2\sigma+\dim \mathcal{P}_m)}$ to optimize the upper bound.

5.4. *Proof of Theorem 3.2*

By the condition of Theorem 3.2, (46) reduces to

$$\mathbb{E}\left\|f_X^n - f_X\right\|^2 \ll n^{-1}\, T^{\dim \mathcal{P}_m/2} \exp\left(\gamma^{-1}T^{\beta}\right) + T^{-\sigma}$$

as $T \to \infty$. Thus, we choose $T \asymp (\log n)^{1/\beta}$ to optimize the upper bound.

5.5. *Proof of Lemma 4.1*

Let $s_j = -(1/2) + i\sigma_j$ for $j = 1, \ldots, m-1$ and $s_m = ((m-1)/4) + i\sigma_m$, where $i^2 = -1$ and $\sigma_1, \ldots, \sigma_m \in \mathbb{R}$. Define $s_j^* = s_{m-j}$ for $j = 1, \ldots, m-1$ and $s_m^* = -(s_1 + \cdots + s_m)$. Consequently,

$$s_j^* + \cdots + s_m^* + \frac{N-j+1}{2} = \frac{2N-m+1}{4} - i\left(\sigma_{m-j+1} + \cdots + \sigma_m\right). \quad (50)$$

By Stirling's approximation,

$$\Gamma(z) \sim z^{z-1/2} e^{-z} \sqrt{2\pi}, \quad (51)$$

for $z \in \mathbb{C}$ with $0 < a \leq \mathrm{Re}(z) < b$, as $|z| \to \infty$; therefore

$$\begin{aligned} &\Gamma\Big(s_j^* + \cdots + s_m^* + \frac{N-j+1}{2}\Big) \\ &\quad \sim \sqrt{2\pi}\Big(\frac{2N-m+1}{4} - i(\sigma_{m-j+1} + \cdots + \sigma_m)\Big)^{\frac{2N-m-1}{4} - i(\sigma_{m-j+1}+\cdots+\sigma_m)} \\ &\quad\quad \times \exp\left\{\frac{2N-m+1}{4} - i(\sigma_{m-j+1} + \cdots + \sigma_m)\right\} \end{aligned}$$

as $|\sigma_{m-j+1} + \cdots + \sigma_m| \to \infty$ for $j = 1, \ldots, m$. Consequently,

$$\begin{aligned} &\left|\Gamma\Big(s_j^* + \cdots + s_m^* + \frac{2N-m+1}{4}\Big)\right|^2 \\ &\quad \sim 2\pi\left(\Big(\frac{2N-m+1}{4}\Big)^2 + (\sigma_{m-j+1} + \cdots + \sigma_m)^2\right)^{(2N-m-1)/2} \\ &\quad\quad \times \exp\left(2(\sigma_{m-j+1} + \cdots + \sigma_m)\arctan\Big(\frac{4(\sigma_{m-j+1} + \cdots + \sigma_m)}{2N-m+1}\Big)\right) \\ &\quad\quad \times \exp\Big(\frac{2N-m+1}{2}\Big) \end{aligned} \quad (52)$$

as $|\sigma_{m-j+1} + \cdots + \sigma_m| \to \infty$ for $j = 1, \ldots, m$.

By the condition that $\lambda_s < T$, it follows that there exists a positive constant C, independent of $\sigma_1, \ldots, \sigma_m$, such that

$$\left|\Gamma_m(s_1^*, \ldots, s_{m-1}^*, s_m^* + N/2)\right|^{-2} \leq C \exp(\pi T^{1/2}) \quad (53)$$

as $T \to \infty$. Since $|h_s(\frac{1}{2}\mathbf{I}_m)|^2 = 1$, it follows that as $T \to \infty$,

$$|\widehat{f}_\varepsilon(s)|^{-2} \ll \exp(\pi T^{1/2}).$$

Acknowledgments

Portions of this research were done during fall 2006 while D. R. was a Research Fellow at the Statistical and Applied Mathematical Sciences Institute, Research Triangle Park, NC. We are grateful to the entire SAMSI community for the warm and generous hospitality accorded to him during his stay. The research of P. K. was supported in part by the Natural Sciences and Engineering Research Council of Canada, grant OGP46204. The research of D. R. was supported in part by the U.S. National Science Foundation, grants DMS-0112069 and DMS-0705210.

References

Basu, S., Fletcher, T. and Whitaker, S. (2006). Rician noise removal in diffusion tensor MRI, in *Medical Image Computing and Computer-Assisted Intervention: Proc. 9th international conference, Copenhagen, Denmark, October 1-6, 2006, Lecture Notes in Computer Science*, Vol. 4190 (Springer, New York), pp. 117–125.

Damon, B. M., Ding, Z., Anderson, A. W., Freyer, A. S. and Gore, J. C. (2002). Validation of diffusion tensor MRI-based muscle fiber tracking, *Magn. Reson. Med.* **48**, pp. 97–104.

Diggle, P. J. and Hall, P. (1993). A fourier approach to nonparametric deconvolution of a density estimate, *J. Roy. Statist. Soc., Ser. B* **55**, pp. 523–531.

Fan, J. (1992). Global behaviour of deconvolution kernel estimates, *Statist. Sinica* **1**, pp. 541–551.

Feuerverger, A. and Mureika, R. A. (1977). The empirical characteristic function and its applications, *Annals of Statistics* **5**, pp. 88–97.

Graczyk, P. (1992). A central limit theorem on the space of positive definite symmetric matrices, *Ann. Inst. Fourier* **42**, pp. 857–874.

Grenander, U. and Miller, M. (1998). Computational anatomy: An emerging discipline, *Quarterly of applied mathematics* **56**, 4, pp. 617–694.

Haff, L. R. (1979). Estimation of the inverse covariance matrix: Random mixtures of the inverse wishart matrix and the identity, *Annals of Statistics* **7**, pp. 1264–1276.

Haff, L. R. (1980). Empirical bayes estimation of the multivariate normal covariance matrix, *Annals of Statistics* **8**, pp. 586–597.

Haff, L. R. (1991). The variational form of certain bayes estimators, *Annals of Statistics* **19**, pp. 1163–1190.

Hasan, K. M., Basser, P. J., Parker, D. L. and Alexander, A. L. (2001). Analytical computation of the eigenvalues and eigenvectors in DT-MRI, *J. Magn. Res.* **152**, pp. 41–47.

Healy, D. M., Hendriks, H. and Kim, P. T. (1998). Spherical deconvolution, *Journal of Multivariate Analysis* **67**, pp. 1–22.

Helgason, S. (1984). *Groups and Geometric Analysis* (Academic Press, Orlando).

Hendriks, H. (1990). Nonparametric estimation of a probability density on a riemannian manifold using fourier expansions, *Annals of Statistics* **18**, pp. 832–849.

Johnstone, I. M. (2001). On the distribution of the largest eigenvalue in principal components analysis, *Annals of Statistics* **29**, pp. 295–327.

Kim, P. T. (1998). Deconvolution density estimation on SO(n), *Annals of Statistics* **26**, pp. 1083–1102.

Kim, P. T. and Koo, J.-Y. (2000). Directional mixture models and optimal estimation of the mixing density, *Canad. J. Statist.* **28**, pp. 383–398.

Kim, P. T. and Koo, J.-Y. (2002). Optimal spherical deconvolution, *Journal of Multivariate Analysis* **80**, pp. 21–42.

Kim, P. T. and Richards, D. St. P. (2001). Deconvolution density estimation on compact lie groups, *Contemp. Math.* **287**, pp. 155–171.

Kim, P. T. and Richards, D. St. P. (2008). Diffusion tensor imaging and deconvolution on spaces of positive definite symmetric matrices, in *Mathematical Foundations of Computational Anatomy: Geometrical and Statistical Methods for Biological Shape Variability Modeling*, pp. 140 149, URL http://www.inria.fr/sophia/asclepios/events/MFCA08/Proceedings/MFCA08_Proceedings.pdf.

Koltchinskii, V., Sakhanenko, L. and Cao, S. (2007). Integral curves of noisy vector fields and statistical problems in diffusion tensor imaging: Nonparametric kernel estimation and hypotheses testing, *Annals of Statistics* **35**, pp. 1576–1607.

Koo, J.-Y. (1993). Optimal rates of convergence for nonparametric statistical inverse problems, *Annals of Statistics* **21**, pp. 590–599.

Le Bihan, D. (1995). *Diffusion and Perfusion Magnetic Resonance Imaging: Applications to Functional MRI* (Raven Press, New York).

Leporé, N., A., B. C., Chou, Y. Y., Chiang, M. C., Dutton, R. A., Hayashi, M., K, Luders, E., Lopez, O. L., Aizenstein, H. J., Toga, A. W., Becker, J. T., and Thompson, P. M. (2008). Generalized tensor-based morphometry of hiv/aids using multivariate statistics on deformation tensors, *IEEE Trans. Medical Imaging* **27**, pp. 129–141.

Letac, G. and Massam, H. (2004). All invariant moments of the wishart distribution, *Scand. J. Statist.* **31**, pp. 295–318.

Mair, B. A. and Ruymgaart, F. H. (1996). Statistical inverse estimation in hilbert scales, *SIAM J. Appl. Math.* **56**, pp. 1424–1444.

Matthews, P. M. and Arnold, D. L. (2001). Magnetic resonance imaging of multiple sclerosis: new insights linking pathology to clinical evolution, *Current Opinion in Neurology* **14**, pp. 279–287.

Muirhead, R. J. (1982). *Aspects of Multivariate Statistics* (Wiley, New York).

Neumann-Haefelin, T., Moseley, M. E. and Albers, G. W. (2000). New magnetic resonance imaging methods for cerebrovascular disease, *Ann. Neurol.* **47**, pp. 559–570.

Pesenson, I. (2006). Deconvolution of band-limited functions on non-compact symmetric spaces, *Houston J. Math.* **32**, pp. 183–204.

Pomara, N., Crandalla, D. T., Choia, S. J., Johnson, G. and Lim, K. O. (2001). White-matter abnormalities in HIV-1 infection: A diffusion tensor imaging study, *Psychiatry Research: Neuroimaging* **106**, pp. 15–24.

Richards, D. St. P. (1985). Applications of invariant differential operators to multivariate distribution theory, *SIAM J. Appl. Math.* **45**, pp. 280–288.

Richards, D. St. P. (1989). The central limit theorem on spaces of positive definite matrices, *Journal of Multivariate Analysis* **29**, pp. 326–332.

Rosenbloom, M., Sullivan, J. V. and Pfefferbaum, A. (2003). Using magnetic resonance imaging and diffusion tensor imaging to assess brain damage in alcoholics, *Alcohol Research and Health* **27**, pp. 146–152.

Schwartzman, A., Dougherty, R. F. and Taylor, J. E. (2008a). False discovery rate analysis of brain diffusion direction maps, *Ann. Appl. Statist.* **2**, pp. 153–175.

Schwartzman, A., Dougherty, R. F. and Taylor, J. E. (2009). Group comparison of eigenvalues and eigenvectors of diffusion tensors, Tech. rep., Harvard University Biostatistics, URL `http://www.bepress.com/harvardbiostat/paper96`.

Schwartzman, A., Mascarenhas, W. and Taylor, J. E. (2008b). Inference for eigenvalues and eigenvectors of gaussian symmetric matrices, *Annals of Statistics* **36**, pp. 2886–2919.

Takemura, A. and Sheena, Y. (2005). Distribution of eigenvalues and eigenvectors of wishart matrix when the population eigenvalues are infinitely dispersed and its application to minimax estimation of covariance matrix, *Journal of Multivariate Analysis* **94**, pp. 271–299.

Terras, A. (1985). *Harmonic Analysis on Symmetric Spaces and Applications, I.* (Springer-Verlag, New York).

Terras, A. (1988). *Harmonic Analysis on Symmetric Spaces and Applications, II.* (Springer-Verlag, New York).

Zhu, H., Zhang, H., Ibrahim, J. G. and Peterson, B. G. (2007). Statistical analysis of diffusion tensors in diffusion-weighted magnetic image resonance data (with discussion), *Journal of the American Statistical Association* **102**, pp. 1085–1102.

Recent History Functional Linear Models

Kion Kim[†], Damla Şentürk[‡] and Runze Li[*]

Department of Statistics, Pennsylvania State University
University Park, PA 16802, USA
[†] *E-mail: kion@psu.edu*
[‡] *E-mail: dsenturk@stat.psu.edu*
[*] *E-mail: rli@stat.psu.edu*

We propose a variant of historical functional linear models for cases where the current response is affected by the predictor process in a window into the past. Different from the rectangular support of functional linear models, the triangular support of the historical functional linear models and the point-wise support of varying coefficient models, the current model has a sliding window support into the past. This idea leads to models that bridge the gap between varying coefficient models and functional linear (historic) models. By utilizing one-dimensional basis expansions and one-dimensional smoothing procedures, the proposed estimation algorithm is shown to have better performance and to be faster than the estimation procedures proposed for historical functional linear models.

Keywords: Basis expansion; Historical functional linear models; Prediction; Smoothing; Two-step estimator; Varying coefficient models

1. Introduction

Functional data are becoming more frequent and functional data analysis is more relevant as it gets easier to collect functional data on a dense grid. Functional linear models are used widely in the literature to investigate the relationship between functional predictors and functional responses (see Ramsay and Silverman, 2005). Let $x_i(s), s \in [0, S]$ be the predictor function and $y_i(t), t \in [0, T]$ be the response function for $i = 1, \ldots, n$ subjects. Assuming that the predictor function affects the response function through a linear model, Bosq (2000), Cardot *et al.* (1999), James (2002), and Ramsay and Silverman (2005) considered the functional linear model given by

$$y_i(t) = \alpha(t) + \int_0^T \beta(s, t) x_i(s) ds + \varepsilon_i(t), \quad s \in [0, T],\ t \in [0, T], \quad (1)$$

where $\alpha(t)$ is the intercept function, $\beta(s,t)$ is the bivariate regression function, and $\varepsilon_i(t)$ is the error function with $E(\varepsilon_i(t)) = 0$ and $\text{cov}(\varepsilon_i(t), \varepsilon_j(t')) = \sigma_{tt'}$ for $i = j$ and 0 otherwise. The model in (1) is given for the specific case of the same support for the predictor and the response. In model (1), the response at the current time t can be affected by the predictor process from any point in time, including the future values.

For predictive purposes, Malfait and Ramsay (2003) tailored the model in (1) to be applicable to situations where the response function $y(t)$ depends only on the predictor $x(s)$ at time $s \leq t$ as

$$y_i(t) = \alpha(t) + \int_{s_0(t)}^{t} \beta(s,t)x_i(s)ds + \varepsilon_i(t), \quad s \in [s_0(t), t], \ t \in [0,T], \quad (2)$$

where $s_0(t) = \max(0, t-\delta)$ and $0 \leq \delta \leq t$. Here, δ denotes the time lag back in time where the predictor function starts to affect the response function. In this model, only the past of the predictor function affects the response at the current time t. Yet another dependence structure was proposed by Hastie and Tibshirani (1993) and Staniswalis and Lee (1998) via the varying coefficient models,

$$y_i(t) = \alpha(t) + x_i(t)\beta(t) + \varepsilon_i(t), \quad t \in [0,T].$$

Unlike in models (1) and (2), in the varying coefficient models, the current response $y_i(t)$ is affected only by the current value of the predictor $x_i(t)$. This model can be thought of as a special case of the historical functional linear model.

In this paper, we focus on a variant of the historical functional linear model that generalizes the varying coefficient model. In this new model, we allow the current response values to be affected only by the recent past of the predictor process,

$$y_i(t) = \alpha(t) + \int_{t-\delta_1}^{t-\delta_2} x_i(s)\beta(s,t)ds + \varepsilon_i(t), \quad t \in [\delta_1, T], \ 0 \leq \delta_2 \leq \delta_1. \quad (3)$$

Here δ_2, particularly useful for prediction models, allows for a lag for the predictor process to start affecting the response, while δ_1 denotes another lag, beyond which the predictor function does not affect the response. We call this model the recent history functional linear model. This particular model is different from the functional linear model with a rectangular support (1), the general historical functional linear model with a triangular support considering the whole past (2), and the varying coefficient model with a point support. Instead, the model in (3) has a sliding window support, in which the predictor process has an effect on the current response.

To illustrate the cases where model (3) would be reasonable, we analyze the relationship between the measurements of rainfall and the flow level of Curlew creek in Florida. A plausible way to model the fluctuations of the flow level would be to assume that the measurements of rainfall from the recent past are affecting the flow level of the river at a given time point. The second lag in model (3), δ_2, would allow for a delay in the way the rainfall from the recent past starting from $t - \delta_1$ affects the flow level at the current time. Note that, for this data set, it would not make sense to think that rainfall from the future affects the current flow level, nor does one need to consider the whole past of rainfall. It is also not plausible to assume that the current rainfall affects the river flow immediately and hence the proposed model provides a nice generalization and sharpening of the previously proposed models.

For the estimation of the functional linear models, a tensor product basis function is used for the two-dimensional basis expansion of the regression function due to the rectangular support property in the literature. For the historical functional linear models, Malfait and Ramsay (2003) proposed an estimation method that is based on a two-dimensional basis expansion over the triangular support via the finite element method. In this paper, we propose an estimation procedure that is based on a one-dimensional basis expansion instead of the two-dimensional ones proposed for the functional linear models and the historical functional linear models. While this simplification results in significant gain in computational costs, it is also easy to implement.

This paper is organized as follows. In Section 2, we introduce the proposed estimation procedure for the recent history functional linear models and in Section 3, the proposed model parameter selection method is discussed. In Section 4, we compare, via simulations, the performance of the proposed estimation method with that of Malfait and Ramsay (2003) proposed for the historical functional linear models. Application to the flow level data is given in Section 5. We conclude with a remarks section.

2. Estimation in Recent History Functional Linear Models

For a sample of n subjects, the response process is modeled via

$$y_i(t) = \alpha(t) + \int_{t-\delta_1}^{t-\delta_2} x_i(s)\beta(s,t)ds + \varepsilon_i(t), \quad t \in [\delta_1, T],\ 0 \le \delta_2 \le \delta_1. \quad (4)$$

The proposed estimation procedure makes use of the observation that the regression function, $\beta(s,t)$, is a univariate function in s for a fixed time point t. Fixing the time point t, we can expand the regression function with respect to s using the basis functions, $\phi_k(\cdot)$, $k = 1, \ldots, K$, resulting in $\beta(s,t) \approx \sum_{k=1}^{K} b_k \phi_k(s)$, where $s \in [t-\delta_1, t-\delta_2]$. Here we assume a fixed known K for each time point t, which yields reasonable approximations at all the time points. In sections that follow, we will discuss criteria to choose K from the data in practice. Repeating this univariate expansion for different t leads us to the two-dimensional expansion

$$\beta(s,t) \approx \sum_{k=1}^{K} b_k(t)\phi_{k,t}(s), \quad s \in [t-\delta_1, t-\delta_2], \tag{5}$$

where $b_k(t)$ denotes the coefficient of the kth basis function at time t. Using the expansion in (5), we can rewrite the regression model in (4) as

$$y_i(t) = \alpha(t) + \sum_{k=1}^{K} b_k(t) \int_{t-\delta_1}^{t-\delta_2} x_i(s)\phi_{k,t}(s)ds + \varepsilon_i'(t), \tag{6}$$

where $\varepsilon_i'(t)$ is the sum of the intrinsic and the approximation errors. Defining $\psi_{i,k}(t)$ as $\psi_{i,k}(t) = \int_{t-\delta_1}^{t-\delta_2} x_i(s)\phi_{k,t}(s)ds$, the proposed model in (4) simplifies to

$$y_i(t) = \alpha(t) + \sum_{k=1}^{K} \psi_{i,k}(t)b_k(t) + \varepsilon_i'(t), \quad t \in [\delta_1, T]. \tag{7}$$

Equation (7) is a varying coefficient model where $\psi_{i,k}(t)$ are the time dependent predictors and $\alpha(t)$, $b_k(t), k = 1, \ldots, K$ are the varying coefficient functions to be estimated.

An outline of the proposed estimation algorithm is as follows. We start by identifying an appropriate basis system $\phi(\cdot)$ for the expansion of the bivariate regression function. After a basis system is chosen, we estimate $\psi_{i,k}(t)$ using numerical integration methods. Using the estimated $\psi_{i,k}(t)$ as the time-varying predictors, the varying coefficient functions in (7) are estimated. Finally, estimates of the bivariate regression function $\hat{\beta}(s,t)$ are obtained using the expansions in (5).

Common choices of basis systems are the spline basis, Fourier basis, wavelet basis, and truncated power basis. The B-spline basis is one of the most widely used spline bases since it is numerically more stable than the truncated power basis and faster than the Fourier basis in approximating functions. For more details about the comparison of the B-spline basis with others, see Fan and Gijbels (1996). The performance of the B-spline basis largely depends on the location and the number of knots. The knot sequence must be determined from the data since the location and the number of knots are closely related with the complexity of the function and the sparsity of data. Nevertheless, for the densely recorded functional data considered here, we will use an equally spaced knot sequence, assuming that the smoothness of the whole process is uniform. The number of knots, however, still matters and this will be determined from the data. For a fixed time t, if there are k equally spaced interior knots over the range $[t-\delta_1, t-\delta_2]$, then there are $K = (k+4)$ B-spline basis functions provided that we use a cubic B-spline basis.

For the numerical integration, we can apply the first order approximation in the case of densely measured functional data. Suppose, without loss of generality, that the predictor function and the response function are measured on T equally spaced time points inside the time interval $[0,1]$ so that we have measurements on $t_j = \frac{j}{T}$, where $j = 1, \ldots, T$. Then $\psi_{i,k}(t)$ is approximated by

$$\begin{aligned}\hat{\psi}_{i,k}(t) &= \int_{t-\delta_1}^{t-\delta_2} x_i(s)\phi_{i,t}(s)ds \\ &\approx \frac{1}{\lfloor(\delta_1-\delta_2)T\rfloor} \sum_{t_j\in[t-\delta_1,t-\delta_2]} x_i\left(\frac{j}{T}\right)\phi_{k,t}\left(\frac{j}{T}\right), \end{aligned} \tag{8}$$

where $\lfloor a \rfloor$ denotes the largest integer smaller than or equal to a. Approximation in (8) is a variation of the usual rectangular rule using the function values evaluated on the knots instead of those evaluated between the knots.

After estimating $\psi_{i,k}(t)$, we next target the coefficient functions $\alpha(t)$ and $b_k(t)$ in (7). For the estimation of varying coefficient models in (7), we adopt the two-step estimation procedure of Fan and Zhang (2000), which rests on the observation that in varying coefficient models a different linear regression model holds between the response and the predictor at each fixed time point t. Based on this observation, they target the raw varying coefficient function estimates in the first step via ordinary least squares. The raw estimates are refined in the second step by one-dimensional smoothing procedures separately for each varying coefficient function. The method is shown to be fast and quite flexible, allowing different degrees of smoothness for different varying coefficient functions in the refinement step. In this paper, we use local linear fits for the refinement step. More specifically, let $\hat{\mathbf{\Psi}}(t)$ be the $n \times K$ matrix that has $\hat{\psi}_{i,k}(t)$ as its $(i,k)^{th}$ element and $\tilde{\mathbf{\Psi}}(t) = [\mathbf{1}, \hat{\mathbf{\Psi}}(t)]$. If we define $\boldsymbol{y}(t)$ as an n-dimensional vector that contains the n realizations of the response variable at time t, then, in the first step, we can get the raw estimates, $\tilde{\alpha}(t)$ and $\tilde{\boldsymbol{b}}(t) = [\tilde{b}_1(t), \ldots, \tilde{b}_K(t)]$, as

$$\begin{pmatrix} \tilde{\alpha}(t) \\ \tilde{\boldsymbol{b}}(t) \end{pmatrix} = \left[\tilde{\boldsymbol{\Psi}}(t)^T \tilde{\boldsymbol{\Psi}}(t)\right]^{-1} \tilde{\boldsymbol{\Psi}}(t)^T \boldsymbol{y}(t), \quad t \in [\delta_1, 1]. \tag{9}$$

Let J be the collection of indices j that satisfy $t_j \in [\delta_1, 1]$ and let τ be the number of time points in J. A linear smoother for $b_k(t)$ at time point t is given by

$$\hat{b}_k(t) = \sum_{j=1}^{\tau} w_k(t_j, t)\tilde{b}_k(t_j),$$

where the weights $w_k(t_j, t)$ can be constructed by various smoothing techniques.

For the local linear fit, let us suppose that the varying coefficient function $b_k(t)$ is two times continuously differentiable. Then the weights are constructed by

$$w_k(t_j, t) = \frac{K_h(t_j - t)\,\{S_{n,2} - (t_j - t)S_{n,1}\}}{S_{n,0}S_{n,2} - S_{n,1}^2}, \tag{10}$$

where $S_{n,j} = \sum_{i=1}^{n} K_h(t_i - t)(t_i - t)^j$ and $K_h(x) = h^{-1}K(x/h)$. To recover the regression function $\beta(s,t)$, we plug the respective estimates obtained above into the expansion in (5).

3. The Choice of Model Parameters: The Number of Basis Functions and Lags

In the proposed model, we need to select two time lags, δ_1 and δ_2, and the number of basis functions used, K. The choice of δ_1 and δ_2 determines the width of the window back in time where the predictor function has an impact on the response variable and the choice of K controls the precision of the approximation of $\beta(s,t)$. Following Malfait and Ramsay (2003), we employ a two step selection algorithm in choosing the two sets of parameters determining window size and number of basis functions. We first choose the number of basis functions for the maximum window size of interest, determining the resolution needed for a reasonable fit. Determining the number of basis functions is equivalent to determining the number of interior knots, denoted as k_0. Once the number of knots k_0 for the largest window is determined, the number of knots used to approximate the regression function for shorter windows is taken to be proportional to the window's relative length compared to the largest window size considered. For example, if 5 knots for the largest window size, say 10, is considered to be reasonable for the approximation of the regression function, then we take 3 knots for a shorter window of size 6.

In the proposed method, for the selection of k_0 at the largest window size, we adopt root squared prediction error (RSPE) from Müller and Zhang (2005). They use RSPE to decide the number of basis functions for the approximation of the two-dimensional regression function. A modification of the RSPE criterion applicable to the proposed model at a given time point t can be written as

$$RSPE(t) = \left[\frac{1}{n}\sum_{i=1}^{n}\left\{\hat{y}_i^{(-i)}(t) - y_i(t)\right\}^2\right]^{1/2},$$

where $\hat{y}_i^{(-i)}(t)$ denotes the estimated response value for the ith subject measured at time t from the data excluding the ith subject.

The size of the window in the proposed estimation procedure utilizing the connection to varying coefficient models determines the total number of predictors in the varying coefficient models in (7). Nevertheless, the choice of the window size is not a typical variable selection problem in that, for different δ_1 and δ_2 combinations, we have different basis functions, hence different predictors. Therefore, the choice of δ_1 and δ_2 is a model selection problem in varying coefficient models rather than a variable selection problem and we adopt a similar selection criterion to Fan *et al.* (2003). For a model selection criterion of varying coefficient models at a given time point t, they use a pointwise AIC criterion utilizing the fact that the varying coefficient model reduces to a linear model at a fixed time point. In the adoption of their idea to our model, we employ the pointwise AIC at time t, $AIC(t)$, defined by

$$AIC(t) = n\left[\log(2\pi) + \log(RSS(t)) + 1\right] + 2K,$$

where $RSS(t) = \sum_{i=1}^{n}(\tilde{y}_i(t) - y_i(t))^2$, $\tilde{y}_i(t) = \tilde{\alpha}(t) + \sum_{k=1}^{K}\phi_k(t)\tilde{b}_k(t)$, and $\tilde{\alpha}(t)$ and $\tilde{b}_k(t)$ are the raw estimates of the intercept and the kth varying coefficient function, respectively. Here K represents the number of basis functions used, hence the number of predictors. In practice, an initial set of δ_1 and δ_2 combinations is considered and the combination that results in the minimum $AIC(t)$ for most of the time points considered is picked.

4. Simulation Study

Goals of the simulation are to study the finite sample properties of the proposed estimator in comparison with those of the estimation method of Malfait and Ramsay (2003), referred to as MR, and to study the performance of the proposed parameter selection criterion. One of the main advantages of the proposed algorithm is that it provides data-driven automatic selection of the window size and the optimal number of knots. On the other hand, the MR method does not provide automatic choice of these parameters. Instead, the parameters of the historical functional linear models are determined by monitoring the changes in R^2, as usually done in linear models. Note also that the MR method does not allow for one of the lags, δ_2. Due to these restrictions, the two goals outlined at the beginning of this section need to be addressed by two separate simulation designs. For the model assessment, we define the normalized integrated error (NIE) criterion, utilizing one-subject leave-out cross-validation, as

$$\text{NIE} = \frac{\int \sqrt{\sum_{i=1}^{n} \{\hat{y}_i^{(-i)}(t) - y_i(t)\}^2 dt}}{\int \sqrt{\sum_{i=1}^{n} \{y_i(t)\}^2 dt}}. \tag{11}$$

For the first part of the simulation, we compare the proposed method with the MR method assuming that the true window size and the model parameters are known. The true window size is fixed at $\delta_1 = 0.1$ and $\delta_2 = 0$. Note that in the MR method, the resolution is controlled by the number of subintervals (denoted N) into which the two-dimensional regression function support is divided, which in turn specifies the location and number of nodes of the two dimensional basis expansion. We consider the performance of the MR method at different resolutions and compare NIEs at 10, 20 and 40 subintervals to those obtained from the proposed method with one interior knot for three different numbers of subjects, $n = 30$, 100, and 300. The computational efficiency of the two algorithms is also compared via the computing time in seconds under the same computing environment. The second part of simulation studies the performance of the model selection criterion proposed for the choice of δ_1 and δ_2 under the true window size of $\delta_1 = 0.1$ and $\delta_2 = 0.05$.

4.1. *Data generation*

Data is generated from the model

$$y_i(t) = \alpha(t) + \int_{t-\delta_1}^{t-\delta_2} \beta(s,t)x_i(s)ds + \varepsilon_i(t), \quad t \in [\delta_1, 1] \tag{12}$$

on 100 equally spaced grid points in [0,1]. The intercept function is taken to be $\alpha(t) = 10\sin(2t)$ for $\delta_1 \le t \le 1$ while the true bivariate regression function, $\beta(s,t)$, is defined to be $\beta(s,t) = 1.5\sin(0.2s - 10) + 0.3\cos(0.2t) - 1$, where $s \in [t - \delta_1, t - \delta_2]$ and $t \in [0, 1]$. For the generation of the predictor function, we use 8 cubic B-spline basis functions with 4 equally spaced interior knots in [0,1] weighted according to 8 randomly generated coefficients from the multivariate normal distribution with mean $\mu = [3, 8, 6, 2, 7, 4, 9, 5]^T$ and variance $\sigma^2_{coef}\boldsymbol{I}$, where $\sigma^2_{coef} = 3$. The variance of the coefficients is chosen so that the predictor functions do not lose the overall pattern but still have a certain amount of variability. A small amount of independent error generated from a normal distribution with mean 0 and variance 0.1 is added to the predictor functions to prevent singularities. A typical set of the generated predictor functions when $n = 30$ is given in Figure 1(a). The error functions, $\varepsilon_i(t), i = 1, \ldots, n$, are generated by the basis expansion, as in the predictor function case, with $\mu = [0, 0, 0, 0, 0, 0, 0, 0]^T$ and $\Sigma_\varepsilon = 5 \cdot \boldsymbol{I}$. The response functions are displayed in Figure 1(b).

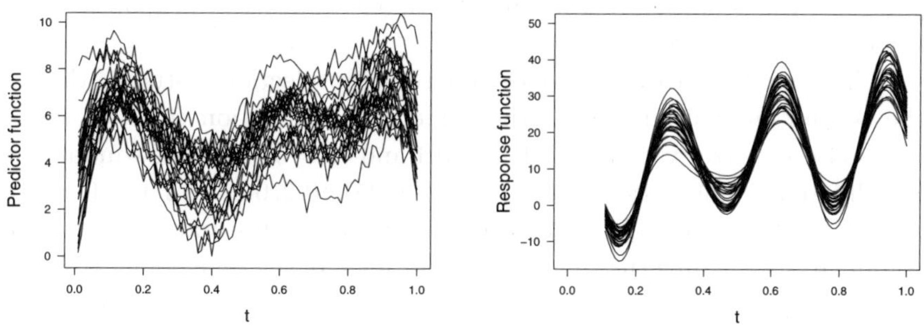

(a) Predictor functions (b) Response functions

Fig. 1. A typical set of predictor and response functions when $n = 30$.

4.2. *Simulation results*

Due to considerations of computational complexity, we consider a modified NIE using 10-fold cross validation instead of one-subject leave-out. For the first comparison simulation, the average NIEs from 500 Monte Carlo runs are reported in Table 1. Note that, in Table 1, N denotes the number of subintervals for the MR method determining precision and the standard deviation of the NIEs are given in parantheses. This table suggests that the proposed estimation method outperforms the MR method in terms of NIE. We also compare the average computing times, in seconds, for the two methods under the same computing environment. Note that computing time is

Table 1. Mean NIE from 500 runs for the proposed and the MR method (with standard deviations in parentheses).

		$n = 30$	$n = 100$	$n = 300$
Proposed method		0.1192 (0.0069)	0.1154(0.0038)	0.114(0.002)
MR method	$N = 10$	0.1464 (0.0111)	0.1384(0.0058)	0.1357(0.0031)
	$N = 20$	0.1526(0.0121)	0.142(0.0062)	0.1388(0.0035)
	$N = 40$	0.1545(0.0142)	0.1457(0.0062)	0.142(0.0034)

defined as the total time to obtain NIEs, which includes multiple model fittings. As the number of subjects or the number of subintervals increases, the average computing time increases. In the simulation with 30 subjects, the MR method with 10 subintervals takes about six times more than the proposed method. For 300 subjects, the MR method with 40 subintervals takes about fifty times more time than the proposed method. This result is expected since while the MR method is based on two-dimensional basis expansions, the proposed method uses only one-dimensional expansions and smoothing procedures.

The performance of the proposed parameter selection is studied through the second simulation. Under the same data-generation scheme, we use $\sigma_\varepsilon^2 = 5$ for the error variance and $\delta_1 = 0.1$ and $\delta_2 = 0.05$ for the window size. Six different combinations, $(0.2, 0), (0.2, 0.05), (0.2, 0.1), (0.1, 0), (0.1, 0.05)$ and $(0.05, 0)$, of (δ_1, δ_2) are considered as candidates and one of the six combinations yielding the smallest AIC for most of the time points is chosen. As the number of subjects increases, the proportion of times the correct choice is made also increases. Starting from 0.727 for $n = 30$, the proportion of correct selections estimated via 500 Monte Carlo runs increases to 0.837 and 0.988 for $n = 100$ and $n = 300$, respectively.

5. Data Analysis

We apply the proposed model to investigate the relationship between the flow level of Curlew creek in Florida and precipitation. From June 7th to October 21st of 2008, both precipitation and flow level are recorded every 15 minutes in inches. We select raining periods for our analysis in which the rain lasts for more than 45 minutes and we consider the 9-hour-long time period from the start of the rain for both the precipitation and flow level. Between the above dates, there are 33 periods that satisfy this description. Since the flow level goes back to normal quickly after the raining period and the precipitation in one raining period does not seem to affect the flow level of the following raining period, each raining period can be thought of as an independent subject. The collected data are presented in Figure 2.

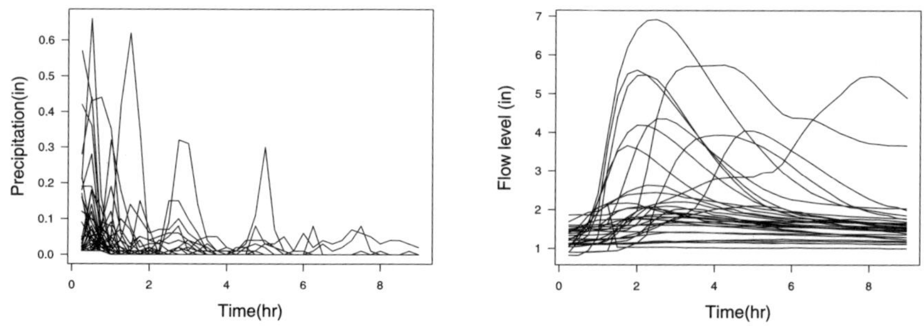

(a) Precipitation data from 33 raining periods

(b) Flow level of Curlew creek, Florida from 33 raining periods

Fig. 2. Precipitation and Flow level of Curlew creek, Florida.

Note that, for the expansion of the regression function, a linear B-spline basis is used instead of a cubic B-spline basis since it yields a better fit in terms of NIE. This implies that the complexity of the regression function can be captured well through the linear approximation. The maximum window size that we consider in this study is $(\delta_1, \delta_2) = (20, 0)$, that is, the flow level of the river at a given time point is assumed to be affected by the patterns of precipitation from the past five hours. The optimal number of knots for the maximum window size is chosen to be 1 and the corresponding average RSPE is 0.6642. Therefore, for smaller window sizes considered, we also apply one interior knot for the basis approximation. We tried 8 different sets of δ_1 and δ_2 combinations: $(20, 15)$, $(20, 10)$, $(20, 5)$, $(20, 0)$, $(15, 10)$, $(15, 5)$, $(15, 0)$, $(10, 5)$, $(10, 0)$ and $(5, 0)$. The optimal combination of δ_1 and δ_2 obtained by the proposed criterion is $(20, 5)$, implying that precipitation from the past 5 to 1.25 hours has an effect on the flow level. For the selected combination, the resulting NIE is 0.258, which implies that about 74% of total variation in the response function is explained by the fitted model. Figure 3 displays flow levels for four randomly selected subjects along with their predicted trajectories.

The estimated regression surface is given in Figure 4. Displayed in Figure 4(b) are 15 profiles of the bivariate regression surface $\hat{\beta}(t - l, t)$, for $l \in \{1.25, 1.5, 1.75, 2, 2.25, 2.5, 2.75, 3, 3.25, 3.5, 3.75, 4, 4.25, 4.5, 4.75\}$ and $t \in \{5.25, 5.5, 5.75, 6, 6.25, 6.5, 6.75, 7, 7.25, 7.5, 7.75, 8, 8.25, 8.5, 8.75, 9\}$

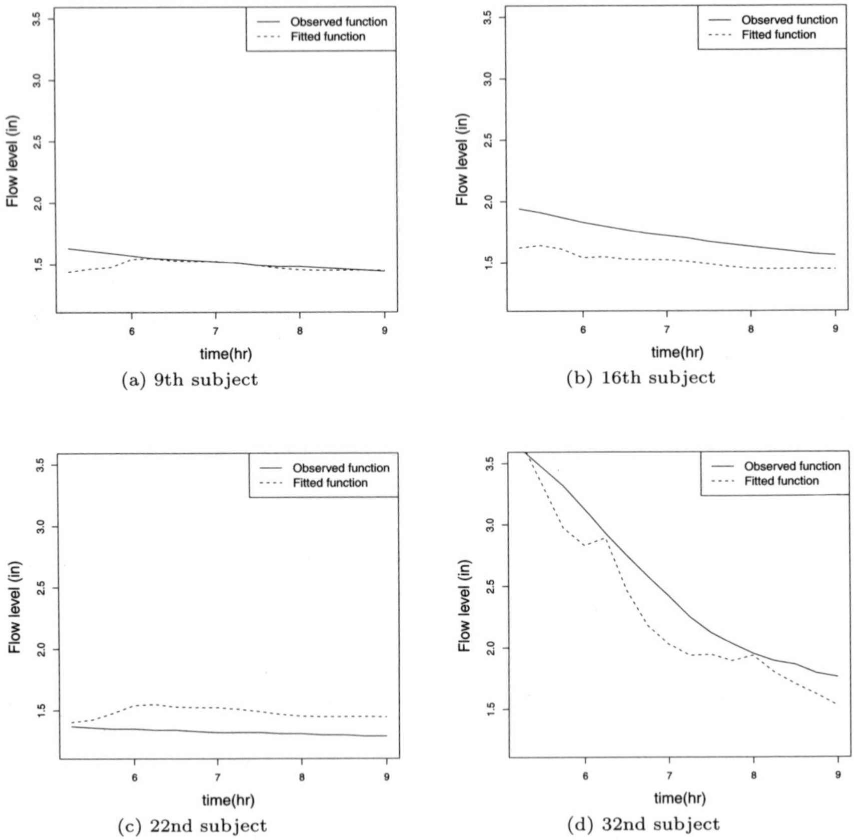

(a) 9th subject

(b) 16th subject

(c) 22nd subject

(d) 32nd subject

Fig. 3. The prediction results for 4 selected subjects. The solid line represents the observed response function and the dashed line denotes the predicted response function.

hours, where profiles are the slices of the regression function in Figure 4(a) parallel to the diagonal and getting further from it by l hours. The profiles are thinner and weaker as they are further from the diagonal. These profiles represent the influence of the predictor function on the response function at l amount back in time from the current time point t. In this plot, the dominating profile (thick solid line) corresponds to $l = 1.25$ hr and the magnitude of the profile decreases as the size of the lag increases. This implies that the current value of the response function is affected most by the predictor function in the closest time point.

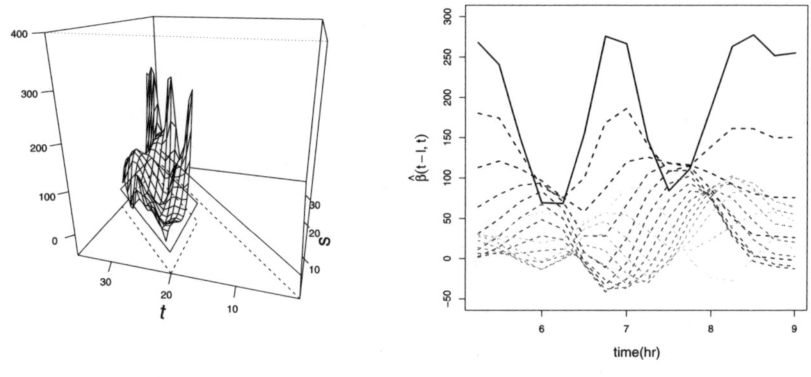

(a) The estimated regression function

(b) The sliced regression function in the direction of the diagonal

Fig. 4. Estimated regression function and profiles parallel to the line $s = t$.

6. Concluding Remarks

In this paper, we propose a variant of the historical functional linear model that models the historical influence of the predictor function on the response function flexibly via the use of two lag parameters. For the estimation of this new class of models, we propose a new method based on a one-dimensional expansion of the regression function utilizing the connection between functional linear models and varying coefficient models. The proposed estimation method assumes densely measured functional data. This assumption plays a key role especially in approximating integrals over the repeated measurements of each subject, such as estimating the $\psi_{i,k}(t)$. In future research, we will consider alternative estimation algorithms that can also accommodate sparse longitudinal data with irregular measurement times and small numbers of observations per subject.

Acknowledgments

Li's research was supported by a National Institute on Drug Abuse (NIDA) grant P50 DA10075. The content is solely the responsibility of the authors and does not necessarily represent the official views of the NIDA or the National Institutes of Health.

References

Bosq, D. (2000). *Linear Processes in Function Spaces: Theory and Applications. Lecture Notes in Statistics 149* (Springer, New York).

Cardot, H., Ferraty, F. and Sarda, P. (1999). Functional linear model, *Statistics and Probability Letters* **45**, pp. 11–22.

Fan, J. and Gijbels, I. (1996). *Local Polynomial Modelling and Its Applications* (Chapman & Hall, London).

Fan, J., Yao, Q. and Cai, Z. (2003). Adaptive varying-coefficient linear models, *Journal of the Royal Statistical Society, Series B* **65**, pp. 57–80.

Fan, J. and Zhang, J. (2000). Two-step estimation of functional linear models with applications to longitudinal data, *Journal of the Royal Statistical Society, Series B* **62**, pp. 303–322.

Hastie, T. J. and Tibshirani, R. J. (1993). Varying-coefficient models, *Journal of the Royal Statistical Society, Series B* **55**, pp. 757–796.

James, G. M. (2002). Generalized linear models with functional predictors, *Journal of the Royal Statistical Society, Series B* **64**, pp. 411–432.

Malfait, N. and Ramsay, J. O. (2003). The historical functional linear model, *The Canadian Journal of Statistics* **31**, pp. 115–128.

Müller, H. G. and Zhang, Y. (2005). Time-varying functional regression for predicting remaining lifetime distributions from longitudinal trajectories, *Biometrics* **61**, pp. 1064–1075.

Ramsay, J. O. and Silverman, B. W. (2005). *Functional Data Analysis* (Springer, New York).

Staniswalis, J. G. and Lee, J. J. (1998). Nonparametric regression analysis of longitudinal data, *Journal of the American Statistical Association* **93**, pp. 1403–1418.

Rank-Based Estimation for Arnold-Transformed Data

John D. Kloke

Center for Research on Health Care, University of Pittsburgh
Pittsburgh, PA 15213, USA
E-mail: klokejd@upmc.edu

Joseph W. McKean

Department of Statistics, Western Michigan University
Kalamazoo, MI 49008, USA
E-mail: joseph.mckean@wmich.edu

Rank-Based methods for linear models for independent and identically distributed data have been developed over the past 30 years. However, little work has been done in the area of mixed models. In this paper, we discuss a transformation approach to modeling a particular mixed model: one with an arbitrary number of fixed effects and covariates but only one random effect. Discussion of the asymptotic theory is given and the results of a simulation study verify the theory. These models are used to estimate the fixed effects from an experiment that uses a randomized block design.

Keywords: Hodges-Lehmann estimate; Mixed model; Nonparametrics; R estimates; Randomized block design; Rank regression scores; Studentized residuals; Wilcoxon methods

1. Introduction

Arnold (1981) developed a transformation approach, based on a Helmert transformation, for the traditional analysis (least squares or maximum likelihood under normal errors) of certain balanced block designs. For the transformed model, the fixed-effects part of the linear model is free of the random block effect. Here, we take "balanced" to mean an equal number of experimental units per block. In this paper we extend Arnold's results to include unbalanced designs. Also, it is easy to see that his estimators are asymptotically normal. This means that for large samples, his normality assumption is not required. Our main focus, however, is on the development of rank-based (R) estimators based on Arnold's transformation.

There has been some previous work on rank estimators for these types of models. Kloke *et al.* (2009) developed inference for these types of models based on estimators that minimize the dispersion function of Jaeckel (1972) (joint rankings). Rashid *et al.* (2010) developed inference for these models based on a many rankings (MR) dispersion (Friedman-type rankings).

In Section 2, we briefly review rank-based estimation for linear models. As we show, the transformation may result in regression through the origin. So for this case, we also discuss an appropriate adjustment for the rank-based estimates. In Section 3, we develop rank-based estimation based on transformed data using Arnold's transformation. We present asymptotic representations from which the theory follows. Standard errors are easily estimated in order to conduct rank-based inference. Our development is for general score functions. The only conditions on the score function are that it is nondecreasing, bounded, and square-integrable. Hence, scores appropriate for skewed error distributions as well as symmetric distributions can be used. In Section 4, we propose Studentization of the residuals based on both the least squares and rank-based procedures. These residuals adjust for both error variation and location in factor space. In Section 5, we present the results of a small Monte Carlo investigation that verify the theoretical results and show the robustness of the rank-based approach over the least squares approach in terms of empirical efficiency. Also, the empirical study shows the validity of confidence intervals based on estimated standard errors. We conclude with an example in Section 6.

2. Rank Estimation in Linear Models

In this section, we establish some notation and briefly discuss the least squares (LS) and rank-based fits of a linear model. Let $\boldsymbol{Y} = (y_1, \ldots, y_n)$ be a vector of observed responses and let $\boldsymbol{x}_i = (x_{i1}, \ldots, x_{ip})'$ be a vector of explanatory variables, which may be covariates, predictors, or indicator variables. Assume that $\boldsymbol{Y}$ follows the linear model

$$y_i = \alpha + \boldsymbol{\beta}'\boldsymbol{x}_i + e_i, \quad i = 1, \ldots, n, \tag{1}$$

where $e_1, \ldots, e_n$ are stochastic errors. In this section, we have assumed an intercept parameter, α. Let $\tilde{x}_{ij}$'s denote the centered predictors; i.e., for each j, $\sum_{i=1}^{n} \tilde{x}_{ij} = 0$. Let $\tilde{\boldsymbol{X}} = [\tilde{x}_{ij}]$ and let $\boldsymbol{X} = [\boldsymbol{1}\ \tilde{\boldsymbol{X}}]$. Assume without loss of generality that $\boldsymbol{X}$ has full column rank. For a matrix $\boldsymbol{W}$ of full column rank, let $\boldsymbol{H}_{\boldsymbol{W}} = \boldsymbol{W}(\boldsymbol{W}^\top\boldsymbol{W})^{-1}\boldsymbol{W}^\top$ denote the projection matrix onto the column space of $\boldsymbol{W}$.

The traditional fit is based on the least squares (LS) estimator,

$$\widehat{\boldsymbol{\beta}}_{LS} = \arg\min_{\boldsymbol{\beta}} \|\boldsymbol{Y} - \tilde{\boldsymbol{X}}\boldsymbol{\beta}\|_2^2, \tag{2}$$

where $\|\cdot\|_2$ is the Euclidean norm. The LS estimator of α is the average of the residuals, i.e., $\widehat{\alpha}_{LS} = \text{Ave}\{\boldsymbol{Y} - \tilde{\boldsymbol{X}}\widehat{\boldsymbol{\beta}}_{LS}\}$.

Let $\varphi(u)$ be a nondecreasing, square-integrable function defined on the interval $(0,1)$. Without loss of generality, assume that $\int \varphi(u)\,du = 0$ and $\int \varphi^2(u)\,du = 1$. The function $\varphi(u)$ is called a score function that generates the rank regression scores as $a(i) = \varphi[i/(n+1)]$, $i = 1, 2, \ldots, n$. Popular score functions are the Wilcoxon, $\varphi(u) = \sqrt{12}[u - (1/2)]$, and the sign, $\varphi(u) = \text{sign}[u - (1/2)]$.

For a specified score function $\varphi(u)$, the rank-based estimate of $\boldsymbol{\beta}$ is given by

$$\widehat{\boldsymbol{\beta}}_{\varphi} = \arg\min_{\boldsymbol{\beta}} \|\boldsymbol{Y} - \tilde{\boldsymbol{X}}\boldsymbol{\beta}\|_{\varphi}, \tag{3}$$

where $\|\cdot\|_{\varphi}$ is the norm given by $\|\boldsymbol{v}\|_{\varphi} = \sum_{i=1}^{n} a[R(v_i)]v_i$ and $R(v_i)$ denotes the rank of v_i among $v_1, \ldots, v_n$; see Chapter 3 of Hettmansperger and McKean (1998).

Because the ranks are invariant to a constant shift, the intercept cannot be estimated using Expression (3). In order to estimate the intercept, let $\hat{\boldsymbol{e}}_{\varphi} = \boldsymbol{Y} - \tilde{\boldsymbol{X}}\widehat{\boldsymbol{\beta}}_{\varphi}$ denote the rank-based residuals. The estimate of the intercept is a location estimator based on these residuals. In this paper, we consider two estimates of the intercept. For the first, we estimate the intercept by the median of the residuals, i.e.,

$$\widehat{\alpha}_S = \text{median}\{\boldsymbol{Y} - \tilde{\boldsymbol{X}}\widehat{\boldsymbol{\beta}}_{\varphi}\}. \tag{4}$$

For the second estimator, suppose $\varphi(u)$ is odd about $1/2$, i.e., $\varphi(1-u) = -\varphi(u)$. Both the Wilcoxon and sign score functions satisfy this property. For such a score function, the corresponding signed-rank score function is given by $\varphi^+(u) = \varphi[(u+1)/2]$; see Chapter 3 of Hettmansperger and McKean (1998) for discussion. The estimate of the intercept is then the signed-rank location estimator based on the residuals, i.e., the estimator $\widehat{\alpha}^+$ that solves the equation

$$\sum_{i=1}^{n} \varphi^+\left[\frac{R|\hat{e}_{\varphi,i} - \alpha|}{n+1}\right] \text{sign}\,[\hat{e}_{\varphi,i} - \alpha] \doteq 0. \tag{5}$$

Simple search algorithms can be used to compute these estimators; see Chapter 3 of Hettmansperger and McKean (1998). If the Wilcoxon scores are used, then Expression (5) simplifies and the resulting estimator is the

Hodges-Lehmann estimator of α, i.e., the median of the Walsh averages of the residuals, given by

$$\widehat{\alpha}^{+} = \text{median}_{i \leq j} \left\{ \frac{\hat{e}_{\varphi,i} + \hat{e}_{\varphi,j}}{2} \right\}. \tag{6}$$

There are advantages and disadvantages for estimators (4) and (6). As for the rank-based estimator $\widehat{\boldsymbol{\beta}}_{\varphi}$ of the regression coefficients, the asymptotic theory for the median of the residuals does not require symmetry of the error distribution. This makes the combination $(\widehat{\alpha}_S, \widehat{\boldsymbol{\beta}}_{\varphi}^{\top})^{\top}$ useful for skewed error distributions as well as symmetric error distributions. However, if the errors are normally distributed, the efficiency of the intercept estimator $\widehat{\alpha}_S$ is quite low, 0.63. In the case of normal errors, the efficiency of the signed-rank estimator of the intercept can be quite high, for example, 0.95 if the signed-rank Wilcoxon scores are used. At times, the high efficiency of a signed-rank estimator outweighs its disadvantage for skewed error distributions; see Freidlin and Gastwirth (2000).

The estimators $(\widehat{\alpha}_S, \widehat{\boldsymbol{\beta}}_{\varphi}^{\top})^{\top}$ or $(\widehat{\alpha}^{+}, \widehat{\boldsymbol{\beta}}_{\varphi}^{\top})^{\top}$ are commonly referred to as **rank-based** regression estimators. A Newton-type algorithm to compute these estimators is discussed in Chapter 3 of Hettmansperger and McKean (1998). For Wilcoxon scores, a set of R functions (R Development Core Team, 2010) to compute these estimates is given in Terpstra and McKean (2005). An R package for general scores (Kloke and McKean, 2010) is currently under development now and will be made available from the authors.

Ties in the responses play a role in rank testing. However, in rank-based estimation, the residuals (not the responses) are ranked and tied residuals are of no consequence to the estimation. To see this, suppose two residuals are equal, say, $e_i = e_j$. Their contribution to the objective function, the norm in Expression (3), is $\{a[R(e_i)] + a[R(e_j)]\}(e_i + e_j)$. Because the score function is nondecreasing, the scores are adjacent to one another, so it does not matter which score either residual obtains because this contribution remains exactly the same. Furthermore, one would never replace the scores by average scores, because then the convexity of the objective function may not hold.

2.1. *Regression through the origin*

Recall that the matrix $\tilde{\boldsymbol{X}}$ is the centered design matrix based on $\boldsymbol{X}$. Whether $\tilde{\boldsymbol{X}}$ or $\boldsymbol{X}$ is used on the right side of Expression (3), the norm distance is the same; hence, for rank-based estimation based on Expression

(3), the design matrix is $\tilde{\boldsymbol{X}}$. In this paper, though, we will first be transforming the data. In this case, there might not be a column of ones in the transformed design matrix. Thus we do not want to center the data but, instead, we want to do regression through the origin. Hence, we briefly discuss an adjustment of the rank-based estimator that allows us to do regression through the origin, which was proposed by Dixon and McKean (1996); see also problem 3.12.39 of Hettmansperger and McKean (1998).

2.2. *Asymptotic results*

In this subsection, we present some asymptotic results; so, for this section only, we assume that the random errors in Model (1) are independent and identically distributed with common density $f(t)$ and distribution function $F(t)$. Also, without loss of generality, assume that the median of e_i is zero. Define the scale parameters τ and τ_S, respectively, by

$$\tau = \left\{ \int \varphi(u) \left[-\frac{f'[F^{-1}(u)]}{f[F^{-1}(u)]} \right] du \right\}^{-1} \tag{7}$$

and

$$\tau_S = 1/[2f(0)]. \tag{8}$$

Let $\boldsymbol{X}^*$ represent an uncentered design matrix and $\tilde{\boldsymbol{X}}^* = (\boldsymbol{I} - \boldsymbol{H}_1)\boldsymbol{X}^*$ represent the centered one. Assume that the usual design conditions hold (see for example Chapter 3 of Hettmansperger and McKean, 1998). First, the model

$$\boldsymbol{y} = [\mathbf{1}\tilde{\boldsymbol{X}}^*] \begin{bmatrix} \alpha_1 \\ \boldsymbol{\beta}_1 \end{bmatrix} + \boldsymbol{e}$$

is fit using the rank-based estimator of Expression (3). This results in the fitted values $\hat{\boldsymbol{y}}_1 = \mathbf{1}\hat{\alpha}_1 + \tilde{\boldsymbol{X}}^*\hat{\boldsymbol{\beta}}_1$. Then the fit is projected to the correct space

$$\hat{\boldsymbol{y}} = \boldsymbol{H}_{\boldsymbol{X}^*}\hat{\boldsymbol{y}}_1,$$

which leads immediately to the estimator

$$\hat{\boldsymbol{\beta}}_R = (\boldsymbol{X}^{*\top}\boldsymbol{X}^*)^{-1}\boldsymbol{X}^{*\top}\hat{\boldsymbol{y}}_1.$$

As discussed in Section 2, common estimates of α are based on signed-rank or sign scores. If the signed-rank estimate of the intercept is used, then the first order approximation of the estimator $\hat{\boldsymbol{\beta}}$ is

$$\hat{\boldsymbol{\beta}} \doteq \tau(\boldsymbol{X}^{*\top}\boldsymbol{X}^*)^{-1}\boldsymbol{X}^{*\top}\varphi(F(\boldsymbol{e})), \tag{9}$$

where $\varphi[F(\boldsymbol{e})]$ is the random vector $(\varphi[F(e_1)],\ldots,\varphi[F(e_n)])^\top$. Likewise, if the median of the residuals is used as the estimate of intercept, the first-order approximation of the the estimate is

$$\hat{\boldsymbol{\beta}} \doteq \tau_s(\boldsymbol{X}^{*\top}\boldsymbol{X}^*)^{-1}\boldsymbol{X}^{*\top}\boldsymbol{H}_{\mathbf{1}_{N^*}}\operatorname{sgn}(\boldsymbol{e}) \\ +\tau_\varphi(\boldsymbol{X}^{*\top}\boldsymbol{X}^*)^{-1}\boldsymbol{X}^{*\top}\boldsymbol{H}_{\tilde{\boldsymbol{X}}^*}\varphi(F(\boldsymbol{e})). \tag{10}$$

See Chapter 3 of Hettmansperger and McKean (1998).

From these representations, it follows immediately that the asymptotic distributions are given by

$$\hat{\boldsymbol{\beta}}^*_R \text{ is approximately } N[\boldsymbol{\beta},\tau^2(\boldsymbol{X}^{*\top}\boldsymbol{X}^*)^{-1}]$$

using signed-rank to estimate the intercept and

$$\hat{\boldsymbol{\beta}}^*_R \text{ is approximately } N(\boldsymbol{\beta},\boldsymbol{V}_s)$$

using the median of the residuals to estimate intercept, where

$$\boldsymbol{V}_s = \tau_s^2(\boldsymbol{X}^{*\top}\boldsymbol{X}^*)^{-1}\boldsymbol{X}^{*\top}\boldsymbol{H}_{\mathbf{1}_{N^*}}\boldsymbol{X}^*(\boldsymbol{X}^{*\top}\boldsymbol{X}^*)^{-1} \\ +\tau_\varphi^2(\boldsymbol{X}^{*\top}\boldsymbol{X}^*)^{-1}\boldsymbol{X}^{*\top}\boldsymbol{H}_{\tilde{\boldsymbol{X}}^*}\boldsymbol{X}^*(\boldsymbol{X}^{*\top}\boldsymbol{X}^*)^{-1}. \tag{11}$$

We will make use of these adjusted rank-based estimates in the next section.

3. Estimation Based on Arnold-Transformed Data

3.1. *The model and the Arnold transformation*

Consider an experiment done over m blocks. Within block k there are n_k observations. Denote the model for block k as

$$\boldsymbol{y}_k = \alpha\mathbf{1}_{n_k} + \boldsymbol{X}_k\boldsymbol{\beta} + b_k\mathbf{1}_{n_k} + \boldsymbol{\epsilon}_k \text{ for } k=1,\ldots,m, \tag{12}$$

where b_k is a random block effect, α is the intercept parameter, $\boldsymbol{y}_k$ is an $n_k\times 1$ vector of responses, $\boldsymbol{X}_k$ is an $n_k\times p$ design matrix, $\mathbf{1}_{n_k}$ is an $n_k\times 1$ vector of ones, and $\boldsymbol{\epsilon}_k$ the $n_k\times 1$ vector of independent and identically distributed errors. As usual in these types of models, we assume that b_k and the random vector $\boldsymbol{\epsilon}_k$ are independent and that the blocks are independent. It will be convenient to work with the error vectors $\boldsymbol{e}_k = b_k\mathbf{1}_{n_k} + \boldsymbol{\epsilon}_k$. We may rewrite (12) as

$$\boldsymbol{y}_k = \alpha\mathbf{1}_{n_k} + \boldsymbol{X}_k\boldsymbol{\beta} + \boldsymbol{e}_k \text{ for } k=1,\ldots,m. \tag{13}$$

If variances exist, $\operatorname{var}[b_k] = \sigma_b^2$ and $\operatorname{var}[\epsilon_{kj}] = \sigma_\epsilon^2$, then the variance-covariance structure for the kth block is compound symmetric, which we denote as

$$\operatorname{var}[\boldsymbol{e}_k] = \sigma^2\boldsymbol{A}_k(\rho) = \sigma^2[(1-\rho)\boldsymbol{I}_{n_k} + \rho\boldsymbol{J}_{n_k}],$$

where $\sigma^2 = \sigma_\epsilon^2 + \sigma_b^2$, $\rho = \sigma_b^2/(\sigma_b^2 + \sigma_\epsilon^2)$, $\boldsymbol{I}_{n_k}$ is an $n_k \times n_k$ identity matrix, and $\boldsymbol{J}_{n_k}$ is an $n_k \times n_k$ matrix of ones.

Let $N = \sum_{k=1}^m n_k$ denote the total sample size. Rewrite (13) as a single linear model:

$$\boldsymbol{y} = \alpha \mathbf{1}_N + \boldsymbol{X}\boldsymbol{\beta} + \boldsymbol{e},$$

where $\boldsymbol{y}^\top = [\boldsymbol{y}_1^\top, \ldots \boldsymbol{y}_m^\top]$, $\boldsymbol{X}^\top = [\boldsymbol{X}_1^\top, \ldots, \boldsymbol{X}_m^\top]$, and $\boldsymbol{e}^\top = [\boldsymbol{e}_1^\top, \ldots, \boldsymbol{e}_m^\top]$. Since there is an intercept parameter in the model, we may assume that $\boldsymbol{X}$ is centered.

Arnold (1981), in Chapters 14 and 15, discusses a Helmert transformation for these types of models for traditional (least squares) analyses for balanced designs (i.e., $n_k = n$). In what follows, we generalize Arnold's results to unbalanced designs.

Consider the $n_k \times n_k$ orthogonal matrix

$$\boldsymbol{\Gamma}_k = \begin{bmatrix} \frac{1}{\sqrt{n_k}} \mathbf{1}_{n_k}^T \\ \boldsymbol{C}_k^T \end{bmatrix},$$

where the columns of $\boldsymbol{C}_k$ are an orthonormal basis for $\mathbf{1}_{n_k}^\perp$ ($\boldsymbol{C}_k^T \mathbf{1}_{n_k} = \mathbf{0}$). Now, apply *Arnold's Transformation* of size n_k to the response vector for the kth block

$$\boldsymbol{y}_k^* = \boldsymbol{\Gamma}_k \boldsymbol{y}_k = \begin{bmatrix} y_{k1}^* \\ \boldsymbol{y}_{k2}^* \end{bmatrix},$$

where the *mean component* is $y_{k1}^* = \alpha^* + b_k^* + \sqrt{n_k}\bar{\boldsymbol{x}}_k^\top \boldsymbol{\beta} + e_{k1}^*$ and the *contrast component* is $\boldsymbol{y}_{k2}^* = \boldsymbol{X}_k^* \boldsymbol{\beta} + \boldsymbol{e}_{k2}^*$. Note that the contrast component contains, as a linear model, the fixed effects of interest and, further, it is free of the random block effect.

Although they are evident based on the transformation, for clarification, we state the other transformed quantities:

$$\begin{aligned}
\bar{\boldsymbol{x}}_k^\top &= \frac{1}{n_k} \mathbf{1}_{n_k}^\top \boldsymbol{X}_k \\
e_{k1}^* &= \frac{1}{\sqrt{n_k}} \mathbf{1}_{n_k}^\top \boldsymbol{e}_k \\
\boldsymbol{X}_k^* &= \boldsymbol{C}_k \boldsymbol{X}_k \\
\boldsymbol{e}_{k2}^* &= \boldsymbol{C}_k^\top \boldsymbol{e}_k = b_k \boldsymbol{C}_k^\top \mathbf{1}_{n_k} + \boldsymbol{C}_k^\top \boldsymbol{\epsilon}_k = \boldsymbol{C}_k^\top \boldsymbol{\epsilon}_k.
\end{aligned}$$

Furthermore, notice that all the information on $\boldsymbol{\beta}$ is in the contrast component if $\bar{\boldsymbol{x}} = \mathbf{0}$. This occurs when the experimental design is replicated at least once in each of the blocks. Also, all of the information on $\boldsymbol{\beta}$ is in the mean component if the covariates are constant within a block. More often, however, there is information on $\boldsymbol{\beta}$ in both of the components. If this is the case, then for balanced designs, one can put both pieces back together and obtain an estimator using all of the information. For unbalanced designs this is not possible. The approach we take is to ignore the information in the mean component and use the contrast component for inference.

Let $N^* = N - m$. Then the full Arnold transformation (AT) is $\boldsymbol{y}_2^* = \boldsymbol{C}^\top \boldsymbol{y}$, where $\boldsymbol{C}^\top = \text{diag}[\boldsymbol{C}_1^\top, \ldots, \boldsymbol{C}_m^\top]$. So we can model $\boldsymbol{y}_2^*$ as

$$\boldsymbol{y}_2^* = \boldsymbol{X}^*\boldsymbol{\beta} + \boldsymbol{e}_2^*, \tag{14}$$

where $\boldsymbol{e}_2^* = \boldsymbol{C}^\top \boldsymbol{e}$, $\text{var}[\boldsymbol{e}_2^*] = \sigma_2^2 I_{N^*}$, $\sigma_2^2 = \sigma^2(1-\rho)$, and $\boldsymbol{X}^* = \boldsymbol{C}^\top \boldsymbol{X}$.

3.2. *Least Squares Estimates Based on the Transformed Data*

In this subsection, we offer an extension of the results of Arnold (1981) to unbalanced cases. The least squares solution to (14) is

$$\hat{\boldsymbol{\beta}}_{ATLS} = \arg\min_{\boldsymbol{\beta}} \|\boldsymbol{y}_2^* - \boldsymbol{X}^*\boldsymbol{\beta}\|_{LS}.$$

That is, $\hat{\boldsymbol{\beta}}_{ATLS} = (\boldsymbol{X}^{*\top}\boldsymbol{X}^*)^{-1}\boldsymbol{X}^{*\top}\boldsymbol{y}_2^*$.

As usual, estimate the intercept based on the mean of the residuals,

$$\begin{aligned}\hat{\alpha}_{LS} &= \frac{1}{N}\mathbf{1}^\top(\boldsymbol{y} - \hat{\boldsymbol{y}}) \\ &= \frac{1}{N}\mathbf{1}^\top(\boldsymbol{I}_N - \boldsymbol{X}(\boldsymbol{X}^{*\top}\boldsymbol{X}^*)^{-1}\boldsymbol{X}^*\boldsymbol{C}^\top)\boldsymbol{y} \\ &= \bar{y}.\end{aligned}$$

It is easy to see that the asymptotic distribution is

$$\begin{bmatrix}\hat{\alpha}_{LS} \\ \hat{\boldsymbol{\beta}}_{ATLS}\end{bmatrix} \dot{\sim} N_{p+1}\left(\begin{bmatrix}\alpha \\ \boldsymbol{\beta}\end{bmatrix}, \begin{bmatrix}\sigma_1^2 & \mathbf{0}^\top \\ \mathbf{0} & \sigma_2^2(\boldsymbol{X}^{*\top}\boldsymbol{X}^*)^{-1}\end{bmatrix}\right), \tag{15}$$

where $\sigma_1^2 = \sigma^2/N^2 \sum_{k=1}^m [(1-\rho)n_k + n_k^2]$ and $\sigma_2^2 = \sigma^2(1-\rho)$. Notice that if inference is to be on $\boldsymbol{\beta}$ then we avoid explicit estimation of ρ. To estimate σ_2^2, we may use $\hat{\sigma}_2^2 = \sum_{k=1}^m \sum_{j=1}^{n_k} \hat{e}_{kj}^{*2}/(N^* - p)$, where $\hat{e}_{kj}^* = y_{kj}^* - \boldsymbol{x}_{kj}^{*\top}\hat{\boldsymbol{\beta}}$.

3.3. *Rank-Based Estimates Based on the Transformed Data*

In this section, we outline a robust, rank-based analogue of Arnold's approach to analyzing randomized block designs. This approach was considered for special cases in Kloke *et al.* (2005) and Kloke and McKean (2004).

Assume, as in Section 2, that we have selected a score function $\varphi(u)$. Estimation of parameters then follows from the geometry of Section 2.1. We define the Arnold's transformation rank-based (ATR) estimator of $\boldsymbol{\beta}$ as the regression through the origin rank estimator of Model (14).

Suppose the median of the residuals is used as the estimate of the intercept ($\hat{\alpha}_S = \text{med}\{\hat{e}_{kj}\}$ where $\hat{e}_{kj} = y_{kj} - \boldsymbol{x}_{kj}^\top \hat{\boldsymbol{\beta}}_{ATR}$). If we also estimate α_1 (in the regression through the origin step) using the median for the residuals then, from Kloke *et al.* (2009), the asymptotic distribution of the regression parameters is

$$\begin{bmatrix} \hat{\alpha}_S \\ \hat{\boldsymbol{\beta}}_\varphi \end{bmatrix} \dot{\sim} N_{p+1}\left(\begin{bmatrix} \alpha \\ \boldsymbol{\beta} \end{bmatrix}, \begin{bmatrix} \sigma_s^2 \tau_{s,e}^2/N & \mathbf{0}^\top \\ \mathbf{0} & \boldsymbol{V} \end{bmatrix} \right), \tag{16}$$

where $\boldsymbol{V}$ is given by (11), $\sigma_s^2 = 1 + n^*\rho_s$, $n^* = \sum_{k=1}^m n_k(n_k - 1)$, and $\rho_s = \text{cov}[\text{sgn}(e_{11}), \text{sgn}(e_{12})]$.

Next assume that the score function $\varphi(u)$ is odd about 1/2; i.e., $\varphi(1 - u) = -\varphi(u)$. Then as discussed in Section 2, Equation (5), suppose we use the signed-rank estimate of the intercept to estimate α_1. Then, similarly to Kloke *et al.* (2009), the asymptotic distribution of the rank-based estimator is

$$\begin{bmatrix} \hat{\alpha} \\ \hat{\boldsymbol{\beta}}_\varphi \end{bmatrix} \dot{\sim} N_{p+1}\left(\begin{bmatrix} \alpha \\ \boldsymbol{\beta} \end{bmatrix}, \begin{bmatrix} \sigma_s^2 \tau_{s,e}^2/N & \mathbf{0}^\top \\ \mathbf{0} & \boldsymbol{V} \end{bmatrix} \right), \tag{17}$$

where $\boldsymbol{V} = \tau^2(\boldsymbol{X}^{*\top}\boldsymbol{X}^*)^{-1}$. In comparing Expressions (15) and (17), we see that the asymptotic relative efficiency (ARE) between the ATLS and the ATR estimators is the same as that between the LS and R estimates in ordinary linear models. In particular, for normal errors, the ARE between the ATLS and ATR (Wilcoxon) is the usual 0.95. Hence, the ATR estimators are efficiently robust.

The scale parameters τ and τ_s are based on the distribution of e_{2kj}^* and can be estimated as discussed in Chapter 3 of Hettmansperger and McKean (1998). From this, an inference is readily formed for the parameters of the model. Validity of the resulting confidence intervals is confirmed in the simulation study of Section 5.

As a final note, upon transformation we did assume working independence. If, however, upon transformation the errors are uncorrelated but not independent, the results of Kloke *et al.* (2009) ensure that the estimates are consistent and asymptotically normal.

4. Studentized Residuals

Studentized residuals are one of the most useful diagnostic tools. This standardization adjusts the residuals for both underlying variance structure of the random errors and their location in factor space. Further, they allow the simple benchmark ± 2 to flag potential outliers. In this section, we develop Studentized residuals for our rank-based AT fit as well as for the AT fit based on LS estimates.

4.1. *AT Least Squares Studentized Residuals*

Define the Arnold's Transformation Least Squares residuals as

$$\begin{aligned}\hat{\boldsymbol{e}}_{ATLS} &= \boldsymbol{y} - \hat{\alpha}_{LS}\mathbf{1}_N - \boldsymbol{X}\hat{\boldsymbol{\beta}}_{ATLS} \\ &= (\boldsymbol{I} - \boldsymbol{H}_1 - \boldsymbol{X}(\boldsymbol{X}^*\boldsymbol{X}^*)^{-1}\boldsymbol{X}^*\boldsymbol{C}^\top)\boldsymbol{y}.\end{aligned}$$

It is easy to see that the variance of $\hat{\boldsymbol{e}}_{ATLS}$ is

$$\begin{aligned}\text{var}[\hat{\boldsymbol{e}}_{ATLS}] = &(\boldsymbol{I} - \boldsymbol{H}_1 - \boldsymbol{X}(\boldsymbol{X}^{*\top}\boldsymbol{X}^*)^{-1}\boldsymbol{X}^*\boldsymbol{C}^\top) \\ &\times\sigma^2\boldsymbol{A}(\rho)(\boldsymbol{I} - \boldsymbol{H}_1 - \boldsymbol{X}(\boldsymbol{X}^{*\top}\boldsymbol{X}^*)^{-1}\boldsymbol{X}^*\boldsymbol{C}^\top).\end{aligned}$$

To estimate the Studentized residuals we use the MSEs from the two fits:

$$\widehat{\sigma^2(1-\rho)} = \text{MSE}_1 = \frac{1}{N^* - p}\sum_{k=1}^{m}\sum_{j=1}^{n_k - 1}\hat{e}_{kj}^{*2},$$

$$\widehat{\sigma^2} = \text{MSE}_2 = \frac{1}{N - p}\sum_{k=1}^{m}\sum_{j=1}^{n_k}\hat{e}_{kj}^{2}.$$

From here we estimate ρ with

$$\hat{\rho} = \max\left\{0, 1 - \frac{\text{MSE}_1}{\text{MSE}_2}\right\}.$$

4.2. *AT Rank-Based Studentized Residuals*

For the linear model with independent and identically distributed errors, McKean *et al.* (1990, 1993) developed Studentized residuals. Our approach follows similarly. There are two Studentizations for the AT rank-based residuals, depending on which estimate of the intercept is used. We begin with the Studentization based on the signed-rank intercept estimator. Let $F(t)$ denote the cdf of the AT random error $e^*_{2,ki}$.

4.2.1. *Signed-Rank Estimate of* α_1

Define the residuals as

$$\hat{\boldsymbol{e}} = \boldsymbol{y} - \hat{\alpha}_s \mathbf{1}_N - \boldsymbol{X}\hat{\boldsymbol{\beta}}_{\varphi}.$$

If we use signed-rank to estimate α_1, then the first-order approximation for the fitted values is

$$\begin{aligned}\hat{\boldsymbol{y}} &= \boldsymbol{X}(\boldsymbol{X}^{*\top}\boldsymbol{X}^*)^{-1}\boldsymbol{X}^*\hat{\boldsymbol{y}}_2^* \\ &\doteq \tau_s \boldsymbol{H}_{\mathbf{1}_N}\mathrm{sgn}(\boldsymbol{e}) + \tau_{\varphi}\boldsymbol{P}\varphi[F(\boldsymbol{e}_2^*)],\end{aligned}$$

where $\boldsymbol{P} = \boldsymbol{X}(\boldsymbol{X}^{*\top}\boldsymbol{X}^*)^{-1}\boldsymbol{X}^{*\top}$. Define $E[e_{kj}] = \mu$. Then $E[\hat{\boldsymbol{e}}] = E[\boldsymbol{e}] = \mu\mathbf{1}$. Therefore, the variance of $\hat{\boldsymbol{e}}$ is

$$\begin{aligned}\mathrm{var}[\hat{\boldsymbol{e}}] &= E[(\hat{\boldsymbol{e}} - \mu\mathbf{1})(\hat{\boldsymbol{e}} - \mu\mathbf{1})^\top] \\ &= E[(\boldsymbol{e} - \tau_s\boldsymbol{H}_{\mathbf{1}_N}\mathrm{sgn}(\boldsymbol{e}) - \tau_{\varphi}\boldsymbol{P}\varphi(F(\boldsymbol{e}_2^*)) - \mu\mathbf{1}) \\ &\quad \times (\boldsymbol{e} - \tau_s\boldsymbol{H}_{\mathbf{1}_N}\mathrm{sgn}(\boldsymbol{e}) - \tau_{\varphi}\boldsymbol{P}\varphi(F(\boldsymbol{e}_2^*)) - \mu\mathbf{1})^\top] \\ &= E[(\boldsymbol{e} - \mu\mathbf{1})(\boldsymbol{e} - \mu\mathbf{1})^\top] - \tau_s E[\boldsymbol{e}\,\mathrm{sgn}(\boldsymbol{e})^\top]\boldsymbol{H}_{\mathbf{1}} \\ &\quad - \tau_{\varphi}E[\boldsymbol{e}\,\varphi(F(\boldsymbol{e}_2^*))^\top]\boldsymbol{P}^\top - \tau_s\boldsymbol{H}_{\mathbf{1}}E[\mathrm{sgn}(\boldsymbol{e})\boldsymbol{e}^\top] \\ &\quad + \tau_s^2\boldsymbol{H}_{\mathbf{1}}E[\mathrm{sgn}(\boldsymbol{e})\mathrm{sgn}(\boldsymbol{e})^\top]\boldsymbol{H}_{\mathbf{1}} + \tau_s\tau_{\varphi}\boldsymbol{H}_{\mathbf{1}}E[\mathrm{sgn}(\boldsymbol{e})\varphi(F(\boldsymbol{e}_2^*))^\top]\boldsymbol{P}^\top \\ &\quad - \tau_{\varphi}\boldsymbol{P}E[\varphi(F(\boldsymbol{e}_2^*))\boldsymbol{e}^\top] + \tau_s\tau_{\varphi}\boldsymbol{P}E[\varphi(F(\boldsymbol{e}_2^*))\mathrm{sgn}(\boldsymbol{e})^\top]\boldsymbol{H}_{\mathbf{1}} \\ &\quad + \tau_{\varphi}^2\boldsymbol{P}E[\varphi(F(\boldsymbol{e}_2^*))\varphi(F(\boldsymbol{e}_2^*))^\top]\boldsymbol{P}^\top.\end{aligned}$$

In practice, of course, estimates of these terms are needed. Estimates of some of the terms are straightforward. We can estimate τ and τ_S as discussed in Chapter 3 of Hettmansperger and McKean (1998). For matrices such as $E[\varphi(F(\boldsymbol{e}_2^*))\varphi(F(\boldsymbol{e}_2^*))^\top]$, we can use the residuals $\hat{e}_i$ from the joint ranks fit (see Kloke *et al.*, 2009) of the original model to obtain the corresponding empirical distribution function, say $\hat{F}_n$. This can be substituted for F in this term. Then, assuming working independence, this reduces to a simple moment estimate times $\boldsymbol{I}_n$. On the other hand, the mixed terms involving $\boldsymbol{e}$ and $\boldsymbol{e}_2^*$ are more complicated. We know that $\mathrm{cov}(\boldsymbol{e}_k, \boldsymbol{e}_{2k}^*) = \sigma_2^2\boldsymbol{C}_k$, so that the correlation between $\boldsymbol{e}_k$ and a function of $\boldsymbol{e}_{2k}^*$ is not straightforward. We are investigating estimates of such terms. For now, we recommend carrying out a bootstrap where the blocks of residuals are sampled with replacement. Then within the block we sample without replacement. An example for the balanced case is discussed in Section 6.

4.2.2. *Median of Residuals, Estimate of* α_1

The first-order expression for $\hat{\mathbf{e}}$ is

$$\hat{\mathbf{e}}_R \doteq \mathbf{e} - \frac{\tau_{s,e}}{N}\mathbf{J}_N \mathrm{sgn}(\mathbf{e}) - \tau_s \mathbf{P}_1 \mathrm{sgn}(\mathbf{e}_2^*) - \tau_\varphi \mathbf{P}_2 \varphi[F(\mathbf{e}_2^*)],$$

where

$$\mathbf{P}_1 = \mathbf{X}(\mathbf{X}^{*\top}\mathbf{X}^*)^{-1}\mathbf{X}^{*\top}\mathbf{H}_{\mathbf{1}_{N^*}}$$

and

$$\mathbf{P}_2 = \mathbf{X}(\mathbf{X}^{*\top}\mathbf{X}^*)^{-1}\mathbf{X}^{*\top}\mathbf{H}_{\tilde{\mathbf{X}}^*}.$$

Define $\mu = E[e_{kj}]$. Then $E[\hat{\mathbf{e}}] = E[\mathbf{e}] = \mu\mathbf{1}$, so that the variance of $\hat{\mathbf{e}}$ is $\mathrm{var}[\hat{\mathbf{e}}] = E[(\hat{\mathbf{e}} - \mu\mathbf{1})(\hat{\mathbf{e}} - \mu\mathbf{1})^\top]$. Substituting $\hat{\mathbf{e}}$ and expanding gives the following sixteen terms:

$$\begin{aligned}
\mathrm{var}[\hat{\mathbf{e}}] = {} & E[(\mathbf{e} - \mu\mathbf{1})(\mathbf{e} - \mu\mathbf{1})^\top] - \frac{\tau_{s,e}}{N} E[\mathbf{e}\,\mathrm{sgn}(\mathbf{e})^\top]\mathbf{J}_N \\
& - \tau_\varphi E[\mathbf{e}\varphi[F(\mathbf{e}_2^*)]^\top]\mathbf{P}_2^\top - \tau_s E[\mathbf{e}\,\mathrm{sgn}(\mathbf{e}_2^*)^\top]\mathbf{1}_{N^*}\mathbf{P}_1^\top \\
& - \frac{\tau_{s,e}}{N}\mathbf{J}_N E[\mathrm{sgn}(\mathbf{e})\mathbf{e}^\top] + \frac{\tau_{s,e}^2}{N^2}\mathbf{J}_N E[\mathrm{sgn}(\mathbf{e})\,\mathrm{sgn}(\mathbf{e})^\top]\mathbf{J}_N \\
& + \tau_\varphi \frac{\tau_{s,e}}{N}\mathbf{J}_N E[\mathrm{sgn}(\mathbf{e})\varphi[F(\mathbf{e}_2^*)]^\top]\mathbf{P}_2^\top \\
& + \tau_s \frac{\tau_{s,e}}{N}\mathbf{J}_N E[\mathrm{sgn}(\mathbf{e})\mathrm{sgn}(\mathbf{e}_2^*)^\top]\mathbf{P}_1^\top - \tau_\varphi \mathbf{P}_2 E[\varphi[F(\mathbf{e}_2^*)]\mathbf{e}^\top)] \\
& - \frac{\tau_{s,e}}{N}\tau_\varphi \mathbf{P}_2 E[\varphi[F(\mathbf{e}_2^*)]\,\mathrm{sgn}(\mathbf{e})^\top]\mathbf{J}_N \\
& - \tau_\varphi^2 \mathbf{P}_2 E[\varphi[F(\mathbf{e}_2^*)]\varphi[F(\mathbf{e}_2^*)]^\top]\mathbf{P}_2^\top \\
& - \tau_s\tau_\varphi \mathbf{P}_2 E[\varphi[F(\mathbf{e}_2^*)]\,\mathrm{sgn}(\mathbf{e}_2^*)^\top]\mathbf{P}_1^\top - \tau_s \mathbf{P}_1 E[\mathrm{sgn}(\mathbf{e}_2^*)\mathbf{e}^\top] \\
& + \frac{\tau_{s,e}}{N}\tau_s \mathbf{P}_1 E[\mathrm{sgn}(\mathbf{e}_2^*)\mathrm{sgn}(\mathbf{e})]\mathbf{J}_N \\
& + \tau_s\tau_\varphi \mathbf{P}_1 E[\mathrm{sgn}(\mathbf{e}_2^*)\varphi[F(\mathbf{e}_2^*)]^\top]\mathbf{P}_2^\top \\
& + \tau_s^2 \mathbf{P}_1 E[\mathrm{sgn}(\mathbf{e}_2^*)\,\mathrm{sgn}(\mathbf{e}_2^*)^\top]\mathbf{P}_1^\top.
\end{aligned}$$

The expectations in the 11th term and the 16th term are equal to the identity matrix. The expectations in the 12th term and the 15th term are equal to a multiple of the identity matrix. All other expectations result in block diagonal matrices.

Similar comments as in the last section pertain to the estimation of these terms. As there, we currently recommend a bootstrap.

5. Simulation Study

To compare the estimators discussed in this paper, we ran a small simulation study. In addition to the estimators proposed in this paper, we included the ordinary rank and least squares estimators. Thus, the procedures we consider are: ordinary least squares (OLS); ordinary rank-based estimates (OR); least squares estimates based on the Arnold transformed data (ATLS); rank-based estimates based on the Arnold transformed data with the median of the residuals as the estimator of the intercept (ATR(s)); and rank-based estimates based on the Arnold transformed data with the signed-rank estimator of the intercept (ATR(sr)). Note that OR and OLS procedures ignore the dependence in the model. All rank-based estimates used the Wilcoxon scores. We ran normal and contaminated normal errors. The contaminated normal errors were obtained with 10% contamination and ratio of scale five to one (contaminated to good). For intraclass correlation coefficient of the uncontaminated errors we set $\rho = 0, \frac{1}{3}, \frac{2}{3}$, and $\frac{19}{20}$. There were $m = 10$ blocks, two of each of the block sizes $4, 5, 6, 7$, and 8. The analysis of covariance model used was

$$y_{kj} = \alpha + \Delta w_{kj} + \boldsymbol{\beta} x_{kj} + b_j + e_{kj},$$

where w_{ij} is a dummy variable alternating between the values of 0 and 1, and the covariate $x_{kj} \sim U(0,1)$. The simulation size was set at 10000 repetitions for each situation.

In the tables below, we display the empirical mean square errors (MSE) for each procedure's estimates of the parameters Δ and β. For the second summary, we display the empirical confidence coefficients for nominal 95% confidence intervals for these parameters. These confidence intervals are based on estimates of the asymptotic variance-covariance matrix of the estimators. Hence, for the ATR procedures, these results are pertinent to a practical inference based on the the rank-based AT estimators. Tables 1 and 2 display the summaries for the normal and contaminated normal situations, respectively.

From Table 1, for $\rho = 0$, OLS performs best in terms of MSE, as it should. Even at $\rho = 1/3$, though, the ATLS estimate has lower MSE than OLS. For these normal situations, in terms of MSE, the ATLS dominates the ATR estimators although not by much. Selecting the MSEs for Δ at $\rho = 2/3$, the empirical efficiency between ATR(sr) and ATLS is $0.0693/0.0730 = 0.95$, about the same as the relative efficiency between Wilcoxon and LS procedures. There is little difference between the ATR(s) and ATR(sr) MSEs. All the empirical confidences for the ATLS and ATR procedures were quite close to the nominal 0.95.

Table 1. Empirical MSEs and Confidence Coefficients for the Normal Error Situations.

			MSEs for Δ		
ρ	OLS	OR	ATLS	ATR(s)	ATR(sr)
0	0.0668	0.0711	0.0678	0.0722	0.0720
1/3	0.0688	0.0742	0.0673	0.0720	0.0718
2/3	0.0794	0.0867	0.0693	0.0730	0.0727
19/20	0.1699	0.1611	0.0689	0.0743	0.0741
			MSEs for β		
ρ	OLS	OR	ATLS	ATR(s)	ATR(sr)
0	0.2146	0.2265	0.2547	0.2741	0.2712
1/3	0.3026	0.3221	0.2558	0.2728	0.2698
2/3	0.5593	0.5908	0.2519	0.2678	0.2662
19/20	3.5928	3.0974	0.2480	0.2664	0.2639
			Empirical Confidences for Δ		
ρ	OLS	OR	ATLS	ATR(s)	ATR(sr)
0	0.9527	0.9578	0.9526	0.9596	0.9590
1/3	0.9781	0.9793	0.9546	0.9585	0.9581
2/3	0.9937	0.9947	0.9528	0.9579	0.9571
19/20	1.0000	1.0000	0.9491	0.9572	0.9555
			Empirical Confidences for β		
ρ	OLS	OR	ATLS	ATR(s)	ATR(sr)
0	0.9474	0.9521	0.9483	0.9575	0.9551
1/3	0.9533	0.9552	0.9512	0.9565	0.9554
2/3	0.9526	0.9576	0.9494	0.9563	0.9544
19/20	0.9513	0.9595	0.9533	0.9607	0.9581

For the contaminated normal situations, in terms of MSE, the clear winners are the two ATR estimators. They are much more efficient than their counterpart, the ATLS estimator. As with the normal situation, selecting the MSEs for Δ at $\rho = 2/3$, the empirical efficiency between ATR(sr) and ATLS is $0.2314/0.1301 = 1.78$; i.e., the ATR(sr) estimator is over 150% more efficient than the ATLS estimator. Hence, for these normal and contaminated normal situations, the ATR estimator is highly efficient. Again, there is little difference between the ATR(s) and the ATR(sr) estimators. All the empirical confidences for the ATLS and ATR procedures are quite close to the nominal 0.95, which is another indication of the practicality of our inference based on the ATR estimators.

Table 2. Empirical MSEs and Confidence Coefficients for the Contaminated Normal Situations.

			MSEs for Δ		
ρ	OLS	OR	ATLS	ATR(s)	ATR(sr)
0	0.2315	0.0989	0.2345	0.1305	0.1303
1/3	0.2310	0.1124	0.2319	0.1299	0.1296
2/3	0.2374	0.1375	0.2314	0.1303	0.1301
19/20	0.3321	0.2633	0.2323	0.1276	0.1273
			MSEs for β		
ρ	OLS	OR	ATLS	ATR(s)	ATR(sr)
0	0.7200	0.3187	0.8511	0.4972	0.4956
1/3	0.8122	0.4368	0.8588	0.4844	0.4830
2/3	1.0803	0.7742	0.8534	0.4813	0.4796
19/20	4.1900	3.5932	0.8501	0.4923	0.4908
		Empirical Confidences for Δ			
ρ	OLS	OR	ATLS	ATR(s)	ATR(sr)
0	0.9555	0.9560	0.9548	0.9611	0.9608
1/3	0.9711	0.9753	0.9567	0.9622	0.9614
2/3	0.9852	0.9905	0.9584	0.9598	0.9589
19/20	0.9996	0.9994	0.9571	0.9624	0.9617
		Empirical Confidences for β			
ρ	OLS	OR	ATLS	ATR(s)	ATR(sr)
0	0.9502	0.9563	0.9506	0.9593	0.9568
1/3	0.9513	0.9549	0.9529	0.9621	0.9601
2/3	0.9514	0.9543	0.9516	0.9610	0.9577
19/20	0.9464	0.9524	0.9512	0.9590	0.9580

6. Example

This example, though simple, demonstrates some of the nice features of Arnold's Transformation, particularly for balanced data. The data in Table 3 are from Table 11.1 on page 260 of Milliken and Johnson (2002). Each row represents a block. There is one covariate and each of the responses is measurements on different treatments. The model for these data is

$$\boldsymbol{y}_k = \alpha \mathbf{1}_2 + \Delta \begin{bmatrix} -0.5 \\ 0.5 \end{bmatrix} + \beta x_k \mathbf{1}_2 + \boldsymbol{\epsilon}_k.$$

Arnold's Transformation for this model is

$$\boldsymbol{\Gamma}_k = \frac{1}{\sqrt{2}} \begin{bmatrix} 1 & 1 \\ 1 & -1 \end{bmatrix}.$$

Table 3. Dataset from Milliken and Johnson (2002).

x	y_1	y_2
23.2	60.4	76.0
26.9	59.9	76.3
29.4	64.4	77.8
22.7	63.5	75.6
30.6	80.6	94.6
36.9	75.9	96.1
17.6	53.7	62.3
28.5	66.3	81.6

The transformed responses are $\boldsymbol{y}_k^* = \boldsymbol{\Gamma}_k \boldsymbol{y}_k = [y_{k1}^*, y_{k2}^*]^\top$, where

$$\begin{aligned} y_{k1}^* &= \alpha^* + \beta^* x_k + \epsilon_{k1}^*, \\ y_{k2}^* &= \Delta^* + \epsilon_{k2}^*, \end{aligned}$$

$\alpha^* = \sqrt{2}\alpha$, $\beta^* = \sqrt{2}\beta$, $\Delta^* = \frac{1}{\sqrt{2}}\Delta$, and the ϵ_{k1}^* for $k = 1, \ldots, m$ are independent and identically distributed, as are the ϵ_{k2}^* for $k = 1, \ldots, m$. Notice that the first component is a simple linear regression model and the second component is a simple location model. For this example, we use signed-rank to estimate both of the intercept terms. The estimates and standard errors of the parameters are given in Table 4.

Table 4. Estimates and standard errors for data from Milliken and Johnson (2002).

	Wilcoxon		MLE	
	Est	SE	Est	SE
α	70.8	3.54	26.0	8.98
Δ	14.45	1.61	14.45	1.19
β	1.43	0.65	1.46	0.33

6.1. *Studentized Residuals*

To calculate the residuals for this example, we form the response vectors

$$\hat{\boldsymbol{y}}_k^* = \begin{bmatrix} \hat{y}_{k1}^* \\ \hat{y}_{k2}^* \end{bmatrix}. \tag{18}$$

Now transform these back to get the fitted values $\hat{\boldsymbol{y}}_k = \boldsymbol{\Gamma}^\top \hat{\boldsymbol{y}}_k^*$. The residuals are defined as $\hat{\boldsymbol{e}}_k = \boldsymbol{y}_k - \hat{\boldsymbol{y}}_k$. To obtain Studentized residuals, a bootstrap procedure is implemented where blocks of residuals are sampled with replacement. To maintain the correlation structure, the errors within a block are then sampled without replacement to form new responses. For the ith step we calculate

$$\boldsymbol{y}_k^{\dagger(i)} = \hat{\boldsymbol{y}}_k + \boldsymbol{e}_k^{\dagger(i)} \text{ for } k = 1, \ldots, m.$$

The model is fit as in the previous susbsection and the residuals are formed as above, giving $\hat{\boldsymbol{e}}_k^{\dagger(i)}$ for $k = 1, \ldots, m$ and $i = 1, \ldots, B$. From here, the variance

$$\hat{\boldsymbol{V}}_e = \text{var}(\hat{\boldsymbol{e}}^{\dagger(1)}, \ldots, \hat{\boldsymbol{e}}^{\dagger(B)})$$

of the residuals is estimated, where $\hat{\boldsymbol{e}}^{\dagger(i)} = [\hat{\boldsymbol{e}}_1^{\dagger(i)T}, \ldots, \hat{\boldsymbol{e}}_m^{\dagger(i)T}]^\top$ is the stacked vector of residuals. The Studentized residuals are then calculated as

$$r_j = \frac{\hat{e}_j}{\sqrt{\hat{v}_{jj}}} \text{ for } j = 1, \ldots, 2m,$$

where v_{jj} is the jth diagonal element of $\hat{\boldsymbol{V}}_e$.

The bootstrap Studentized residuals with $B = 10000$ for the least squares (top) and Wilcoxon fits are given in Figure 1. Since there are not serious outliers in this dataset, the plots are quite similar.

6.2. *Sensitivity Analysis*

To demonstrate the robustness of our procedure, we conducted a sensitivity analysis based on the changes to the response variable. We set the second data point $y_{12}^{(i)} = y_{11} + \Delta y$, where Δy varied from -30 to 30. We then estimated the parameters $\Delta^{(i)}$ based on the data set with the outlier. In Figure 2, we display the relative change of the estimate,

$$\frac{\hat{\Delta}^{(i)} - \hat{\Delta}}{\hat{\Delta}},$$

as a function of Δy. The relative change in the estimates is small, which indicates that our estimates are robust.

7. Conclusion

In this paper, we have developed rank-based estimators for Arnold Transformed data. This transformation is appropriate for many types of randomized block designs, with or without covariates, that commonly occur in practice. It assumes that each block has a random effect. If variances exist,

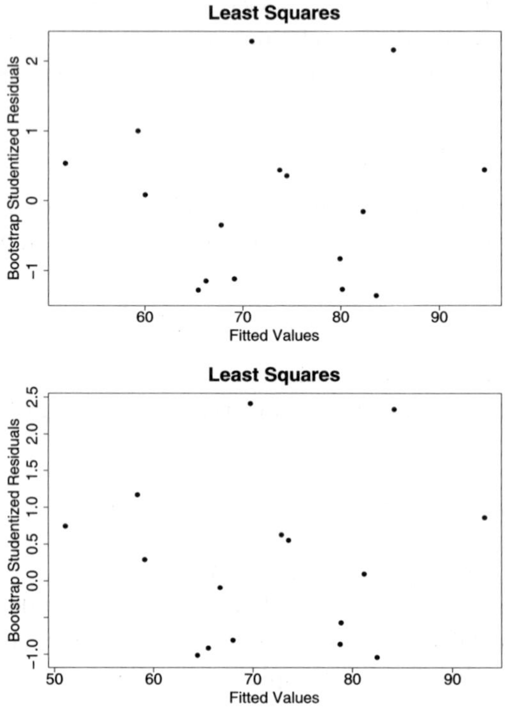

Fig. 1. Bootstrap Studentized residuals

this implies that the covariance structure is compound symmetric. Arnold (1981) proposed applying an orthogonal type of transformation to the data from such designs and then using least squares (LS) estimation techniques for the inference. We offer a simple extension to unbalanced cases, though in practice one would still want to use a maximum likelihood estimator on the untransformed data. This extension is more to set the stage for our rank-based estimation procedure.

Our estimation procedure involves using rank-based estimates (R estimates) on the transformed data. The process includes an adjustment for rank-based fitting through the origin that we discuss in Section 2. In Section 3, we present our estimates and develop their asymptotic distribution. The development is for general rank score functions. In particular, scores appropriate for skewed and well as symmetric error distributions can be used. The resulting covariance structure of the estimates lends itself to straightforward estimation, so besides estimation, we also have a practi-

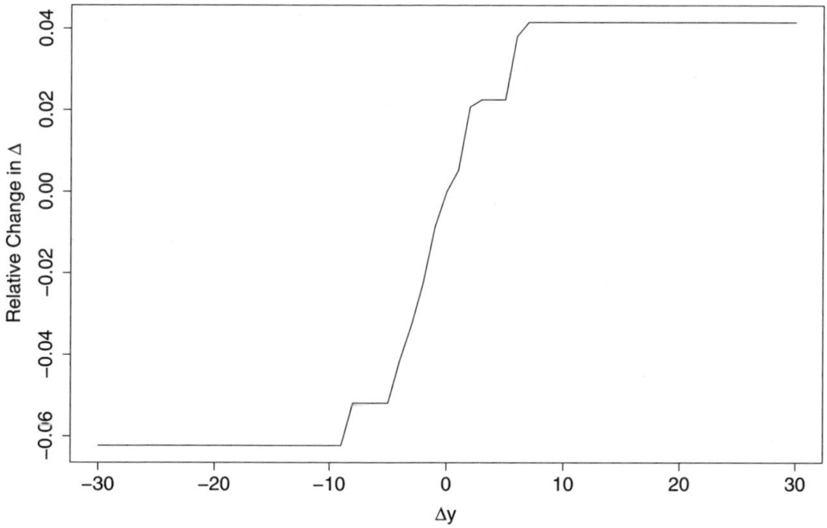

Fig. 2. Relative change of estimate.

cal inference procedure. Furthermore, R code (R Development Core Team, 2010) for many of the procedures is available from the authors.

Based on asymptotic relative efficiency (ARE) results, these rank-based estimates appear to be highly efficient. The simulation study of Section 5 verifies these efficiency results as well as the validity of the associated inference procedures. In Section 6, we use an example to illustrate our procedures and a sensitivity analysis confirms the robustness.

In this paper, we have developed robust estimators for simple mixed (compound symmetric, if second moments exist) models based on the Arnold transformation. Clearly, for mixed models in general, other transformations can be used. The joint ranks estimators discussed in Kloke *et al.* (2009) are for models more general than the compound symmetric models. These estimators, combined with robust estimators of covariance, lend themselves naturally to either a general R estimator based on a robust transformation or a generalized estimating equation-weighted estimator. We are currently investigating such estimators. Occasionally, transformations are used for nonlinear models. We note, however, for the mixed model setting of this paper, that the nonlinear R estimators proposed by Abebe and McKean (2007) can be extended to models of the form (12) with a nonlinear component, i.e.,

$$y_{k,i} = g(\boldsymbol{\theta}, \boldsymbol{x}_{k,i}) + b_k + \epsilon_{k,i} \text{ for } k = 1, \ldots, m, i = 1, \ldots, n_k,$$

where g is a specified function and $\boldsymbol{\theta}$ is a $s \times 1$ vector of parameters.

For this paper we have ignored any information about $\boldsymbol{\beta}$ that might be in the mean component. It is possible, for balanced designs, that we may be able to use the information in both components. We currently have a manuscript in preparation that considers this approach. Also, work is progressing on a diagnostics paper, including bootstrap standard errors for the Studentized residuals.

Acknowledgments

We appreciate the helpful suggestions from the referees on our original manuscript.

References

Abebe, A. and McKean, J. W. (2007). Highly efficient nonlinear regression based on the wilcoxon norm, in D. Umbach (ed.), *Festschrift in Honor of Mir Masoom Ali on the Occasion of his Retirement* (Ball State University, Muncie, IN), pp. 340–357.

Arnold, S. F. (1981). *The Theory of Linear Models and Multivariate Analysis* (John Wiley & Sons, New York).

Dixon, S. L. and McKean, J. W. (1996). Rank-based analysis of the heteroscedastic linear model, *Journal of the American Statistical Association* **91**, pp. 699–712.

Freidlin, B. and Gastwirth, J. L. (2000). Should the median test be retired from general use? *The American Statistician* **54**, pp. 161–164.

Hettmansperger, T. P. and McKean, J. W. (1998). *Robust Nonparametric Statistical Methods* (Arnold, London).

Jaeckel, L. A. (1972). Estimating regression coefficients by minimizing the dispersion of residuals, *Annals of Mathematical Statistics* **43**, pp. 1449–1458.

Kloke, J. D. and McKean, J. W. (2004). Rank analyses of one-way repeated measures models, in *2004 Proceedings of the American Statistical Association, Nonparametrics Section [CD-ROM]* (American Statistical Association, Alexandria, VA), pp. 1980–1985.

Kloke, J. D. and McKean, J. W. (2010). Rfit: An R package for rank-based estimation, In preparation.

Kloke, J. D., McKean, J. W. and Rashid, M. (2009). Rank-based estimation and associated inferences for linear models with cluster correlated errors, *Journal of the American Statistical Association* **104**, pp. 384–390.

Kloke, J. D., Salman, M. and McKean, J. W. (2005). A small sample investigation of rank-based profile analyses, *Journal of Statistical Research* **39**, pp. 163–190.

McKean, J. W., Sheather, S. J. and Hettmansperger, T. P. (1990). Regression diagnostics for rank-based methods, *Journal of the American Statistical Association* **85**, pp. 1018–1028.

McKean, J. W., Sheather, S. J. and Hettmansperger, T. P. (1993). The use and interpretation of residuals based on robust estimation, *Journal of the American Statistical Association* **88**, pp. 1254–1263.

Milliken, G. A. and Johnson, D. E. (2002). *Analysis of Messy Data, Vol. 3: Analysis of Covariance* (Chapman & Hall/CRC, New York).

R Development Core Team (2010). *R: A Language and Environment for Statistical Computing*, R Foundation for Statistical Computing, Vienna, Austria, URL `http://www.R-project.org`, ISBN 3-900051-07-0.

Rashid, M., McKean, J. W. and Kloke, J. D. (2010). Rank-based procedures for multi-center clinical trials, Submitted.

Terpstra, J. and McKean, J. W. (2005). Rank-based analyses of linear models using R, *Journal of Statistical Software* **14**, 7, URL `http://www.jstatsoft.org/v14/i07`.

QQ Plots for Assessing Symmetry Models

John I. Marden

Department of Statistics, University of Illinois at Urbana-Champaign
Champaign, IL 61820, USA
E-mail: jimarden@illinois.edu

Analogous to the univariate plots, multivariate QQ plots can be used to compare two multivariate distributions (empirical or theoretical) by matching a set of quantiles in one distribution with the corresponding set in the other. We base our plots on Chaudhuri's spatial quantiles, which are vectors of the same dimension as the observations. The QQ plots consist of arrows pointing from one distribution's quantiles to the other's. In two dimensions, the arrows can be plotted directly. In higher dimensions, we look at projections. Principal component-like projections are used to find directions in which the two distributions are most different.

We focus on assessing how well certain symmetry models, including spherical symmetry, fit multivariate samples. The QQ plots compare the spatial quantiles of the data to that of a symmetrized version of the data. A randomization technique aids in choosing the number of dimensions that show deviations from the symmetry of interest.

Keywords: QQ plot; Spatial quantile; Spherical symmetry; Invariant normal models

1. Introduction

QQ plots are popular for comparing a univariate sample to a hypothesized model, where the observations are plotted versus the corresponding quantiles of the model. The plots can be used to test visually whether the data fit the hypothesis, and if not, where the discrepancies might lie. In this paper we consider multivariate QQ plots. We are primarily interested in assessing certain structural hypotheses for the models, but the ideas apply more broadly. We use the spatial quantiles introduced by Chaudhuri (1996) and Koltchinskii (1997), which are the same dimension as the data.

We start with a sample of $p \times 1$ vectors,

$$\mathbf{X}_1, \ldots, \mathbf{X}_n \ \text{ independent and identically distributed (iid)} \sim F, \qquad (1)$$

and a (possibly estimated) hypothesized distribution F_0. For each observed vector $\mathbf{x}_i$, we find the hypothetical quantile $\mathbf{q}_i$ whose position among the distribution of $\mathbf{Z} \sim F_0$ is the same as that of $\mathbf{x}_i$ among the observed data. See Section 2 for the precise definition. The $\mathbf{q}_i$ can be thought of as the "fit" of the data to the hypothesized model, and the difference,

$$\mathbf{a}_i = \mathbf{x}_i - \mathbf{q}_i, \tag{2}$$

as the "residual," which we will call an "arrow." These arrows are the basis on which to assess the model. Marden (1998), for the bivariate case, draws an arrow from each hypothesized quantile $\mathbf{q}_i$ to the corresponding data point $\mathbf{x}_i$. Test statistics are based on the lengths of the arrows. Chakraborty (2001) proposes affine equivariant versions of these methods. This article presents some additional strategies based on analyzing the arrows, especially a projection pursuit rotation of the data based on the dimensions for which the arrows are longest. Related work can be found in Serfling (2004), which presents a comprehensive treatment of the use of spatial quantiles in assessing multivariate skewness, kurtosis, peakedness, and tailweight. As for us, the focus is on multidimensional measures and plots.

The next section defines the spatial quantiles. Section 3 describes the symmetry models of interest, which include simple $\pm$ symmetry, certain permutation symmetries, and spherical symmetry. Section 4 contains several examples of varying dimensions. Section 5 compares the QQ-based rotation to two types of principal-component-based rotations, and to a rotation based on PP plots. Some calculations necessary for the spherical symmetry case are relegated to Section 6.

2. Multivariate spatial quantiles

The spatial quantiles are based on multivariate spatial signs and ranks. The **sign** of a vector $\mathbf{x} \in \mathbb{R}^p$ is the unit vector in the direction of $\mathbf{x}$:

$$\text{Sign}(\mathbf{x}) = \begin{cases} \frac{\mathbf{x}}{\|\mathbf{x}\|} & \text{if } \mathbf{x} \neq \mathbf{0} \\ \\ \mathbf{0} & \text{if } \mathbf{x} = \mathbf{0}, \end{cases} \tag{3}$$

where $\|\mathbf{x}\|$ is the usual Euclidean norm and $\mathbf{0}$ is the $p \times 1$ vector of 0's. The **rank** of a vector $\mathbf{x}$ relative to a distribution F on $\mathbb{R}^p$ is

$$\text{Rank}(\mathbf{x}\ ;\ F) = E\left[\text{Sign}(\mathbf{x} - \mathbf{Z})\right], \quad \text{where} \quad \mathbf{Z} \sim F. \tag{4}$$

In one dimension, $\text{Rank}(x\ ;\ F) = 2[F(x) - 1]$, so is essentially the usual rank, shifted and scaled to lie between -1 and 1. In general dimensions, it is the average over $\mathbf{Z} \sim F$ of the unit vectors pointing from $\mathbf{Z}$ to $\mathbf{x}$, representing the position of $\mathbf{x}$ among the $\mathbf{Z}$. The spatial median is the vector with rank $\mathbf{0}$, and as $\mathbf{x}$ strays farther from the center, the length of its rank increases.

The **quantile** is the inverse of the rank. It is indexed by $\mathbf{r}$ in the open p-dimensional unit disk, so that the $\mathbf{r}^{th}$ quantile is the Quantile($\mathbf{r}$; F) that satisfies

$$\text{Rank}[\text{Quantile}(\mathbf{r}\ ;\ F)\ ;\ F] = \mathbf{r}. \tag{5}$$

Such quantile may not exist for all $\mathbf{r}$, so we use the definition given in Chaudhuri (1996) that Quantile($\mathbf{r}$; F) is the argument $\mathbf{q} \in \mathbb{R}^p$ that minimizes

$$E[\|\mathbf{q} - \mathbf{Z}\| + \mathbf{r}'(\mathbf{Z} - \mathbf{q})]. \tag{6}$$

As long as F does not concentrate all its mass on a straight line, the quantile exists and is unique for any $\mathbf{r}$ with $\|\mathbf{r}\| < 1$.

Consider the observations from the iid sample (1), $\mathbf{x}_1, \ldots, \mathbf{x}_n$. We will take the sample quantiles to be the $\mathbf{x}_i$ themselves, though one could take any set desired. The rank of $\mathbf{x}_i$ relative to the sample is, as in (4),

$$\text{Rank}(\mathbf{x}_i\ ;\ \widehat{F}_n) = \frac{1}{n}\sum_{j=1}^{n}\text{Sign}(\mathbf{x}_i - \mathbf{x}_j), \tag{7}$$

where $\widehat{F}_n \equiv \widehat{F}_n(\ \cdot\ ; \mathbf{x}_1, \ldots, \mathbf{x}_n)$ is the empirical distribution function. The corresponding quantile for the hypothesized distribution F_0 is then

$$\mathbf{q}_i = \text{Quantile}[\text{Rank}(\mathbf{x}_i\ ;\ \widehat{F}_n)\ ;\ F_0], \tag{8}$$

so that

$$\text{Rank}(\mathbf{x}_i\ ;\ \widehat{F}_n) = \text{Rank}(\mathbf{q}_i\ ;\ F_0). \tag{9}$$

Thus $\mathbf{q}_i$ is positioned within the F_0 distribution as the $\mathbf{x}_i$ is positioned among the data points. The fit of the data to the model F_0 is assessed using the $\mathbf{x}_i$, $\mathbf{q}_i$, and their corresponding arrows (2).

An alternative approach compares the ranks rather than the quantiles, so that the arrows would be

$$\mathbf{a}_i^* = \text{Rank}(\mathbf{x}_i\ ;\ \widehat{F}_n) - \text{Rank}(\mathbf{x}_i\ ;\ F_0). \tag{10}$$

Plots based on these arrows are multivariate analogs of Probability-Probability (PP) plots. Because QQ plots are more popular than PP plots in univariate situations, in this article we focus on QQ plots. Analysis of PP plots would also be of interest, especially as they may be more robust to outliers. In Section 5 we have some comparisons of the two types of plot.

3. Symmetry models

Symmetry in multivariate data is often of interest, either as an integral part of a model, or as an assumption needed for carrying out certain inference procedures. For example, many multivariate hypothesis testing procedures are based on the multivariate normal distribution but work also for elliptically symmetric distributions. See Kariya and Sinha (1989). Or in repeated measures models, one often assumes an intraclass correlation structure (permutation invariance) within observations.

The symmetry models we consider are defined by subgroups $\mathcal{G}$ of the group of $p \times p$ orthogonal matrices, $\mathcal{O}_p$. Given $\mathcal{G}$, the distribution function F on $\mathbb{R}^p$ is $\mathcal{G}$-invariant if

$$\mathbf{Z} \sim F \implies \mathbf{Z} =^{\mathcal{D}} g\mathbf{Z} \text{ for all } g \in \mathcal{G}, \tag{11}$$

where "$=^{\mathcal{D}}$" means equal in distribution. Andersson (1975) and Andersson *et al.* (1983) provide a thorough treatment of such models when the multivariate normal distribution is assumed, using deep algebraic results to obtain distributions of relevant estimators and test statistics. Popular models include overall $\pm$ symmetry, where $\mathcal{G}$ consists of just $\pm\mathbf{I}_p$; componentwise $\pm$ symmetry, where $\mathcal{G}$ consists of all diagonal matrices with ± 1's on the diagonal; permutation invariance or exchangeability, where $\mathcal{G}$ is the set of $p \times p$ permutation matrices; compound symmetry (Votaw, 1948), where $\mathcal{G}$ is the set of block-diagonal matrices with permutation matrices making up the blocks; and spherical symmetry, where $\mathcal{G} = \mathcal{O}_p$. In the normal case, componentwise $\pm$ symmetry implies independence of the elements (and in any case implies zero correlation, if correlations exist), and permutation invariance plus componentwise $\pm$ symmetry implies spherical symmetry.

If the covariance matrix $\mathbf{\Sigma}$ of $\mathbf{Z}$ exists, then $\mathcal{G}$-invariance implies that

$$g\mathbf{\Sigma}g' = \mathbf{\Sigma} \;\text{ for all }\; g \in \mathcal{G}. \tag{12}$$

Under multivariate normality, the maximum likelihood estimator of $\mathbf{\Sigma}$ given $\mathcal{G}$-invariance is the symmetrized sample covariance matrix,

$$\widehat{\mathbf{\Sigma}}(\mathcal{G}) = E[\mathbf{GSG}'], \quad \mathbf{G} \sim \text{Uniform}(\mathcal{G}), \tag{13}$$

the Haar probability measure on $\mathcal{G}$, where $\mathbf{S} = \sum \mathbf{X}_i\mathbf{X}_i'/n$. Then the likelihood ratio statistic for testing two nested $\mathcal{G}$-invariant models is based on the ratio of the determinants of the MLE's. Because the spatial signs and ranks are equivariant under the orthogonal group, their sample covariance matrices have the same structure as $\mathbf{\Sigma}$ in (12) under $\mathcal{G}$-invariance. Thus we can find robust alternatives to the normal-based procedures. See Marden

(1999), Marden and Gao (2002), and Visuri *et al.* (2000). Note that these procedures are based on moments, whether of the raw data or of the signs or ranks.

In this paper we consider assessing symmetry models using the multivariate QQ plots and the resulting arrows, which are sensitive to departures beyond those that show up in moments, and suggest for which dimensions the models' lack of fit is most pronounced. Section 4 contains illustrative examples. For assessing a particular $\mathcal{G}$-invariant model based on the sample (1), we use the symmetrized distribution function:

$$\widehat{F}_n^{\mathcal{G}}(\mathbf{x}) = E[\widehat{F}_n(\mathbf{x}; \mathbf{G}_1\mathbf{x}_1, \ldots, \mathbf{G}_n\mathbf{x}_n)], \quad \mathbf{G}_1, \ldots, \mathbf{G}_n \text{ are iid Uniform}(\mathcal{G}). \tag{14}$$

Then the formula (8) becomes

$$\mathbf{q}_i = \text{Quantile}[\text{Rank}(\mathbf{x}_i \ ; \ \widehat{F}_n) \ ; \ \widehat{F}_n^{\mathcal{G}}]. \tag{15}$$

If the invariance group is finite, then from (4) and (14), the rank of any $\mathbf{x}$ with respect to the symmetrized empirical distribution function is

$$R(\mathbf{x} \ ; \ \widehat{F}_n^{\mathcal{G}}) = \frac{1}{n}\frac{1}{\#\mathcal{G}} \sum_{i=1}^{n} \sum_{\mathbf{g}\in\mathcal{G}} \text{Sign}(\mathbf{x} - \mathbf{g}\mathbf{x}_i), \tag{16}$$

which is the same as the rank with respect to an augmented data set of $\#\mathcal{G} \times n$ observations consisting of the data plus all their rotations,

$$\{\mathbf{g}\mathbf{x}_i \mid i = 1, \ldots, n, \ \mathbf{g} \in \mathcal{G}\}. \tag{17}$$

For spherical symmetry, the $\mathcal{G}$ is the orthogonal group $\mathcal{O}_p$, hence

$$R(\mathbf{x} \ ; \ \widehat{F}_n^{\mathcal{O}_p}) = \frac{1}{n} \sum_{i=1}^{n} E[\text{Sign}(\mathbf{x} - \mathbf{\Gamma}\mathbf{x}_i)], \quad \mathbf{\Gamma} \sim \text{Uniform}(\mathcal{O}_p), \tag{18}$$

the Haar probability measure on $\mathcal{O}_p$. In Section 6, we show that the rank can be written

$$\text{Rank}(\mathbf{x} \ ; \ \widehat{F}_n^{\mathcal{O}_p}) = \text{Sign}(\mathbf{x}) \ \frac{1}{n} \sum_{i=1}^{n} c_p\left(\frac{\|\mathbf{x}_i\|}{\|\mathbf{x}\|}\right), \tag{19}$$

where for $t \geq 0$,

$$c_p(t) = \int_{-1}^{1} \frac{1 - tu}{1 - 2tu + t^2} \ f_p(u)du \tag{20}$$

and

$$f_p(u) = \frac{\Gamma(\frac{p}{2})}{\sqrt{\pi}\ \Gamma(\frac{p-1}{2})} \ (1 - u^2)^{\frac{p-3}{2}}. \tag{21}$$

In (38) of Section 6, we write this function in terms of hypergeometric functions for $t \neq 1$. For a rank $\mathbf{r}$, the corresponding quantile is in the same direction,

$$\text{Quantile}(\mathbf{r}\ ;\ \widehat{F}_n^{\mathcal{O}_p}) = \text{Sign}(\mathbf{r})\ d(\|\mathbf{r}\|), \tag{22}$$

where for $v > 0$, $d(v)$ is the solution to

$$v = \frac{1}{n} \sum_{i=1}^{n} c_p \left(\frac{\|\mathbf{x}_i\|}{d(v)} \right). \tag{23}$$

The longer the arrows (2), the less well the data fit the hypothesized symmetry, hence basing statistics on the (squared) lengths of the arrows can be useful. In many dimensions, it can be informative to find projections of the data that show the longest arrows. That is, we wish to find the vector $\mathbf{b}$, $\|\mathbf{b}\| = 1$, that maximizes

$$\sum_{i=1}^{n} (\mathbf{b}'\mathbf{a}_i)^2, \tag{24}$$

or the $p \times K$ matrix $\mathbf{B}$ with orthonormal columns to maximize

$$\sum_{i=1}^{n} \|\mathbf{B}'\mathbf{a}_i\|^2. \tag{25}$$

This task is easily accomplished using principal components. Write

$$\sum_{i=1}^{n} \|\mathbf{B}'\mathbf{a}_i\|^2 = \sum_{i=1}^{n} \text{trace}(\mathbf{B}'\mathbf{a}_i\mathbf{a}_i'\mathbf{B}) \tag{26}$$

$$= \text{trace}\left[\mathbf{B}' \left(\sum_{i=1}^{n} \mathbf{a}_i\mathbf{a}_i' \right) \mathbf{B} \right]. \tag{27}$$

Thus the matrix of eigenvectors of $\sum_{i=1}^{n} \mathbf{a}_i\mathbf{a}_i'$ rotate the data so that the first dimension has the largest parts of the arrows, the second dimension has the second-largest parts, etc. By plotting the rotated data, we can see the projections that most contribute to the lengths of the arrows, i.e., the projections that contribute most to the lack of fit to the hypothesized symmetry.

Note that any hypothesized symmetry will give rise to a rotation of the data that may reveal some interesting features. Other principal-component-like approaches are based on finding the eigenvalues of one multivariate scatter matrix to another, e.g., finding the regular principal components after robustly sphering the data. For examples, see the generalized principal component analysis of Caussinus and Ruiz-Gazen (1993), the use of multivariate kurtosis scatter matrices relative to covariance matrices in Oja *et al.* (2006), and the general theory of invariant coordinate systems in Tyler *et al.* (2009). These approaches can be especially effective in independent component analysis. In what follows, we will refer to these methods generically as *generalized principal components.*

We calibrate the lengths of the arrows using randomization, where we compare the observed arrows with those obtained from randomized samples, where a randomized sample is

$$(\mathbf{G}_1\mathbf{x}_1, \ldots, \mathbf{G}_n\mathbf{x}_n) \tag{28}$$

for $\mathbf{G}_1, \ldots, \mathbf{G}_n$ being iid Uniform($\mathcal{G}$).

4. Examples

4.1. *Plus/minus symmetry in two dimensions*

We start with two bivariate ($p = 2$) examples to illustrate the ideas, where we are interested in assessing simple $\pm$ symmetry, $\mathcal{G} = \{+\mathbf{I}_p, -\mathbf{I}_p\}$. The data are from the South African Heart Disease study (Rousseauw *et al.*, 1983). There are ten variables in this data set on $n = 462$ men, measuring various health indicators related to heart disease. We will look at the measures of ldl cholesterol level versus obesity and adiposity versus obesity. For each bivariate sample, we start by taking logs of both variables. Next we sphere the data using Tyler's robust estimate of covariance (Tyler, 1987). That is, we transform the data points via

$$\mathbf{y}_i = \widehat{A}\mathbf{x}_i \tag{29}$$

for $i = 1, \ldots, n$, where $\widehat{A}$ is chosen so that

$$\frac{1}{n}\sum_{i=1}^{n} \text{Sign}(\mathbf{y}_i - \widehat{\boldsymbol{\theta}})\text{Sign}(\mathbf{y}_i - \widehat{\boldsymbol{\theta}})' = \frac{1}{p}\mathbf{I}_p, \tag{30}$$

the $\widehat{\boldsymbol{\theta}}$ being the spatial median of the $\mathbf{y}_i$'s. The algorithm we use is found in Hettmansperger and Randles (2002). Finally, we center the $\mathbf{y}_i$'s by subtracting the coordinatewise Hodges-Lehmann estimate of the median, which for variable j is the median of the $(y_{ij} + y_{kj})$'s for $i \neq k$.

Figure 1 plots the two resulting scatter plots, along the left-hand side. The ldl versus obesity data look relatively symmetric, while the adiposity versus obesity data look definitely asymmetric. In order to find the QQ plots, we need to obtain the symmetrized data sets, which for each plot is simply the union of the observed data and the negatives of the observed

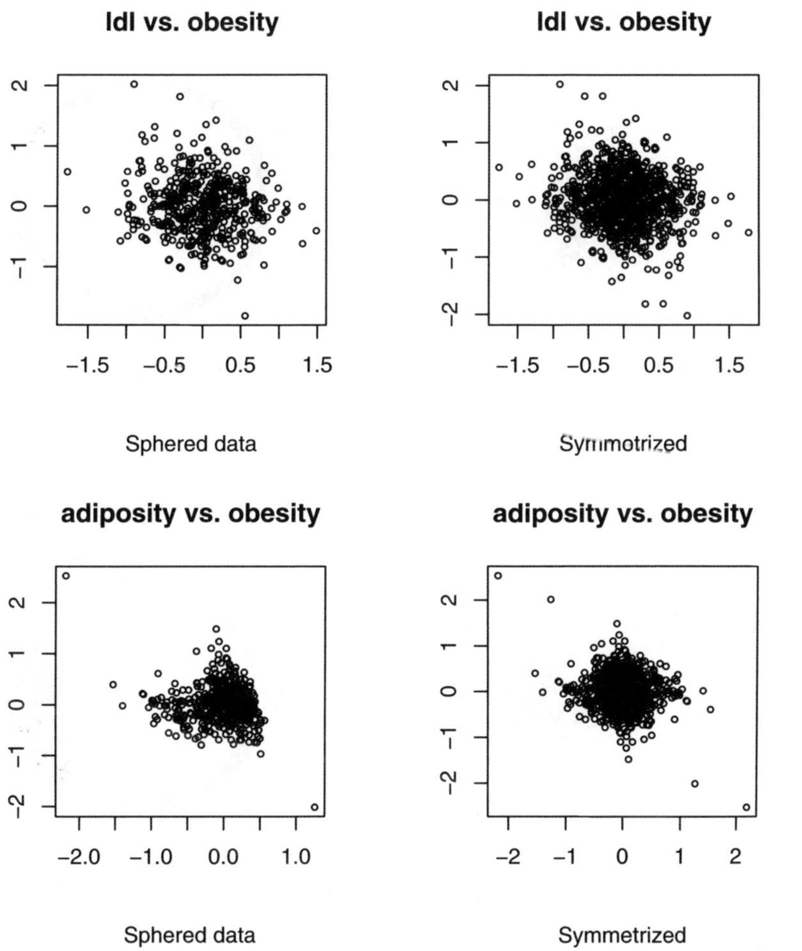

Fig. 1. The left-hand plots show the two scatter plots after preprocessing (i.e., taking logs, sphering, and centering). The right-hand plots contain the symmetrized samples, consisting of the data combined with the negatives of the data.

data. These plots are along the right-hand side in Figure 1. The top two plots are fairly similar, modulo twice as many points on the right, while the bottom two plots are quite different. In the latter, the observed data have more observations skewed to the north and west than one would expect with symmetry.

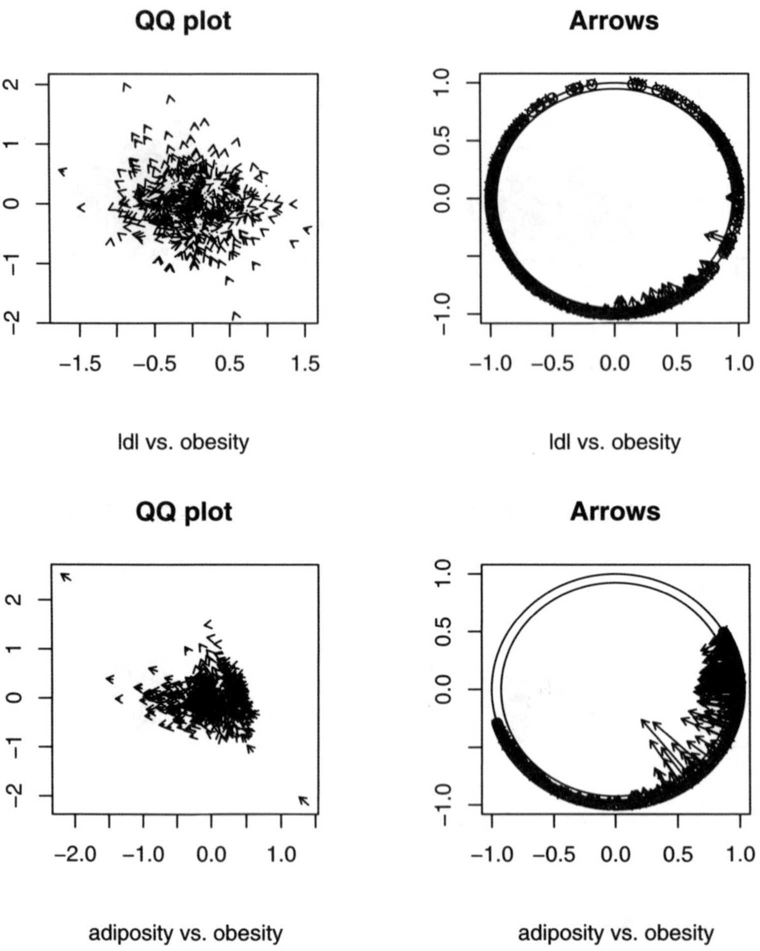

Fig. 2. The left-hand plots show the QQ plots for testing ± symmetry for the two scatter plots. The right-hand plots arrange the arrows around the unit circle.

Figure 2 contains the QQ plots. The top left-hand QQ plot is for the ldl versus obesity data. The tips of the arrows represent the observed data, and the bases are the quantiles under symmetry. The arrows are fairly small in length, but there does appear to be a trend towards pointing to the northwest. Most of the arrows are concentrated in the middle, so are difficult to see. The "Arrows" plot arranges the arrows so that they start at the edge of the unit circle and point towards the origin, keeping their orientations the same. They are scaled for visual appeal. We see more clearly that the largest arrows are pointing towards the northwest, and otherwise the arrows are reasonably well scattered about the circle. To calibrate their lengths, we performed 100 randomizations, each taking the data and randomly multiplying each $\mathbf{y}_i$ with a ± 1. We recentered the result, then calculated the arrows. The inner circle on the arrows plot represents the 95th percentile of the lengths of these random arrows. About 13% of the observed arrows exceeded this percentile, suggesting some lack of symmetry in the data not easily evident by eye. We also used the mean of the arrows' squared lengths as a test statistic, and the means of the randomization samples' arrows' squared lengths as a benchmark, to obtain an approximate randomized p-value of 6%. (It is not an exact randomization value because of the sphering and centering of the data.) Using the mean lengths of the arrows, the randomization p-value was 3%. Thus if the data are not symmetric, they are not far off.

We performed the same steps for the adiposity versus obesity data, obtaining the QQ plot and arrows plot in the second row of Figure 2. The QQ plot shows some relatively long arrows, and many pointing to the northwest, but again most arrows are packed together. The arrows plot (scaled with the same factor as that for the ldl versus obesity plot) exhibits many longer arrows, some pointing northwest, and a large group pointing west. The inner circle is again based on 100 randomizations of these data, where about 41% of the observed arrows are longer than the 95th percentile among the randomized ones. Using the average arrow squared length as a test statistic, the randomization p-value was 0. In fact, the observed mean of 0.0023 was more than four times the maximum of the 100 randomization means, 0.0005. We can see that the QQ plot indeed emphasizes the lack of symmetry for these two variables, and that the arrows plot clearly reveals the direction of the skewness.

4.2. *Compound symmetry*

Compound symmetry as presented by Votaw (1948) is characterized by permutation invariance within groups of variables. We illustrate with a data set on exam scores in a statistics class. There are $n = 191$ students and four variables: the scores on three midterm exams (X_1, X_2, and X_3) and one final exam (X_4). The symmetry of interest is that the three midterms are permutation-invariant (after centering), so that $\mathcal{G}$ is the set of permutation matrices of the form

$$\mathbf{P} = \begin{pmatrix} \mathbf{P}^* & 0 \\ 0 & 1 \end{pmatrix}, \tag{31}$$

where $\mathbf{P}^*$ is a 3×3 permutation matrix. A $\mathcal{G}$-invariant covariance matrix (12) has structure

$$\mathbf{\Sigma} = \begin{pmatrix} a & b & b & c \\ b & a & b & c \\ b & b & a & c \\ c & c & c & d \end{pmatrix}. \tag{32}$$

The sample covariance matrices of the raw data and the spatial ranks are, respectively,

$\widehat{\mathbf{\Sigma}}$:

	Exam1	Exam2	Exam3	Final Exam
Exam1	80.99	52.21	55.80	58.86
Exam2	52.21	123.27	67.73	74.77
Exam3	55.80	67.73	174.02	112.48
Final Exam	58.86	74.77	112.48	197.88

and

$100 \times \widehat{\mathbf{\Sigma}}_R$:

	Exam1	Exam2	Exam3	Final Exam
Exam1	7.55	2.60	3.17	2.84
Exam2	2.60	11.59	3.68	3.87
Exam3	3.17	3.68	10.97	4.70
Final Exam	2.84	3.87	4.70	13.05

. (33)

The raw covariance matrix could possibly fit the symmetry, though the three variances for the midterm exams may be too disparate. The more robust variance of the ranks looks better, with possibly the first variable having relatively low variance. As in (16), we find the quantiles with respect to the symmetrized data (after subtracting variable means) by replicating the data six times, where each replication has its first three variables permuted with a different permutation of $\{1, 2, 3\}$. Figure 3 is a histogram of the lengths of the resulting arrows. Note that there is one large outlier, while the rest of the arrows seem to follow a reasonably tight distribution. This outlier received a '0' on the third exam. We redo the analysis without that individual. There are other values that could be considered outliers, but they do not appear to be outliers with respect to compound symmetry, so we will keep them.

The covariance matrices in (33) calculated without the outlier do not change much, except that the variance of Exam3 for the raw data is now 138.04, producing a better fit to the compound symmetry. The arrows for the modified data do not appear to contain any outliers. Using the mean squared length of the arrows as a test statistic, the estimated p-value using

100 randomizations is 8%, which fits marginally well. Thus it is plausible that the first three exams, centered, are exchangeable.

We also tried the symmetry induced by permuting Exam1 and Exam2, and permuting Exam3 and the Final Exam. The randomized p-value here is 47%, giving us no reason to doubt the symmetry. Thus it looks like rather than the three midterms being exchangeable, it is that the first two exams are exchangeable (these exams are taken before covering the topic of inference in the course), and the last two exams are exchangeable.

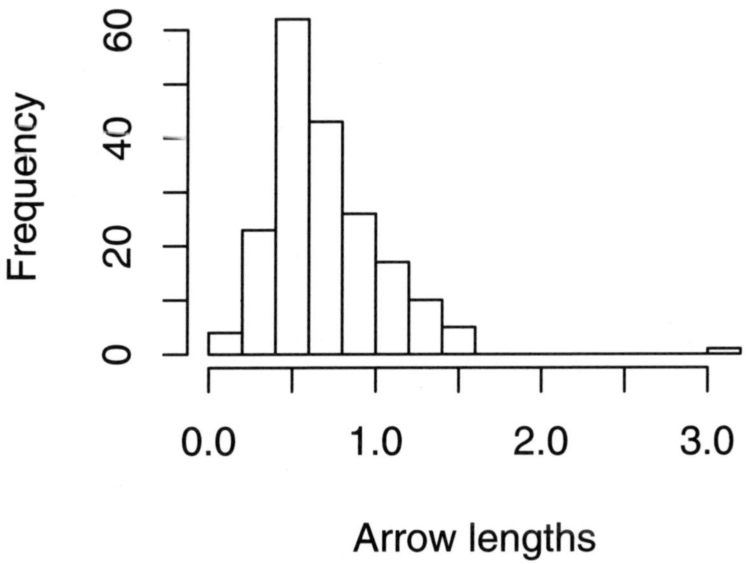

Fig. 3. Histogram for the lengths of the arrows for testing compound symmetry using the entire exams data set.

4.3. *Spherical symmetry*

Recall that the distribution F is spherically symmetric if (11) holds with $\mathcal{G} = \mathcal{O}_p$. The distribution of $\mathbf{X}$ is called elliptically symmetric if $\mathbf{X} = \mathbf{AZ}+\boldsymbol{\theta}$ for some $p\times p$ matrix $\mathbf{A}$ and $p\times 1$ vector $\boldsymbol{\theta}$, where $\mathbf{Z}$ is spherically symmetric. Testing for spherical symmetry and elliptical symmetry has a long history. Mauchly (1940) derived the likelihood ratio test for the null hypothesis of spherical symmetry under the assumption of multivariate normality, and Kariya and Eaton (1977) and King (1980) extend consideration to general elliptically symmetric situations. These articles assume elliptical symmetry under the alternative hypothesis, in which case testing the null hypothesis of spherical symmetry is equivalent to testing that the center $\boldsymbol{\theta} = \mathbf{0}$ and the scale matrix $\mathbf{AA}' = c\mathbf{I}_p$ for some constant c. The corresponding tests may not have good power when the alternative is more general than elliptical symmetry. For work on tests for more general alternatives, see Romano (1989), Baringhaus (1991), Koltchinskii and Li (1998), Hallin and Paindaveine (2006), and references therein.

In the example here, we are more interested in assessing elliptical symmetry, which involves first shifting and sphering the data before finding the QQ plots relative to spherical symmetry. See Beran (1979) and Koltchinskii and Sakhanenko (2000) for tests of the null hypothesis of elliptical symmetry.

Consumers' Union (1990) provides data on $n = 111$ automobiles with $p = 11$ size measurements: Length, length of wheel-base, width, height, front headroom, rear headroom, front legroom, rear seating room, front shoulder room, rear shoulder room, and luggage room. The data can be found in the S-Plus (Insightful Corporation, 1988-2005) data frame *cu.dimensions*. One problem arises from the fact that some automobiles do not have a delineated luggage area, either because they are too small to have trunks (sports cars), or the luggage space and passenger space comprise one large area (minivans). SUV's were not prevalent yet. These observations had values coded "-2" or "-3" for their luggage room.

Before addressing spherical or elliptical symmetry, we consider the simpler $\pm$ symmetry as in Section 4.1. The data are first sphered using Tyler's transformation as in (29) and (30) and centered using the spatial median. We then find the quantiles with respect to the symmetrized data, as in (22). Rather than trying to plot the arrows, we find the two-dimensional projection that maximizes the sum of squares of the projected arrows, as in (27). Figure 4 contains the result. The idea is that this plot exhibits the two dimensions that contribute most to lack of symmetry.

The main features of this plot are easy to describe. There is a very tight cluster consisting of most of the data (96 of the 111 points) in the upper right. At the upper left, there is a small cluster of five automobiles: Chevrolet Corvette, Honda Civic CRX, Mazda MX5 Miata, Mazda RX7, and Nissan 300ZX. These are the small sports cars. At the lower right is a more spread out cluster of ten minivans: Chevrolet Astro, Chevrolet Lumina APV, Dodge Caravan, Dodge Grand Caravan, Ford Aerostar, Mazda MPV, Mitsubishi Wagon, Nissan Axxess, Nissan Van, and Volkswagen Vanagon.

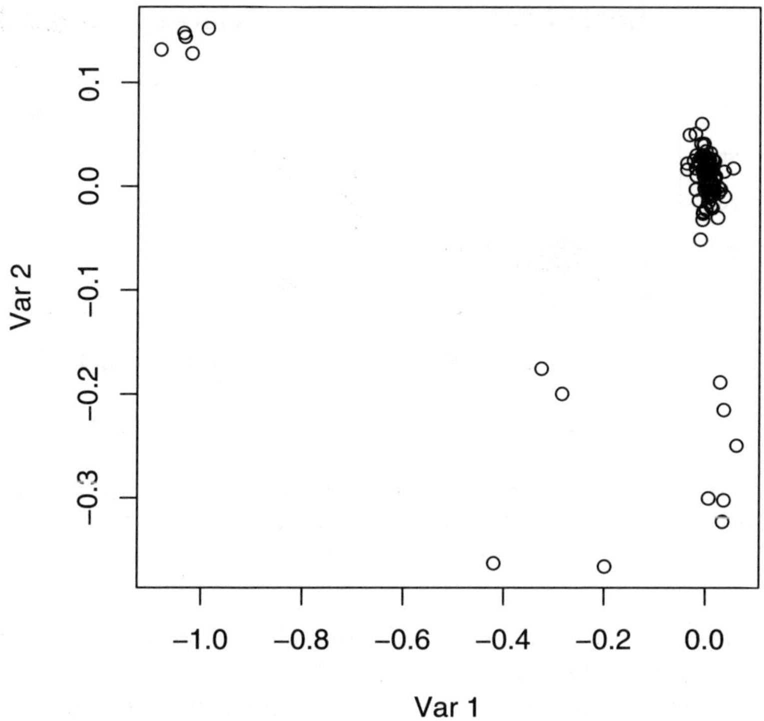

Fig. 4. The first two components after rotating the automobile data based on the squared lengths of the arrows for testing ± symmetry.

Thus assessing symmetry leads directly to the first steps for analyzing the data. In Figure 5 we plot the first two variables for other principal component decompositions. The first plot uses regular principal components on the raw data where the variables are scaled to have equal sample variance (but not sphered), and the second uses the generalized principal components idea in Oja *et al.* (2006) of taking the usual principal components of the Tyler-sphered data (i.e., the principal components of the $\mathbf{y}_i$'s in equations 29 and 30). Note that the latter is quite close to what we see in Figure 4, while the regular principal components plot shows some clustering but does not reveal the features so starkly.

We turn to assessing spherical symmetry, after sphering using the sample covariance matrix, so that we are actually testing for elliptical symmetry of the original data. That is, we transform each observation $\mathbf{x}_i$ to $\mathbf{y}_i = \mathbf{S}^{-1/2}(\mathbf{x}_i - \overline{\mathbf{x}})$, where $\mathbf{S}$ is the sample covariance matrix and $\overline{\mathbf{x}}$ is the sample

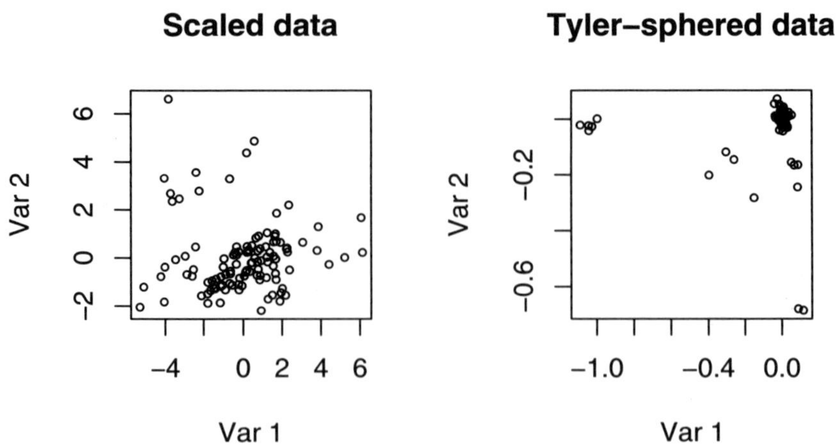

Fig. 5. The first two principal components for the automobile data. The left-hand plot is based on the data scaled so that the variances are equal. The right-hand plot is based on the data sphered using Tyler's transformation.

mean. The goal is to rotate the data so that all the non-sphericity is in the first k components for some k.

Our approach is stepwise. We start by finding the arrows $\mathbf{a}_i$ for testing spherical symmetry on the $\mathbf{y}_i$'s using all p variables, and finding the eigenvalues and eigenvectors of $\sum_{i=1}^n \mathbf{a}_i \mathbf{a}_i'$. Let λ_1 be the largest eigenvalue and γ_1 the corresponding eigenvector, and let $\mathbf{\Gamma}^*$ be the $p \times (p-1)$ matrix whose columns are the remaining eigenvectors. The first component of the desired rotation consists of the $\gamma_1' \mathbf{y}_i$'s.

The second step considers the $(p-1)$-dimensional data consisting of the $\mathbf{\Gamma}^{*'} \mathbf{y}_i$'s. We find the arrows, say $\mathbf{a}_i^*$, for testing spherical symmetry for these data and the corresponding eigenvalues and eigenvectors of the matrix $\sum_{i=1}^n \mathbf{a}_i^* \mathbf{a}_i^{*'}$. Let λ_2 be the largest of these eigenvalues and γ_2 the corresponding eigenvector, so that the second component of interest consists of the $\gamma_2' \mathbf{y}_i^*$'s.

We continue until obtaining p components and corresponding eigenvalues. Then λ_k measures the lack of spherical symmetry in the final $p-k+1$ components, smaller values indicating better fit to spherical symmetry. To assess smallness, we perform 100 randomization trials of the entire process, recording the resulting λ_k's.

Somewhat analogous statistics found in the literature include the Kolmogorov-Smirnov-type statistic in Koltchinskii and Li (1998) and the Cramér-von-Mises-type statistic in Baringhaus (1991) for testing spherical symmetry, as well as the approach in Koltchinskii and Sakhanenko (2000) for testing elliptical symmetry. Our test is based on the sizes of the arrows (2) in the QQ plots, while the others are based on the arrows (10) in the PP plots, or similar quantities. We use the randomization technique suggested in Romano (1989), but because we randomize after centering and sphering the data, our approach is only an approximate randomization. We do not claim any optimality for our statistic. Rather, we are tailoring it to the graphics, which are our prime interest.

Figure 6 plots the observed eigenvalues, along with the boxplots for the randomizations, for the entire data set and for the reduced data set eliminating the sports cars and minivans. For the complete data, the eigenvalues are not plausibly small until $k = 5$, while for the reduced data set, even the second eigenvalue is not very large. Comparing the sizes of eigenvalues for the two plots, we see that those for the complete data are generally much larger than for the reduced data set. Figure 7 plots the first four components (one versus two and three versus four) from the reduced data. There does not appear to be much remarkable, although there is skewing in the first variable and the second plot may show some clustering on variable three. Generally, the spherical symmetry assumption for the reduced data is reasonable.

5. Comparing rotations

There are many possible ways to rotate the data to find interesting projections. Different approaches will lead to different rotations, so it is reasonable to consider a number of different rotations. Here we do a brief empirical comparison of four approaches: regular principal components on data with the variables scaled to have equal variance ("PC"), the rotations suggested by the arrows for QQ plots or PP plots when assessing spherical symmetry based on the data sphered using the sample covariance matrix ("QQ" and "PP", respectively), and generalized principal components using the usual principal components on the data sphered by Tyler's transformation ("GPC"). We are not trying to find the "best" rotation, but rather determine how distinct the approaches are.

To compare two rotations, for each $k = 1, \ldots, p$, we find R^2/k, where R^2 is the trace of the multiple correlation matrix between the first k com-

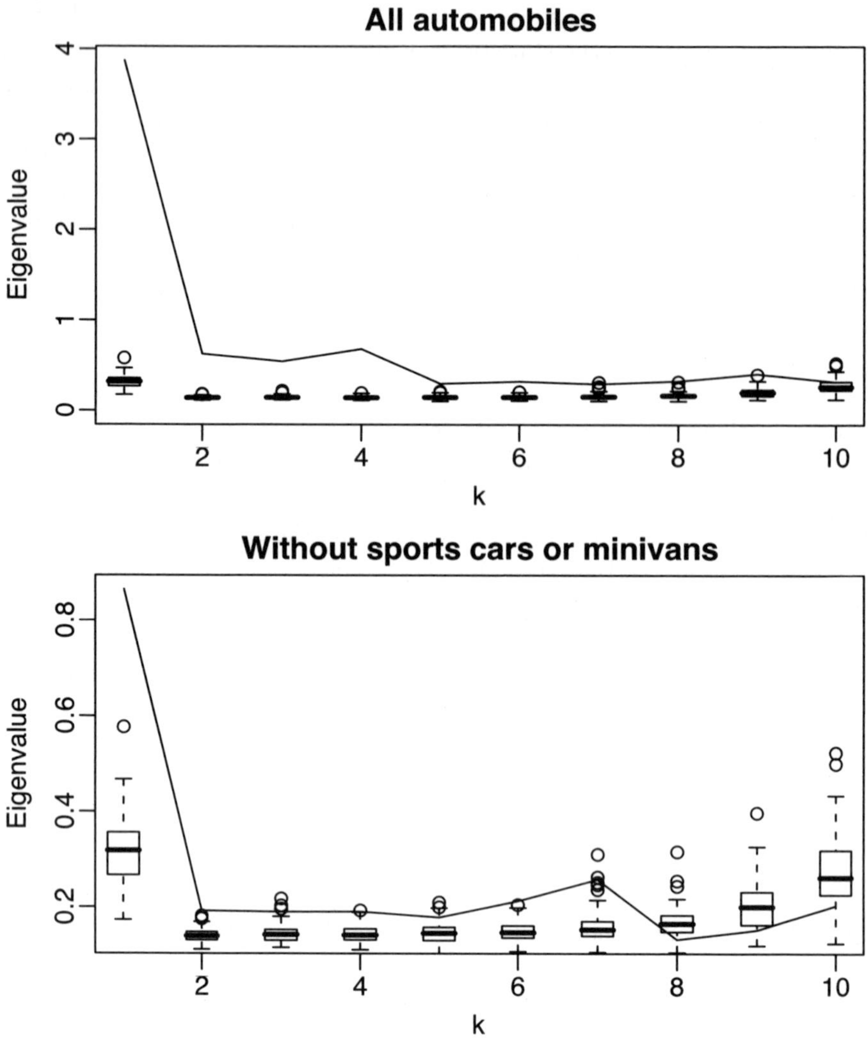

Fig. 6. Eigenvalues for the sum of squares of arrows for the cars' data. The top graph shows the results for all the data, the bottom for all but the sports cars and minivans. The boxplots are constructed from 100 randomization samples.

ponents resulting from each rotation. Thus, $R^2/k = 1$ for a given k means that the two rotations have the same k-dimensional projection.

Figure 8 shows the results for the complete automobile data, the reduced automobile data, and the Boston Housing data (Newman *et al.*, 1998). The straight solid line represents the expected value of the measure if the two

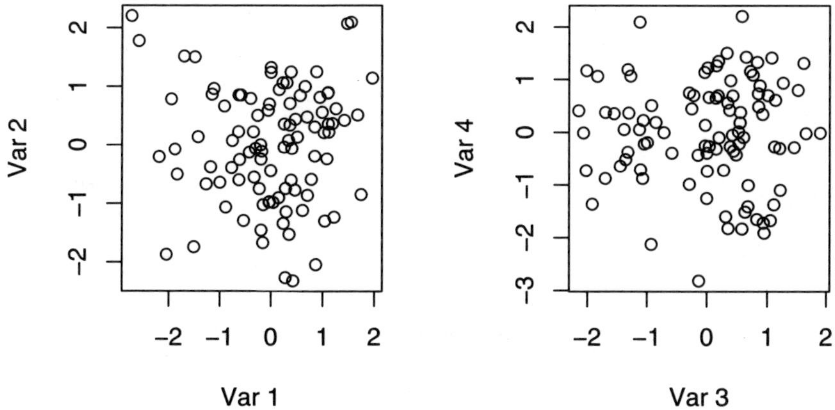

Fig. 7. Scatter plots of the first four components for the reduced cars' data, testing spherical symmetry.

rotations are independent and uniform. Note that for the most part, the graphs are very close to the solid line, suggesting that the rotations are exhibiting distinct characteristics of the data. One exception is for the complete automobile data, where, at $k = 4$, the $R^2/k \approx 0.87$ for the QQ-based and generalized principal components approaches. Also, note that for the latter two plots, the QQ and PP rotations are quite similar, whereas for the first plot, QQ and PP are similar only for $k = 1$. A possible explanation is that the PP plot is less sensitive to outliers, of which the complete automobile data has several. In this case, the QQ plots reveal the outliers more clearly.

6. Calculating the ranks for symmetrized distributions

The definition of the rank of $\mathbf{x}$ relative to the spherically symmetrized data is given in (18). Focus on the integral for the ith summand. If $\mathbf{x} = \mathbf{0}$, then $E[\text{Sign}(\mathbf{\Gamma x}_i)] = \mathbf{0}$ by symmetry of $\mathbf{\Gamma}$'s distribution. Suppose $\mathbf{x} \neq \mathbf{0}$. We can find $\mathbf{g_x}$ and $\mathbf{g}_i$ in $\mathcal{O}_p$ so that

$$\mathbf{x} = \|\mathbf{x}\|\ \mathbf{g_x e}_1 \ \text{ and } \ \mathbf{x}_i = \|\mathbf{x}_i\|\ \mathbf{g}_i \mathbf{e}_1, \tag{34}$$

where $\mathbf{e}_1 = (1, 0, \ldots, 0)'$, the $p \times 1$ vector with 1 as the first element and 0 elsewhere. Also, note that $\text{Sign}(c\mathbf{z}) = \text{Sign}(\mathbf{z})$ for $c > 1$, and for orthogonal

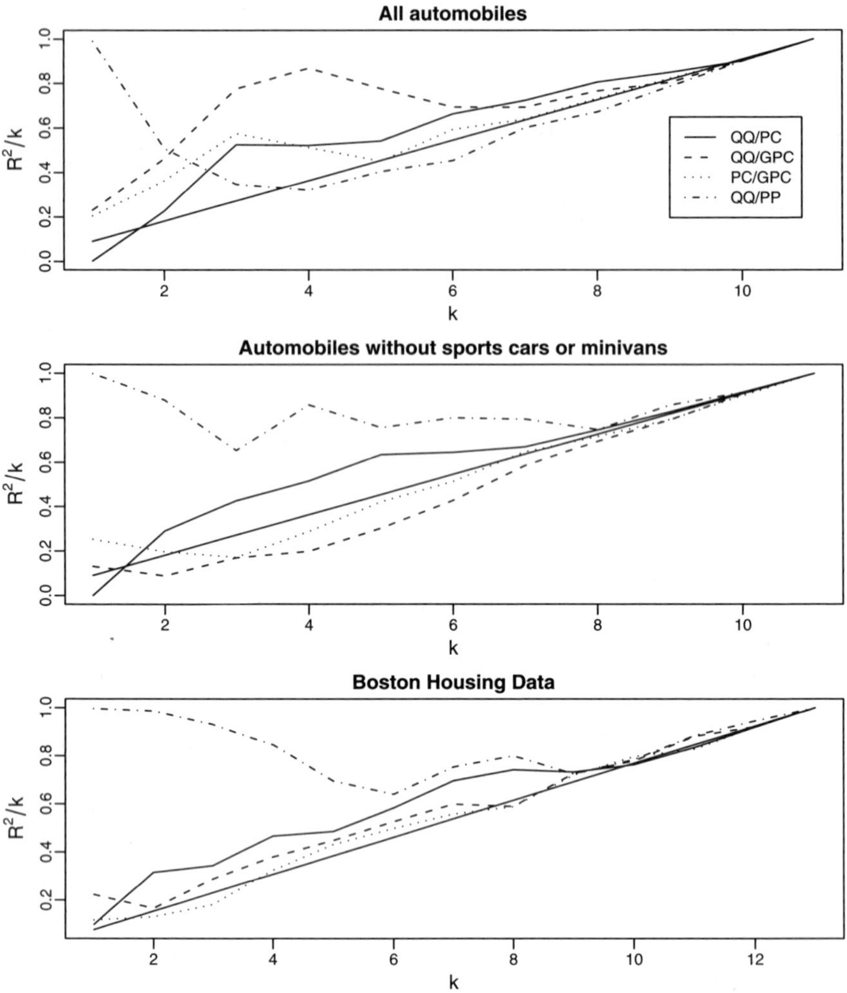

Fig. 8. Comparing four rotations for three data sets. The rotations are regular principal components (PC), the QQ-based plots (QQ), the generalized principal components (GPC), and the PP-based plots (PP).

$\mathbf{g}$, $\text{Sign}(\mathbf{gz}) = \mathbf{g}\ \text{Sign}(\mathbf{z})$. Then by the uniformity of the measure on $\mathbf{\Gamma}$,

$$\begin{aligned} E[\text{Sign}(\mathbf{x} - \mathbf{\Gamma x}_i)] &= E[\text{Sign}(\|\mathbf{x}\|\mathbf{g}_{\mathbf{x}}\mathbf{e}_1 - \|\mathbf{x}_i\|\mathbf{\Gamma g}_i\mathbf{e}_1)] \\ &= \mathbf{g}_{\mathbf{x}}\ E[\text{Sign}(\|\mathbf{x}\|\mathbf{e}_1 - \|\mathbf{x}_i\|\mathbf{g}_{\mathbf{x}}^{-1}\mathbf{\Gamma g}_i\mathbf{e}_1)] \\ &= \mathbf{g}_{\mathbf{x}}\ E[\text{Sign}(\mathbf{e}_1 - (\|\mathbf{x}_i\|/\|\mathbf{x}\|)\mathbf{\Gamma e}_1)] \\ &= \mathbf{g}_{\mathbf{x}}\ E[\text{Sign}(\mathbf{e}_1 - (\|\mathbf{x}_i\|/\|\mathbf{x}\|)\boldsymbol{\gamma})], \end{aligned} \tag{35}$$

where $\boldsymbol{\gamma}$ is the first column of $\boldsymbol{\Gamma}$, which is distributed uniformly over the p-dimensional unit sphere. The distribution of $\boldsymbol{\gamma}$ is invariant under component change of sign, and components 2 through p of the Sign in the last expression in (35) are odd functions of the corresponding components of $\boldsymbol{\gamma}$, hence their expected values are zero. Thus

$$\begin{aligned} \mathbf{g_x}\ E[\text{Sign}(\mathbf{e}_1 - (\|\mathbf{x}_i\|/\|\mathbf{x}\|)\boldsymbol{\gamma}] &= \mathbf{g_x}\mathbf{e}_1\ E\left[\frac{1-(\|\mathbf{x}_i\|/\|\mathbf{x}\|)\gamma_1}{\|\mathbf{e}_1 - (\|\mathbf{x}_i\|/\|\mathbf{x}\|)\boldsymbol{\gamma}\|}\right] \\ &= \text{Sign}(\mathbf{x})\ c_p(\|\mathbf{x}_i\|/\|\mathbf{x}\|), \end{aligned} \tag{36}$$

because by (34), $\mathbf{g_x}\mathbf{e}_1 = \text{Sign}(\mathbf{x})$, where $c_p(t)$ is the function

$$c_p(t) = E\left[\frac{1-t\gamma_1}{\|\mathbf{e}_1 - t\boldsymbol{\gamma}\|}\right] = E\left[\frac{1-t\gamma_1}{\sqrt{1-2t\gamma_1+t^2}}\right]. \tag{37}$$

Because the vector $\boldsymbol{\gamma}$ is uniformly distributed over the p-dimensional unit sphere, it has the same distribution as $\mathbf{Z}/\|\mathbf{Z}\|$, where $\mathbf{Z} \sim N_p(\mathbf{0}, \mathbf{I}_p)$. Thus γ_1 is distributed as $Z_1/\|\mathbf{Z}\|$, hence has density given in (21), verifying the expression (19). (To see this distribution, note that the distribution of γ_1 is symmetric about 0 and $\gamma_1^2 \sim \text{Beta}[1/2, (p-1)/2]$.)

After some manipulations, it can be shown that for $t \neq 1$,

$$c_p(t) = {}_2F_1(\frac{1}{2}, \frac{p-1}{2}, p-1; z_t) - \frac{t}{1+t}\ {}_2F_1(\frac{1}{2}, \frac{p+1}{2}, p; z_t), \tag{38}$$

where $z_t = 4t/(1+t)^2$ and ${}_2F_1$ is the hypergeometric function as in Chapter 15 of Abramowitz and Stegun (1972). Also,

$$c_p(1) = \frac{2^{p-2}\Gamma(\frac{p}{2})^2}{\sqrt{\pi}\ \Gamma(p-\frac{1}{2})}. \tag{39}$$

Acknowledgments

Many thanks go to the organizers of this volume, and the referees for their perceptive comments.

References

Abramowitz, M. and Stegun, I. A. (1972). *Handbook of Mathematical Functions* (U.S. Government Printing Office, Washington, D.C.).

Andersson, S. (1975). Invariant normal models, *Annals of Statistics* **3**, pp. 132–154.

Andersson, S. A., Brons, H. K. and Jensen, S. T. (1983). Distribution of eigenvalues in multivariate statistical analysis, *Annals of Statistics* **11**, 2, pp. 392–415.

Baringhaus, L. (1991). Testing for spherical symmetry of a multivariate distribution, *Annals of Statistics* **19**, 2, pp. 899–917.

Beran, R. (1979). Testing for ellipsoidal symmetry of a multivariate density, *Annals of Statistics* **7**, 1, pp. 150–162.

Caussinus, H. and Ruiz-Gazen, A. (1993). Projection pursuit and generalized principal component analyses, in S. Morgenthaler, E. Ronchetti and W. A. Stahel (eds.), *New Directions in Statistical Data Analysis and Robustness* (Birkhäuser Verlag), pp. 35–46.

Chakraborty, B. (2001). On affine equivariant multivariate quantiles, *Annals of the Institute of Statistical Mathematics* **53**, 2, pp. 380–403.

Chaudhuri, P. (1996). On a geometric notion of quantiles for multivariate data, *Journal of the American Statistical Association* **91**, pp. 862–872.

Consumers' Union (1990). Body dimensions, *Consumer Reports* **April**, pp. 286–288.

Hallin, M. and Paindaveine, D. (2006). Semiparametrically efficient rank-based inference for shape: I. optimal rank-based tests for sphericity, *Annals of Statistics* **34**, 6, pp. 2707–2756.

Hettmansperger, T. P. and Randles, R. H. (2002). A practical affine equivariant multivariate median, *Biometrika* **89**, 4, pp. 851–860.

Insightful Corporation (1988-2005). *S-PLUS version 7.0 for Windows (computer software)* (Insightful Corporation, Seattle, Washington).

Kariya, T. and Eaton, M. L. (1977). Robust tests for spherical symmetry, *Annals of Statistics* **5**, 1, pp. 206–215.

Kariya, T. and Sinha, B. K. (1989). *Robustness of Statistical Tests* (Academic Press).

King, M. L. (1980). Robust tests for spherical symmetry and their application to least squares regression, *Annals of Statistics* **8**, 6, pp. 1265–1271.

Koltchinskii, V. and Sakhanenko, L. (2000). Testing for ellipsoidal symmetry of a multivariate distribution, in D. M. E. Giné and J. Wellner (eds.), *High Dimensional Probability II, Progress in Probability*, Vol. 47 (Birkhäuserauser, Boston), pp. 493–510.

Koltchinskii, V. I. (1997). M-estimation, convexity and quantiles, *Annals of Statistics* **25**, pp. 435–477.

Koltchinskii, V. I. and Li, L. (1998). Testing for spherical symmetry of a multivariate distribution, *Journal of Multivariate Analysis* **65**, pp. 228–244.

Marden, J. and Gao, Y. (2002). Rank-based procedures for structural hypotheses on covariance matrices, *Sankhyā, Series A* **64**, 3, pp. 653–677.

Marden, J. I. (1998). Bivariate QQ-plots and spider web plots, *Statistica Sinica* **8**, pp. 813–826.

Marden, J. I. (1999). Multivariate rank tests, in S. Ghosh (ed.), *Multivariate Analysis, Design of Experiments, and Survey Sampling* (Marcel Dekker, New York), pp. 401–432.

Mauchly, J. W. (1940). Significance test for sphericity of a normal n-variate distribution, *Annals of Mathematical Statistics* **11**, 2, pp. 204–209.

Newman, D., Hettich, S., Blake, C. and Merz, C. (1998). UCI repository of machine learning databases, URL `http://www.ics.uci.edu/~mlearn/`

`MLRepository.html`.

Oja, H., Sirkiä, S. and Eriksson, J. (2006). Scatter matrices and independent component analysis, *Austrian Journal of Statistics* **35**, 2–3, pp. 175–189.

Romano, J. P. (1989). Bootstrap and randomization tests of some nonparametric hypotheses, *Annals of Statistics* **17**, 1, pp. 141–159.

Rousseauw, J., Plessis, J., Benade, A., Jordaan, P., Kotze, J., Jooste, P. and Ferreira, J. (1983). Coronary risk factor screening in three rural communities, *South African Medical Journal* **64**, pp. 430–436.

Serfling, R. (2004). Nonparametric multivariate descriptive measures based on spatial quantiles, *Journal of Statistical Planning and Inference* **123**, 2, pp. 259–278.

Tyler, D. E. (1987). A distribution-free M-estimator of multivariate scatter, *Annals of Statistics* **15**, pp. 234–251.

Tyler, D. E., Critchley, F., Dumbgen, L. and Oja, H. (2009). Invariant co-ordinate selection (with discussion), *Journal of the Royal Statistical Society, Series B* **71**, 3, pp. 549–592.

Visuri, S., Koivunen, V. and Oja, H. (2000). Sign and rank covariance matrices, *Journal of Statistical Planning and Inference* **91**, 2, pp. 557 – 575.

Votaw, D. F. (1948). Testing compound symmetry in a normal multivariate distribution, *Annals of Mathematical Statistics* **19**, pp. 447–473.

A Comparison of Estimators for the Variance of Cross-Validation Estimators of the Generalization Error of Computer Algorithms

Marianthi Markatou[†] and Rositsa Dimova[‡]

Department of Biostatistics, Columbia University
New York, NY 10032, USA
[†] *E-mail: mm168@columbia.edu*
[‡] *E-mail: rbd2107@columbia.edu*

Anshu Sinha

Department of Biomedical Informatics, Columbia University
New York, NY 10032, USA
E-mail: anshu.sinha@dbmi.columbia.edu

This paper studies variance estimators of cross-validation estimators of the generalization error. Three estimators are discussed, and their performance is evaluated in a variety of data models and data sizes. It is shown that the standard error associated with the moment approximation estimator is smaller than that associated with the other two. The effect of training and test set size on these estimators is discussed.

Keywords: Cross-Validation; Generalization Error; Moment Approximation; Variance

1. Introduction

A number of problems can be categorized as either prediction or classification problems. The goal in prediction is to estimate a function $y = f(x;\theta)$ that can predict a value y given an input vector x and a set of estimated parameters θ. In classification, we wish to learn a mapping from a vector of measurements x to a categorical variable y. The variable to be predicted is the class variable taking values in the set $\{c_1, c_2, ..., c_m\}$.

A variety of algorithms is available for prediction and classification. Over the years, many variations of known classification and prediction algorithms have been developed. How does one then choose an appropriate algorithm for a given problem? There are many studies in the literature comparing different algorithms using both real and artificial data sets.

One measure of algorithmic performance is the generalization error. The generalization error is the error an algorithm makes when applied to a new data set, different from the data set on which it is trained. Therefore, the generalization error can be considered as a measure of predictive accuracy, with small generalization error indicating superior performance.

One way therefore to select among the different algorithms is to compute the generalization error on test data and estimate the variance that is associated with it. The choice of a variance estimator of the generalization error is extremely important: a poor choice could lead one to incorrectly report that a new method is better than a previously proposed one. In machine learning, the comparison of different supervised learning algorithms is done through the use of appropriate significance tests (Dietterich, 1998). These tests are often used to support the claim that a new learning algorithm generalizes better than existing algorithms. If the variance estimators used in these tests take into account the variability due to the test examples only, this could lead to serious underestimation of the variance of cross-validation estimators of the generalization error and therefore to incorrect conclusions.

The problem of variance estimation of estimators of the generalization error is not well studied in the literature. Relevant works include McLachlan (1972, 1974, 1976), Nadeau and Bengio (2003), Bengio and Grandvalet (2004), Markatou *et al.* (2005) and Tian (2006). Other related literature is Efron (1983, 1986, 2004), Breiman (1992), and Markatou *et al.* (2008).

Providing a variance estimate of an estimator of the generalization error is a difficult problem, particularly when one would like to take into account all possible sources of variation. These sources of variation include the randomness in the training and test sets as well as the randomness in the data. In this paper, we present an extensive, comparative study of three estimators of the variance of the generalization error: the Nadeau-Bengio biased estimator (NB-Biased) (Nadeau and Bengio, 2003), the Nadeau-Bengio unbiased estimator (NB-Unbiased) (Nadeau and Bengio, 2003), and the moment approximation estimator (MA) (Markatou *et al.*, 2005, 2008). We examine the estimators in the case of both random cross-validation and k-fold cross-validation, with special emphasis on predicting the sample mean. Although prediction and classification in practice is more complex than predicting the mean, this simple case illustrates the difficulties associated with the task of estimating variances. Moreover, it illustrates the role that the different training and test sets play in the problem of variance estimation.

Additionally, the performance of the estimators is studied in terms of bias and variance, in both large samples and small samples. In both cases, the MA estimator outperforms both Nadeau-Bengio estimators. The Nadeau-Bengio biased estimator of variance is easy to compute, but this is at the expense of increased variability of the value of the estimator. On the other hand, the MA estimator does not exhibit high variance in any model or sample size we studied for large samples, and exhibits considerably smaller variance than the Nadeau-Bengio estimators in small samples.

The paper is organized as follows. Section 2 presents the three estimators under discussion. Section 3 presents an analysis of the bias of the estimators. Section 4 presents simulation results and discusses the findings. Finally, section 5 offers conclusions.

2. The Estimators of Variance of Cross-Validation Estimators of the Generalization Error

In this section we discuss the mathematical framework within which the estimators will be derived and present the three estimators under consideration.

2.1. *The Framework and Relevant Definitions*

Let $X_1, X_2, ..., X_n$ be independent and identically distributed random variables that follow a distribution F. Assume that $\Omega = \{X_1, X_2, ..., X_n\}$ constitutes the data universe, and let S be a set of size $n_1 < n$, $S \subset \Omega$. Denote by S^c the complement of S with respect to the data universe Ω.

Definition 2.1. Any subset S of the data universe Ω that is such that $\text{Card}(S) = n_1 < n$ and that is used to construct a statistical or algorithmic rule is called a training set.

Definition 2.2. The data set S^c, the elements of which have the same distribution as the elements of the set S, is called the test set. The test set has cardinality $n - n_1$.

Definition 2.3. A prediction rule or model constructed using the elements of training sets is a measurable function f such that $f : S \to R$.

Definition 2.4. Let $L : R \times R \to R$ be a function, Y a target variable, and $\hat{f}(x)$ a prediction rule. The function $L[Y, \hat{f}(X)]$, $X \in R^p$, that measures the error between target variable and the prediction made by the prediction rule is called a loss function.

The most popular form of a loss function is the square error loss defined as $L[Y, \hat{f}(X)] = [Y - \hat{f}(x)]^2$. Other typical choices include the absolute error loss given as $L[Y, \hat{f}(X)] = |Y - \hat{f}(x)|$ and the negative likelihood function defined as $L[Y, \hat{f}(X)] = -\log f[y; \hat{\theta}(x)]$, where $f[y, \theta(x)]$ is the density of Y indexed by a parameter $\theta(x)$ and $\hat{\theta}$ indicates the estimate of θ based on the training set data.

Definition 2.5. (Hastie *et al.*, 2001) The generalization error or test error of an algorithm is the expected prediction error over an independent test sample and it is given by the quantity

$$E\{L[Y, \hat{f}(X)]\},$$

where both X and Y are drawn randomly from their joint distribution, and the expectation is taken over everything that is random. This includes the randomness in the training sample that produced $\hat{f}(X)$.

A variety of methods are available to estimate the generalization error and can be classified either as analytical (Efron, 1983, 1986) or direct, such as cross-validation or bootstrap. We study the following two forms of cross-validation.

Completely random selection: This form of cross-validation creates J training sets of a given size n_1 that are randomly selected from the universe of available data. The corresponding test sets are formed from the complements of the respective training sets.

Notice that, in this case, the training sets formed, as well as the test sets, have common elements. This introduces dependence, which appears as a covariance term in the estimation of the variance of cross-validation estimators of the generalization error. This dependence must be taken into account and estimated from the data. Efron (1983, 1986, 2004) notes that the general L_2 prediction error for cross-validation is underestimated if it is not adjusted by a covariance penalty. One of the variance estimators that we discuss, the moment approximation estimator, accounts explicitly for this dependence, and estimates the relevant terms from the data.

Nonoverlapping test sets: In this case, the test sets are controlled so that they do not have elements in common. That is, if S_j^c and $S_{j'}^c$ are two different test sets, then $S_j^c \cap S_{j'}^c = \varnothing$.

This form of cross-validation corresponds to k-fold cross-validation, where the data universe is split into k non-overlapping test sets. This way, the dependence due to overlapping of test sets is eliminated. However, the dependence due to overlapping training sets is still present.

The specific cross-validation estimator of the generalization error that we are going to study in terms of its variance is defined as follows.

Let $\{1, 2, 3, ..., n\}$ be a set of distinct integers and A_j be a random set of n_1 distinct numbers from the set $\{1, 2, ..., n\}, n_1 < n$. Let $n_2 = n - n_1$ be the size of the corresponding test set, with n_1 being a fixed number. Select, independently, J random sets $A_1, A_2, ..., A_J$ and denote by $S_j = \{X_i : i \in A_j\}, j = 1, 2, ..., J$, the training sets formed by subsampling Ω according to the index sets A_j. The corresponding test set is then $S_j^c = \{X_i : i \in A_j^c\}$. Define $L(j, i) = L(S_j, X_i)$ to be the loss function defined on S_j and the elements of the test set. The usual "average test set error" measured on S_j^c is then

$$\hat{\mu}_j = \frac{1}{n_2} \sum_{i \in S_j^c} L(j, i). \tag{1}$$

The cross-validation estimator of the generalization error that we will study is given by the expression

$$\hat{\mu}_J = \frac{1}{J} \sum_{j=1}^{J} \hat{\mu}_j. \tag{2}$$

The estimator depends on the value of J, the size of the training and test sets, the loss function and the size of the data universe.

Nadeau and Bengio (2003) perform a theoretical investigation of the variance of the cross-validation estimator of the generalization error using complete random selection. If the only randomness that is taken into account is that of the test set, the variance will be underestimated. Moreover, Nadeau and Bengio (2003) note that there is no general unbiased estimator of the variance that involves the loss function in a linear or quadratic way. In what follows, we will describe two variance estimators introduced by Nadeau and Bengio (2003). We will then present the analysis of Markatou *et al.* (2005).

2.2. *The Variance Estimators*

We first describe the estimators proposed by Nadeau and Bengio (2003). These are estimators of the variance of the cross-validation estimator of the generalization error.

The first estimator, which we denote by NB-biased in our graphs and tables, is given by the expression

$$\left(\frac{1}{J}+\frac{\rho}{1-\rho}\right)S^2_{\hat{\mu}_j},$$

$$S^2_{\hat{\mu}_j}=\frac{1}{J-1}\sum_{j=1}^{J}(\hat{\mu}_j-\hat{\mu}_J)^2,$$

and $\hat{\mu}_j$ is the test set error defined by (1). The quantity ρ indicates the correlation between $\hat{\mu}_j$ and $\hat{\mu}_{j'}$, that is, the correlation between two different estimates of test set errors. If this correlation were known, the above estimator would be an unbiased estimator of the variance of the cross-validation estimator $\hat{\mu}_J$. Nadeau and Bengio (2003) suggest estimating ρ by $\hat{\rho}=n_2/n$, where $n_2=\text{Card}(S_j^c)$, $j=1,2,...,J$. Here, the cardinality of all S_j^c, $j=1,2,...,J$, is exactly the same and equal to n_2, so there is no need to index the estimate of the correlation by j. This choice of $\hat{\rho}$ would result in either underestimation or overestimation of the variance, depending on whether $\hat{\rho}$ overestimates or underestimates the true correlation.

The second estimator, introduced again by Nadeau and Bengio (2003), can be described as follows. Let n be the size of the data universe and assume, without loss of generality, that n is even. Randomly split the data set into two equally-sized data subsets. Compute the cross-validation estimator, as given by (2), on each of the data subsets and denote by $\hat{\mu}_1$ the estimator computed using data from the first subset and by $\hat{\mu}_2$ the estimator computed using data from the second subset. Repeat this process of randomly splitting the data into two equally-sized subsets and computing the corresponding estimators M times. The estimator of the variance is then given by the expression

$$\frac{1}{2M}\sum_{m=1}^{M}(\hat{\mu}_{1,m}-\hat{\mu}_{2,m})^2.$$

Nadeau and Bengio (2003) have established that this estimator is unbiased for $\text{Var}(\hat{\mu}_J)$ when the size of the training set is $n_1'=\lfloor\frac{n}{2}\rfloor-n_2$. They recommend using $M=10$. In our tables and graphs we will denote this estimator by NB-unbiased.

It is instructive to note that Nadeau and Bengio (2003) did not study the variance of the k-fold cross-validation estimators of the generalization error. Bengio and Grandvalet (2004) showed that there does not exist any unbiased, universal estimator of the variance of k-fold cross-validation that is valid under all distributions.

Markatou *et al.* (2005, 2008) address the problem of variance estimation of the cross-validation generalization error as a problem of approximating the moments of the distribution of the cross-validation statistic. They offer a unifying framework, within which both complete random cross-validation and k-fold cross-validation can be studied.

The idea of Markatou *et al.* (2005) is simple. The cross-validation estimator of the generalization error is viewed as a statistic and the needed moments of this statistic are approximated. The approximation illustrates the role of training and test sets in the performance of the algorithm and takes into account the dependence introduced by overlapping training and test sets.

For the case of predicting the mean, the analysis carried out by Markatou *et al.* (2005, 2008) obtains a variance estimator through the formula

$$\mathrm{Var}(\hat{\mu}_J) = \frac{1}{J^2}\sum_{j=1}^{J}\mathrm{Var}(\hat{\mu}_j) + \frac{1}{J^2}\sum\sum_{j\neq j'}\mathrm{Cov}(\hat{\mu}_j, \hat{\mu}_{j'}), \tag{3}$$

where the variance term is given as

$$\begin{aligned}\mathrm{Var}(\hat{\mu}_j) = \frac{1}{n_2}\Big[&\mathrm{Var}[L(\mu, X_i)] + \frac{\sigma^2}{n_1}\Big\{E\Big(\Big[L'(\mu, X_i)\Big]^2\Big) \\ &+ \mathrm{Cov}\Big[L(\mu, X_i), L''(\mu, X_i)\Big]\Big\}\Big] \\ &+ \frac{n_2-1}{n_2}\frac{\sigma^2}{n_1}\Big\{E\Big[L'(\mu, X_i)\Big]\Big\}^2 + O\Big(\frac{1}{n_1^2}\Big),\end{aligned}$$

and the covariance term is given by the expression

$$\begin{aligned}\mathrm{Cov}(\hat{\mu}_j, \hat{\mu}_{j'}) = \Big(1 - \frac{E(Y^*)}{n_2^2}\Big)\Big[&\frac{\sigma^2}{n_1^2}E(Y)\Big\{E\Big[L'(\mu, X_i)\Big]\Big\}^2 \\ &- \frac{\sigma^4}{4n_1^2}\Big\{E\Big[L''(\mu, X_i)\Big]\Big\})^2 + O\Big(\frac{1}{n_1^2}\Big)\Big] \\ &+ \frac{E(Y^*)}{n_2^2}\Big\{\mathrm{Var}\left[L(\mu, X_i)\right] + \frac{\sigma^2}{n_1}\Big\{E\Big[L(\mu, X_i)L''(\mu, X_i)\Big] \\ &- E[L(\mu, X_i)]E\Big[L''(\mu, X_i)\Big]\Big\} + \frac{\sigma^2}{n_1^2}E(Y)E\left[L'(\mu, X_i)^2\right] \\ &- \frac{\sigma^4}{4n_1^2}\left\{E\left[L''(\mu, X_i)\right]\right\}^2 + O\Big(\frac{1}{n_1^2}\Big)\Big\},\end{aligned}$$

where $Y = \mathrm{Card}(S_j \bigcap S_{j'})$, $Y^* = \mathrm{Card}(S_j^c \bigcap S_{j'}^c)$, $\mu = E_\Omega(X_i)$, $\sigma^2 = \mathrm{Var}_\Omega(X_i)$, $L' = \partial L/\partial\mu$, and $L'' = \partial^2 L/\partial\mu^2$. Markatou *et al.* (2005) proved that Y and Y^* each have a hypergeometric distribution, and therefore $E(Y) = n_1^2/n$ and $E(Y^*) = n_2^2/n$.

When the estimator of the generalization error is obtained via k-fold cross-validation, the corresponding expressions are simpler and are given as follows:

$$\begin{aligned}\text{Var}(\hat{\mu}_j) = \frac{k}{n}\Big\{ &\text{Var}[L(\mu, X_i)] + \frac{\sigma^2}{n}\frac{k}{k-1}\Big\{E\left[L'^2(\mu, X_i)\right] \\ &+ \text{Cov}\left[L(\mu, X_i), L''(\mu, X_i)\right]\Big\}\Big\} + \frac{n-k}{n}\frac{\sigma^2}{n}\frac{k}{k-1} \\ &+ \{E\left[L'(\mu, X_i)\right]\}^2 + O\left(\frac{1}{n_1^2}\right)\end{aligned}$$

and

$$\begin{aligned}\text{Cov}(\hat{\mu}_j, \hat{\mu}_{j'}) = &\frac{\sigma^2 k(k-2)}{n(k-1)^2}\left\{\left[E\left[L'(\mu, X_i)\right]\right\}^2\right. \\ &- \frac{\sigma^4 k^2}{4n^2(k-1)^2}\left\{E\left[L''(\mu, X_i)\right]\right\}^2 + O\left(\frac{1}{n_1^2}\right).\end{aligned}$$

Therefore, the variance of the cross-validation estimator of the generalization error is obtained by estimating from the training data the mean and variance included in the aforementioned formulas.

The assumption that the loss function is smooth, that is, differentiable, may be considered as a limitation of the method. However, this is not the case, as the theory developed in Markatou *et al.* (2005) applies to loss functions that can be approximated by continuous differentiable functions. Two such functions, commonly used, are the absolute error loss and the indicator function that is used in classification. For details, see Markatou *et al.* (2005) and Tian (2006). Markatou *et al.* (2005) mathematically analyzed prediction with linear regression models and classification; Tian (2006) studied the case of kernel regression and developed the corresponding theory.

In the next section, we present a detailed study of the three estimators discussed above for predicting the mean. Although this is a relatively easy case to handle, it is important as it illuminates the effects of the various components, such as training and test set size and randomness, on the variance estimators of the generalization error when it is computed by both methods, completely random and k-fold cross-validation. Moreover, it illustrates the performance of these estimators in the small-sample case.

3. Bias Calculation

To address the issue of bias, we derived the exact value of the variance of the cross-validation estimator of the generalization error when predicting the mean of normally distributed data sample.

Let $X_1, \ldots, X_n$ be independently and identically distributed $N(\mu, \sigma^2)$ random variables. The generalization error is estimated by (2), where

$$\hat{\mu}_j = \frac{1}{n_2} \sum_{i \in S_j^c} (\bar{X}_{S_j} - X_i)^2. \tag{4}$$

Recall that (3) can be written as:

$$\mathrm{Var}(\hat{\mu}_J) = \frac{1}{J}\mathrm{Var}(\hat{\mu}_j) + \frac{J-1}{J}\mathrm{Cov}(\hat{\mu}_j, \hat{\mu}_{j'}). \tag{5}$$

We prove the following propositions.

Proposition 3.1. *Let $X_1, X_2, \ldots, X_n$ be independent and identically distributed random variables from a $N(\mu, \sigma^2)$ distribution. Then*

$$Var(\hat{\mu}_j) = \frac{2\sigma^4}{n_1^2 n_2}\left[(n_1+1)^2 + n_2 - 1\right].$$

Proof. We have

$$\begin{aligned}\mathrm{Var}(\hat{\mu}_j) = {} & \frac{1}{n_2}\mathrm{Var}\left[(\bar{X}_{S_j} - X_k)^2\right] \\ & + \frac{1}{n_2^2}\sum_{i \in S_j^c}\sum_{i' \neq i}\mathrm{Cov}\left[(\bar{X}_{S_j} - X_i)^2, (\bar{X}_{S_j} - X_{i'})^2\right],\end{aligned} \tag{6}$$

where $X_k \in S_j^c$. Then $\mathrm{Var}[(\bar{X}_{S_j} - X_k)^2] = (2\sigma^4/n_1^2)(n_1+1)^2$.

To obtain the second term in (6), we first compute $E[(\bar{X}_{S_j} - X_i)^2] = (\sigma^2/n_1)(n_1+1)$ and

$$E\left[(\bar{X}_{S_j} - X_i)^2(\bar{X}_{S_j} - X_{i'})^2\right] = \sigma^4\left(\frac{3}{n_1^2} + \frac{2}{n_1} + 1\right),$$

where $i \neq i' \in S_j^c$. Thus,

$$\mathrm{Cov}\left[(\bar{X}_{S_j} - X_i)^2, (\bar{X}_{S_j} - X_{i'})^2\right] = \frac{2\sigma^4}{n_1^2}.$$

After substituting these into Equation (6), we obtain the statement for the variance. □

Proposition 3.2. *Under the assumption that $X_1, X_2, ..., X_n$ are independent and identically distributed random variables from a $N(\mu, \sigma^2)$ distribution,*

$$Cov(\hat{\mu}_j, \hat{\mu}_{j'}) = \frac{2\sigma^4}{n_1^2 n^2 (n-1)} \left[n_1^2 (n+1)^2 + (n-n_1)^2 - 4n_1 \right].$$

Proof.

$$\begin{aligned} \text{Cov}(\hat{\mu}_j, \hat{\mu}_{j'}) &= E(\hat{\mu}_j \hat{\mu}_{j'}) - E(\hat{\mu}_j) E(\hat{\mu}_{j'}) \\ &= \left(1 - \frac{E\{\text{Card}(S_j^c \cap S_{j'}^c)\}}{n_2^2}\right) \text{Cov}\left\{(\bar{X}_{S_j} - X_i)^2, (\bar{X}_{S_{j'}} - X_{i'})^2\right\} \\ &\quad + \frac{E\{\text{Card}(S_j^c \cap S_{j'}^c)\}}{n_2^2} \text{Cov}\left\{(\bar{X}_{S_j} - X_i)^2, (\bar{X}_{S_{j'}} - X_i)^2\right\}. \end{aligned}$$

The key is the computation of the expectation of the product $(\bar{X}_{S_j} - \mu)^2 (\bar{X}_{S_{j'}} - \mu)^2$.

$$\begin{aligned} \bar{X}_{S_j} - \mu &= \frac{1}{n_1} \sum_{l=1}^{n_1} (X_{l,S_j} - \mu) \\ &= \frac{1}{n_1} \left\{ \sum_{l=1}^{Y} (X_l - \mu) + \sum_{l'=Y+1}^{n_1} (X_{l',S_j} - \mu) \right\}, \end{aligned}$$

where $Y = \text{Card}(S_j \cap S_{j'})$. That is, the first part of the sum corresponds to the common elements of the training sets S_j and $S_{j'}$, and the second part consists of the elements that are training-set specific. Then

$$\begin{aligned} E\left\{(\bar{X}_{S_j} - \mu)^2 (\bar{X}_{S_{j'}} - \mu)^2\right\} &= \frac{1}{n_1^4} E\left\{ \left[\sum_{l=1}^{Y} (X_l - \mu) + \sum_{l'=Y+1}^{n_1} (X_{l',S_j} - \mu) \right]^2 \right. \\ &\quad \left. \times \left[\sum_{l=1}^{Y} (X_l - \mu) + \sum_{k=Y+1}^{n_1} (X_{k,S_{j'}} - \mu) \right]^2 \right\} \\ &= \frac{1}{n_1^4} E\left\{ \left[\sum_{l=1}^{Y} (X_l - \mu) + \sum_{l'=1}^{n_1 - Y} (X_{l',S_j} - \mu) \right]^2 \right. \\ &\quad \left. \times \left[\sum_{l=1}^{Y} (X_l - \mu) + \sum_{k=1}^{n_1 - Y} (X_{k,S_{j'}} - \mu) \right]^2 \right\} \end{aligned}$$

Notice that the above sums are random. Therefore, the expectation is obtained using a double-expectation argument, first by conditioning on Y and then by taking the expectation with respect to the distribution of Y. This completes the proof. □

Proposition 3.3. *Let $X_1, X_2, ..., X_n$ be independent and identically distributed random variables from a $N(\mu, \sigma^2)$ distribution. Then*

$$Var(\hat{\mu}_J) = \frac{1}{J}\frac{2\sigma^4}{n_1^2 n_2}\left[(n_1+1)^2 + n_2 - 1\right] + \frac{J-1}{J}\frac{2\sigma^4}{n_1^2 n^2 (n-1)}\left[n_1^2(n+1)^2 + (n-n_1)^2 - 4n_1\right].$$

Proof. The variance of $\hat{\mu}_J$ is obtained by substituting the expressions for $\mathrm{Var}(\hat{\mu}_j)$ and $\mathrm{Cov}(\hat{\mu}_j, \hat{\mu}_{j'})$ into (5). □

Notice that the variance of $\hat{\mu}_J$ explicitly depends on the variance of the underlying population and the size of the training and test sets.

Similar expressions can be obtained for the case where the underlying population is a mixture of normals. If the distribution is given by $pN(\mu_1, \sigma_1^2) + (1-p)N(\mu_2, \sigma_2^2)$, then computing the exact variance of $\hat{\mu}_J$ under square error loss in the problem of prediction of the mean relies on using the moment generating function for a mixture of normals to obtain higher-order moments, as well as higher-order moments of the distribution of order moments of the distribution of Y and Y^*. The point is that the moments of the underlying population influence the variance of $\hat{\mu}_J$.

4. A Simulation Study

We now describe simulation results that illustrate the behavior of the variance estimators under different distributional models and sample sizes and that exemplify the contribution of each component in the computation of the estimator. The simulation study mainly focuses on prediction of the mean, partly because of its computational simplicity, and partly because the relatively easier mathematical analysis allows a clear elucidation of the contributions to the variance estimator made by the training and test sets. Additionally, we present a smaller study in the regression case to indicate mainly the feasibility of computing the moment approximation (MA) estimator. The loss function we use throughout this section is square error loss. All the experiments were done in R, version 2.9.1.

4.1. *Normal distribution*

We compare the two Nadeau-Bengio estimators and the MA estimator of the variance of the cross-validation estimator of the generalization error when predicting the mean of a normally distributed data sample. Each experiment consists of 100 samples of size $n = 100$, 500, or 1000 generated from $N(0,1)$. We apply the two schemes of cross-validation, completely random and k-fold. In order to identify the effect of the size of the training and test sets on the variance estimator in the case of completely random cross-validation, we select various values for the ratio n_2/n_1, i.e. $\{0.1, 0.2, 0.3, 0.4, 0.5, 0.6, 0.7, 0.8, 0.9, 1\}$, where $n_2 = n - n_1$. The parameter J is chosen to be equal to 15 and the constant M is chosen to be equal to 10, as suggested by Nadeau and Bengio (2003). For k-fold cross-validation, we take the number of folds, k, to be 4, 5, or 10. In these experiments, we also compute the bias of the estimators.

In the case of the completely random MA estimator, we estimated the parameter σ^2 using two different schemes. The first (denoted as MA) used the sample variance computed on the entire data set. The second (denoted as MA-training) computed the sample variance using the training set data S_j, $j = 1, 2, \ldots, J$. The results presented in Table 1 indicate that MA-training has smaller bias and mean squared error (MSE) than the two Nadeau-Bengio estimators, but slightly larger bias and MSE than the MA estimator. Moreover, the variance of MA-training is slightly larger than the variance of the MA estimator. Both MA estimators exhibit smaller overall MSE than the two Nadeau-Bengio estimators.

Figure 1 plots the square root of the variance estimators as functions of the ratio n_2/n_1 for each sample size n. The plot also includes the exact

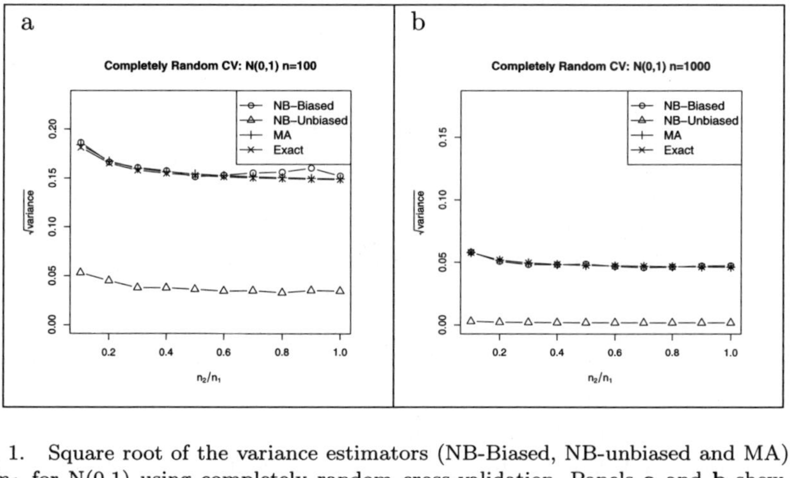

Fig. 1. Square root of the variance estimators (NB-Biased, NB-unbiased and MA) vs. n_2/n_1 for N(0,1) using completely random cross-validation. Panels **a** and **b** show the case for $n = 100$ and 1000, respectively, where n is the universe size, n_1 is the size of the training set, and n_2 is the size of the test set.

Table 1. Results for the NB-biased, NB-unbiased and MA (computed in two ways) estimators of the variance of the complete random CV estimator of the generalization error when data come from N(0,1) and $n = 100$.

		NB-biased			
n_2/n_1	Exact Var	Mean	SE	Bias	MSE
1	0.02206	0.02313	0.01242	0.00107	0.00016
0.8	0.02234	0.02433	0.01426	0.00198	0.00021
0.5	0.02330	0.02288	0.01325	−0.00042	0.00018
0.1	0.03287	0.03468	0.02507	0.00181	0.00063
		NB-Unbiased			
1	0.02206	0.00118	0.00172	−0.02088	0.00044
0.8	0.02234	0.00107	0.00170	−0.02128	0.00046
0.5	0.02330	0.00131	0.00223	−0.02199	0.00049
0.1	0.03287	0.00283	0.00410	−0.03004	0.00092
		MA			
1	0.02206	0.02225	0.00678	0.00019	0.00005
0.8	0.02234	0.02263	0.00689	0.00029	0.00005
0.5	0.02330	0.02374	0.00723	0.00044	0.00005
0.1	0.03287	0.03381	0.01030	0.00094	0.00011
		MA-training			
1	0.02206	0.02255	0.00702	0.00050	0.00005
0.8	0.02234	0.02332	0.00745	0.00098	0.00006
0.5	0.02330	0.02401	0.00757	0.00071	0.00006
0.1	0.03287	0.03393	0.01035	0.00105	0.00011

variance of $\hat{\mu}_J$ obtained in Proposition 3.3. As the ratio n_2/n_1 approaches 1, the size of the test set approaches the size of the training set, and the variance decreases. Notice that when the sample size, n, increases, $\text{Var}(\hat{\mu}_J)$, its estimators, and their bias decrease. It is seen that the two MA estimators (only MA is shown in the figure) have the smallest bias. NB-biased also has small bias and closely follows the exact variance as the sample size increases. However, the NB-biased estimator remains much more variable than the MA estimator, with standard error being two times larger than that of the MA estimator.

Cross-validation error estimation has been quite popular for microarray classification. A key point in microarray classification is that the sample size is typically small. It has been reported in the literature (Braga-Neto and Dougherty, 2004; Fu *et al.*, 2005) that cross-validation estimators are especially problematic in small-sample settings and typically exhibit high variance. To that end, we studied the behavior of the three estimators in the case of small samples. Table 2 contains the results from these experiments for $n = 16$ and $n = 20$.

Table 2. Results for the NB-biased, NB-unbiased, and MA estimators of the variance of the complete random CV estimator of the generalization error when data come from N(0,1).

			NB-biased			
n	n_2/n_1	Exact Var	Mean	SE	Bias	MSE
16	0.23	0.19064	0.22750	0.43804	0.03686	0.19324
16	0.33	0.17841	0.19832	0.24969	0.01991	0.06274
16	0.6	0.16672	0.19100	0.32094	0.02428	0.10359
16	0.78	0.16194	0.20674	0.25117	0.04480	0.06510
16	1	0.16194	0.21709	0.29869	0.05515	0.09226
20	0.25	0.14537	0.15730	0.18877	0.01193	0.03578
20	0.33	0.13818	0.13737	0.17310	−0.00081	0.02997
20	0.43	0.13017	0.15047	0.17052	0.02030	0.02949
20	0.67	0.12611	0.14000	0.15749	0.01390	0.02500
20	1	0.12491	0.15289	0.14316	0.02798	0.02128
			NB-unbiased			
16	0.23	0.19064	1.08905	5.36968	0.89841	29.64060
16	0.33	0.17841	0.61842	3.56404	0.44000	12.89595
16	0.6	0.16672	0.57828	3.09017	0.41157	9.71851
16	0.78	0.16194	0.49261	2.02057	0.33067	4.19203
16	1	0.16194	0.50050	1.68959	0.33855	2.96932
20	0.25	0.14537	0.44629	2.58903	0.30092	6.79364
20	0.33	0.13818	0.37966	2.07095	0.24148	4.34713
20	0.43	0.13017	0.30901	1.79042	0.17884	3.23760
20	0.67	0.12611	0.21503	0.83141	0.08892	0.69914
20	1	0.12491	0.27724	0.94875	0.15233	0.92333
			MA			
16	0.23	0.19064	0.22229	0.20678	0.03165	0.04376
16	0.33	0.17841	0.20540	0.19107	0.02699	0.03724
16	0.6	0.16672	0.18617	0.17318	0.01945	0.03037
16	0.78	0.16194	0.17175	0.15977	0.00980	0.02562
16	1	0.16194	0.17175	0.15977	0.00980	0.02562
20	0.25	0.14537	0.16493	0.12627	0.01956	0.01633
20	0.33	0.13818	0.15547	0.11902	0.01730	0.01447
20	0.43	0.13017	0.14371	0.11002	0.01354	0.01229
20	0.67	0.12611	0.13546	0.10370	0.00935	0.01084
20	1	0.12491	0.13153	0.10070	0.00662	0.01018

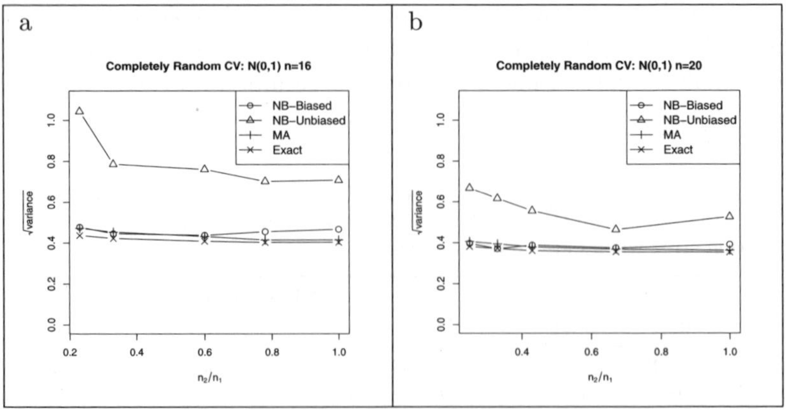

Fig. 2. Square root of the variance estimates (NB-Biased, NB-unbiased, and MA) vs. n_2/n_1 for $N(0,1)$ using completely random cross-validation on small data sets. Panels **a** and **b** show the case for $n = 16$ and 20, respectively, where n is the universe size, n_1 is the size of the training set, and n_2 is the size of the test set.

Observe that the NB-biased and MA estimators behave similarly in terms of bias, with the MA estimator exhibiting better bias performance when the ratio n_2/n_1 becomes large. The MA estimator still exhibits approximately half the standard error exhibited by the NB-biased estimator when $n = 16$, but both standard errors are considerably increased in small samples. Both estimators perform better than the NB-unbiased estimator. A word of explanation is in order here. Nadeau and Bengio established that their second estimator, NB-unbiased, is indeed unbiased when the size of the training set is $\lfloor n/2 \rfloor - n_2$, which is not the case in our computations. However, we kept the name here to indicate the potential of the estimator to be unbiased. Figure 2 illustrates the above results, showing that the MA estimator still closely tracks the exact variance.

We present results relating to the performance of the k-fold cross-validation estimators in Tables 3 and 4. Observe that the value of both Nadeau-Bengio estimators increases with k and decreases as n increases, whereas the value of the MA estimator decreases as both k and n increase. The MA estimator has the smallest bias and this bias is negative, indicating that the estimator slightly underestimates the variance of $\hat{\mu}_J$. Moreover, the MA estimator exhibits the smallest standard error and hence the smallest MSE. These results hold true for the case of the small samples as well (see table 4). Again, the MA estimator exhibits the smallest standard error and bias, and therefore the smallest MSE. However, the variance is much larger than the variance exhibited in the large samples.

Table 3. Mean, bias, variance, and mean squared error (MSE) of the NB-biased, NB-unbiased, and MA estimators for the variance of the k-fold CV estimator ($k = 4$, 5, and 10) of the generalization error when data come from $N(0,1)$.

		NB-biased			
n	k	Mean	SE	Bias	MSE
100	4	0.03685	0.03358	0.01338	0.00131
100	5	0.04845	0.06487	0.02545	0.00486
100	10	0.21608	0.28397	0.19273	0.11778
500	4	0.00591	0.00473	0.00150	0.00002
500	5	0.00629	0.00584	0.00201	0.00004
500	10	0.01142	0.00849	0.00726	0.00012
1000	4	0.00299	0.00255	0.00080	0.00001
1000	5	0.00275	0.00179	0.00063	0.00000
1000	10	0.00388	0.00228	0.00182	0.00001
		NB-unbiased			
100	4	0.05791	0.02781	0.03444	0.00196
100	5	0.05920	0.03264	0.03619	0.00238
100	10	0.07092	0.04055	0.04757	0.00391
500	4	0.00795	0.00375	0.00354	0.00003
500	5	0.00821	0.00400	0.00393	0.00003
500	10	0.00905	0.00443	0.00488	0.00004
1000	4	0.00440	0.00201	0.00222	0.00001
1000	5	0.00443	0.00205	0.00231	0.00001
1000	10	0.00453	0.00206	0.00248	0.00001
		MA			
100	4	0.01845	0.00840	−0.00502	9.576E-05
100	5	0.01813	0.00781	−0.00488	8.485E-05
100	10	0.01510	0.00635	−0.00825	0.000108
500	4	0.00391	0.00065	−0.00050	6.681E-07
500	5	0.00387	0.00064	−0.00041	5.750E-07
500	10	0.00376	0.00064	−0.00040	5.724E-07
1000	4	0.00196	0.00021	−0.00022	9.397E-08
1000	5	0.00196	0.00022	−0.00016	7.267E-08
1000	10	0.00193	0.00021	−0.00012	5.961E-08

Table 4. Mean, bias, variance, and mean squared error (MSE) of the NB-biased, NB-unbiased, and MA estimators for the variance of the k-fold CV estimator (k = 4, 5, and 10) of the generalization error when data come from $N(0,1)$.

		NB-biased			
n	k	Mean	SE	Bias	MSE
16	2	0.32738	0.55674	0.08102	0.31653
16	4	0.52542	0.72855	0.31683	0.63116
20	2	0.35071	0.75263	0.16308	0.59304
20	4	0.29608	0.39977	0.14139	0.17981
20	5	0.62942	0.78053	0.47155	0.83158
		NB-unbiased			
16	2	0.69130	0.78136	0.44495	0.80850
16	4	0.86948	0.85405	0.66089	1.16618
20	2	0.47118	0.59807	0.28355	0.43809
20	4	0.66127	0.60170	0.50658	0.61866
20	5	0.63301	0.82273	0.47514	0.90264
		MA			
16	2	0.10879	0.15050	−0.13757	0.04157
16	4	0.05725	0.05642	−0.15134	0.02609
20	2	0.09130	0.09460	−0.09633	0.01823
20	4	0.06700	0.12593	−0.08770	0.02355
20	5	0.03861	0.04422	−0.11926	0.01618

4.2. *Other distributions*

We now investigate the performance of the NB-biased, NB-unbiased, and MA estimators when data are generated from distributions other than the normal. We generated data from a mixture of normal distributions, $(1-\varepsilon)N(0,1)+\varepsilon N(0,9)$, where $\varepsilon \in \{i \times 0.05, i = 1, \ldots, 10\}$. Data were also generated from a Gamma(1,2) distribution. A hundred different data sets were generated, with the number of samples in each set, n, equal to 100, 500, or 1000. In order to identify the effect of the size of training and test sets on the variance estimator, we used different values for the ratio n_2/n_1. As in the previous section, the constants $J = 15$ and $M = 10$, as suggested by Nadeau and Bengio (2003), were used.

Figure 3 presents plots of the three estimators as a function of the different mixing proportions ε and universe sizes for the completely random cross-validation case. Panels **a, c,** and **e** correspond to universe size $n = 100$;

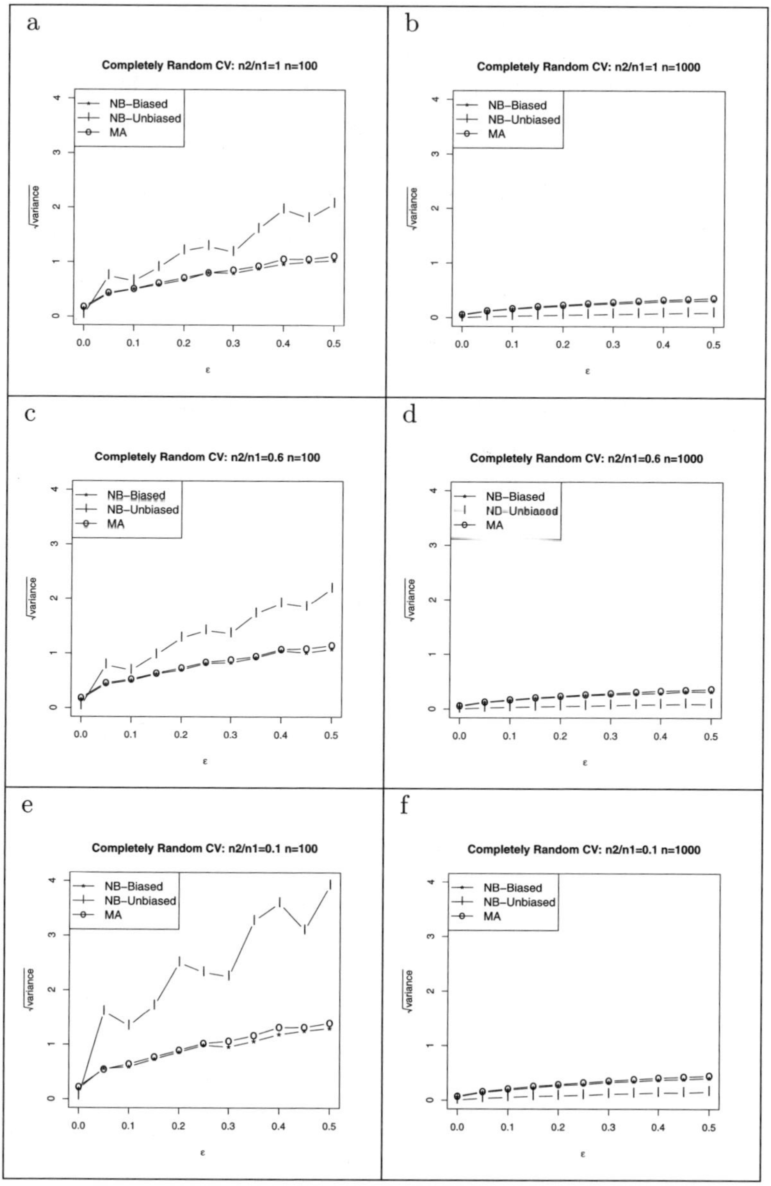

Fig. 3. Square root of the variance estimators (NB-Biased, NB-unbiased, and MA) vs. ε for the data distribution $(1-\varepsilon)N(0,1)+\varepsilon N(0,9)$ using completely random cross-validation. Panels **a** and **b** show the case $n_2/n_1 = 1$ for $n = 100$ and 1000, respectively. Panels **c** and **d** show the case $n_2/n_1 = 0.6$ for $n = 100$ and 1000, respectively. Panels **e** and **f** show the case $n_2/n_1 = 0.1$ for $n = 100$ and 1000, respectively.

in panel **a** the sizes of the training and test sets are equal, while in the other cases the sizes of the test set are $n_2 = 37$ (panel **c**) and $n_2 = 10$ (panel **e**). Notice that the curve of the NB-unbiased estimator is higher than the other two curves, which follow each other closely. As the universe size n becomes larger, the situation changes; the curve that corresponds to the NB-unbiased stays below the other two curves that still follow each other. All values increase with increasing mixing proportions.

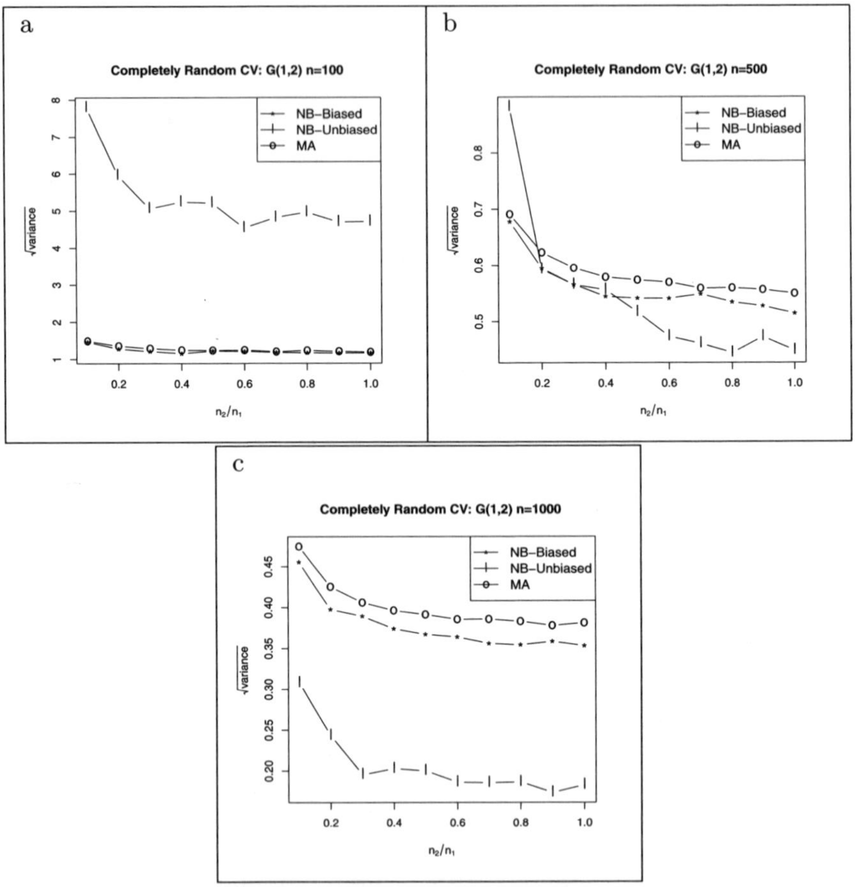

Fig. 4. Square root of the variance estimators (NB-Biased, NB-unbiased, and MA) for Gamma(1,2) using completely random cross-validation. Panels **a**–**c** show the cases for $n = 100$, 500, and 1000, respectively where n is the universe size, n_1 is the size of the training set, and n_2 is the size of the test set.

We also compute the variance estimates when data come from a Gamma(1,2) distribution. Figure 4 presents the variance estimator curves as a function of the ratio n_2/n_1 and the universe size. The results here are similar to those from a normal mean 0, variance 1 dataset, in that the NB-unbiased estimator produces a standard error curve that is consistently below the standard error curves of both the NB-biased and MA estimators.

Figure 5 presents the square roots of the values of these estimates as a function of the mixing proportion ε for universe sizes $n = 100$ and 1000 and for $k = 4$, 5, and 10. In the case of $n = 100$, the value of all estimators increases as the mixing proportion increases. Moreover, when $k = 4$ or 5, clearly the highest curve is provided by the NB-unbiased estimator, followed by the NB-biased curve and the MA curve. When the number of folds equals 10, the NB-unbiased and NB-biased curves are very close to each other, much more so in this case than in either $k = 4$ or 5 folds. The same patterns described for $n = 100$ hold for $n = 500$ and 1000 (results are only shown for $n = 1000$). The MA curves are below the NB-biased and NB-unbiased curves for 4, 5, and 10 folds. The higher values exhibited by the NB-biased estimator as compared with the MA estimator may be due to the fact that $\hat{\rho}$ is overestimating the correlation, while the covariance terms needed to compute MA are directly estimated from the data. The even higher values of the NB-unbiased estimator may be due to the fact that splitting the sample reduces the size of the training and test sets and thus produces a more variable estimator.

Table 5. Variance estimates of k-fold CV estimator ($k = 4$, 5, and 10) of the generalization error. Data distribution is Gamma(1,2).

Size (n)	k	NB-bias (SE)	NB-unb. (SE)	MA (SE)
100	4	2.0433 (2.7361)	3.5061 (4.2573)	1.2723 (1.4386)
100	5	1.6625 (2.1073)	3.4621 (3.9966)	1.1727 (1.3235)
100	10	3.8728 (4.9685)	4.1573 (4.7725)	1.0030 (1.1777)
500	4	0.3406 (0.2845)	0.5781 (0.4475)	0.2629 (0.1656)
500	5	0.3328 (0.3068)	0.5926 (0.4545)	0.2607 (0.1634)
500	10	0.3527 (0.2453)	0.6189 (0.4711)	0.2518 (0.1625)
1000	4	0.1856 (0.1670)	0.2741 (0.1662)	0.1301 (0.0643)
1000	5	0.1529 (0.1297)	0.2743 (0.1655)	0.1295 (0.0647)
1000	10	0.1375 (0.0911)	0.2787 (0.1713)	0.1268 (0.0638)

Table 5 presents results when the underlying data generating mechanism is Gamma(1,2). Similar observations hold here in that the MA estimates

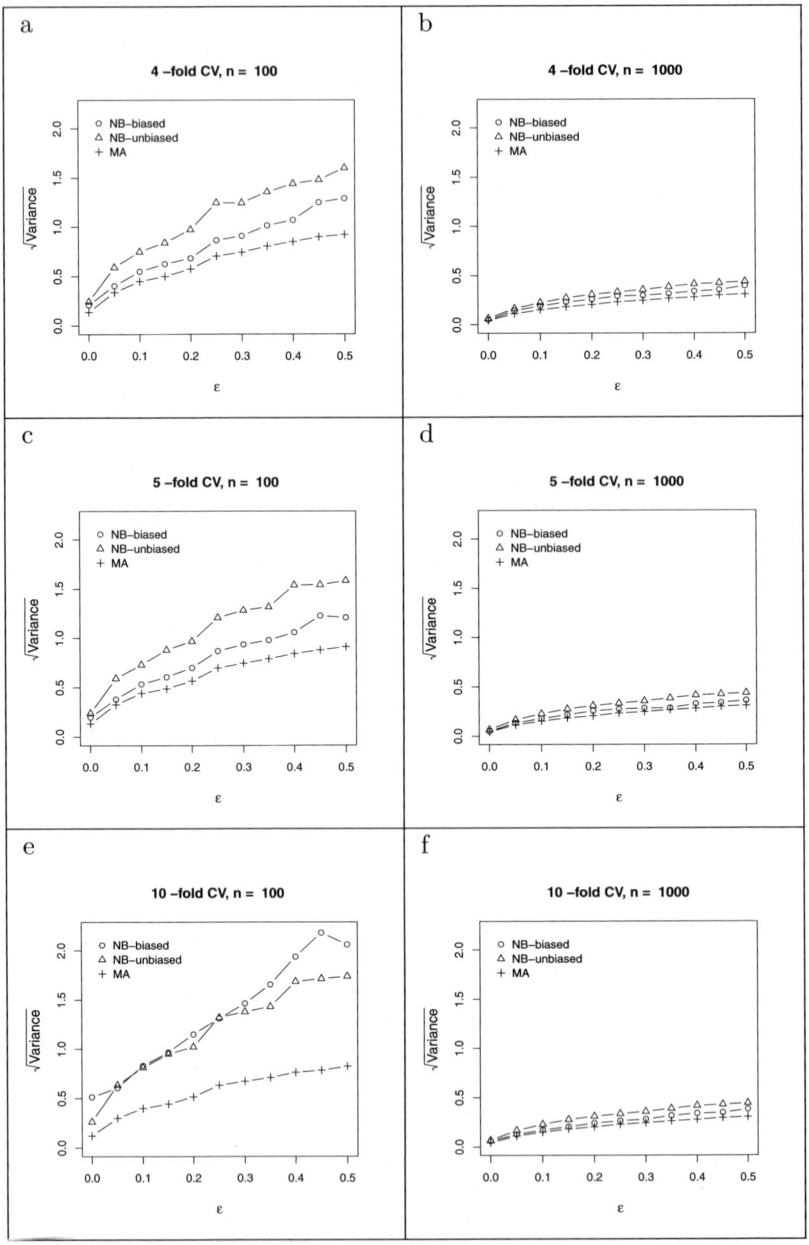

Fig. 5. Square roots of the mean values of the NB-biased, NB-unbiased, and MA estimators of the variance of the k-fold CV estimator ($k = 4$, 5, and 10) of the generalization error when data come from $(1-\varepsilon)N(0,1) + \varepsilon N(0,9)$ and $n = 100$ and 1000.

are smaller than the corresponding NB-biased and NB-unbiased estimates. Moreover, notice that the moment approximation estimates have smaller standard errors than their other counterparts.

4.3. *Linear Regression*

To see whether these results hold for other than prediction of the mean cases, we carried out a small simulation experiment where the prediction model was a linear regression model.

Data were generated in R according to the model $y_i = 2+3x_{1i}+3x_{2i}+\varepsilon_i$, $i = 1, 2, \ldots, n$. The universe sizes were 200, 500, and 1000. The errors, ε_i, were generated independently from a standard normal distribution. The explanatory variables were generated once, first from a bivariate normal distribution with mean vector 0 and identity covariance matrix and then from a bivariate normal with mean vector 0 and covariance given by

$$\Sigma = \begin{pmatrix} 1 & 0.2 \\ 0.2 & 1 \end{pmatrix}.$$

Figure 6 presents the standard deviation curves of the two estimators suggested by Nadeau and Bengio (2003), for both the independent and dependent cases, for different sizes of the universe and as functions of the ratio of the test and training set sizes. The curve of the NB-unbiased estimator does exhibit less smoothness than the curve of the NB-biased estimator, particularly when the universe size is not large. This is because the computation of the estimator uses smaller sizes of training and test sets due to the splitting of the data universe.

Figure 7 presents the MA estimator standard error curve for $n = 200$ and as a function of n_2/n_1. The generated trend observed in the case of prediction of the mean holds here also. That is, as the test set size increases, the variance decreases. This can also be seen from Table 6, which shows the values of the MA estimator and its associated standard error.

The above results indicate the behavior of the three variance estimators in a variety of data generating models and as functions of the sizes of the training and test sets. The advantage of the two Nadeau-Bengio estimators lies in their computational simplicity. On the other hand, the moment approximation estimator, although computationally not as efficient as the other two, has smaller variance and explicitly accounts for the dependence introduced by the overlap in the training and test sets.

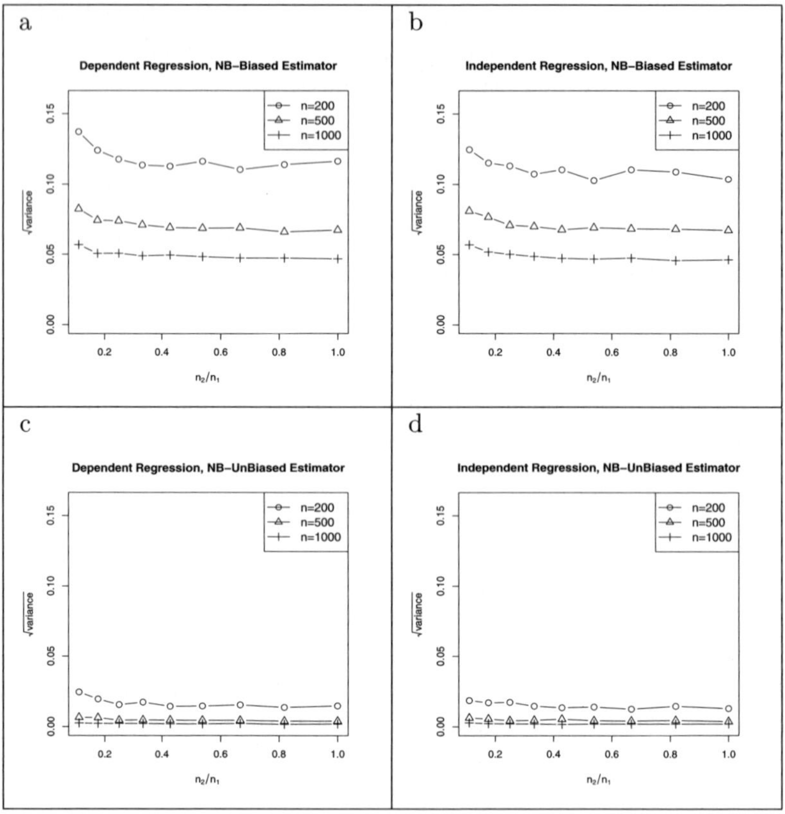

Fig. 6. Square roots of the variance estimators (NB-Biased, NB-unbiased, and MA) vs. n_2/n_1 in the case of linear regression using completely random cross-validation. Values were generated from the function $y = 2 + 3x_1 + 3x_2 + \varepsilon$ with ε being $N(0,1)$.

Table 6. MA variance estimates of the CV estimator of the generalization error in the Linear Regression model.

Size (n)	n_1	n_2	MA	SE(MA)
200	180	20	0.01607	0.00390
200	160	40	0.01281	0.00307
200	150	50	0.01217	0.00294
200	120	80	0.01118	0.00260
200	100	100	0.01087	0.00260

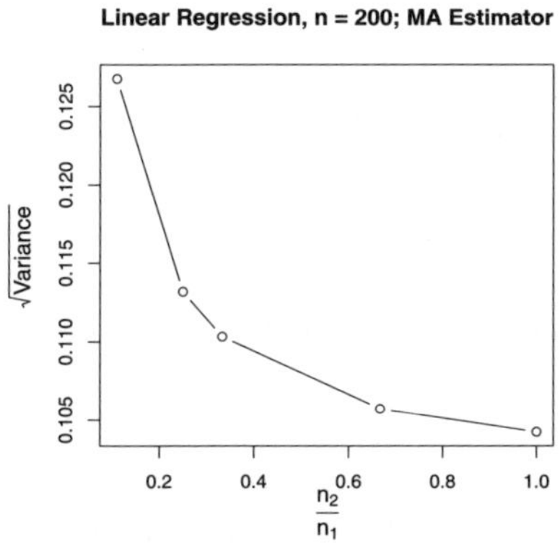

Fig. 7. Square roots of the mean values of the MA estimator in the Linear Regression Model $y = 2 + 3x_1 + 3x_2 + \varepsilon$ with ε being $N(0, 1)$ and with a sample size of 200.

5. Conclusions

At this point it is instructive to note that estimation of the generalization error can also be done analytically. Hastie *et al.* (2001) define the training error as the average loss over the training sample, given as

$$\frac{1}{n_1}\sum_{i=1}^{n_1} L(y_i, \hat{f}(x_i)).$$

The training error indicates how well a statistical or algorithmic rule fits the training data. It is not a good estimate of the generalization error because it consistently decreases when more complex rules are used, and thereby overfitting is observed. Typically, the training error rate is smaller than the generalization error; this is because we use the same data to construct the rules and to assess its error. Efron (1983, 1986) shows that the generalization error is a sum of two components, the training error and the optimism bias. The optimism bias is a penalty term related to the complexity of the model or rule. Efron (1986, 2004) shows that for loss functions that belong in the q-class, the optimism corresponds to a covariance penalty. Efron (2004) shows how to estimate these covariance penalties and illustrates their relationship to cross-validation. This work may facilitate further insights into the problem of variance estimation of the cross-validation estimators of the generalization error and is currently the subject of further work by the authors.

In this paper, we discuss the estimation of variance of cross-validation estimators of the generalization error. We study the performance of three estimators, first focusing on the simple, illuminating case of error in predicting the mean and further extending it to the regression case. An alternative method for comparing classifiers is via the comparison of widths of associated confidence intervals. McLachlan (1975, 1992) derived a method for obtaining a confidence interval for the actual error rate, while Dobbin (2009) derived a method for constructing confidence intervals for the error rate of a prediction rule that can be used when the sample size n is small but the number of predictors p is large ($p \gg n$). In this paper, we do not study confidence intervals for the estimate of the generalization error; instead, we study, in detail, estimators of its variance.

Our results show that the variance curves associated with the NB-unbiased estimator of variance are substantially different than the variance curves associated with the other two estimators when the completely random cross-validation estimator of the generalization error is used. When the k-fold cross-validation estimator is used, the NB-unbiased estimator exhibits this behavior only for smaller universe sizes ($n = 100$), becoming closer to the other two in larger universe sizes. In all cases, the NB-unbiased estimator performed worse than the NB-biased and MA estimators.

The results illustrate the effect of the training and test set size on the estimators of the variance. In particular, the moment approximation estimator clearly accounts for the dependence introduced by the overlap between training and test sets and estimates the relevant covariance terms from the data. The MA estimator also exhibits a substantially smaller standard error in comparison to the other two estimators. Further work on this topic is needed, to fully understand the behavior of cross-validation as a model selection or model assessment tool.

Acknowledgments

M. M. is supported by NSF DMS-0504957; the Center for Biologics Evaluation and Research, Office of Biostatistics and Epidemiology, of the Food and Drug Administration is also gratefully acknowledged. A. S. is supported by the National Library of Medicine, Informatics Training Program (5 T15 LM007079-15).

References

Bengio, Y. and Grandvalet, Y. (2004). No unbiased estimator of the variance of k-fold cross-validation, *Journal of Machine learning Research* **5**, pp. 1089–1105.

Braga-Neto, U. M. and Dougherty, E. R. (2004). Is cross-validation valid for small sample microarray clasification? *Bioinformatics* **20**, pp. 374–380.

Breiman, L. (1992). The little bootstrap and other methods for dimenstionality selection in regression: x-fixed prediction error, *Journal of the American Statistical Association* **87**, pp. 738–754.

Dietterich, J. G. (1998). Approximate statistical tests for comparing supervised classification learning algorithms, *Neural Computations* **10**, pp. 1895–1923.

Dobbin, K. K. (2009). A method for constructing a confidence bound for the actual error rate of a prediction rule in hiogh dimensions, *Biostatistics* **10**, pp. 282–296.

Efron, B. (1983). Estimating the error of a prediction rule: Improvement on cross-validation, *Journal of the American Statistical Association* **78**, pp. 316–331.

Efron, B. (1986). How biased is the apparent error rate of a prediction rule? *Journal of the American Statistical Association* **81**, pp. 461–470.

Efron, B. (2004). The estimation of the prediction error: Covariance penalties and cross validation (with discussion), *Journal of the American Statistical Association* **99**, pp. 619–642.

Fu, W. J., Carroll, R. J. and Wang, S. (2005). Estimating missclassification error with small samples via bootstrap cross-validation, *Bioinformatics* **21**, pp. 1979–1986.

Hastie, T., Tibshirani, R. and Friedman, J. (2001). *The elements of statistical Learning. Data mining, Inference and Prediction*, Springer Series in Statistics (Springer Verlag, New York).

Markatou, M., Dimova, R. and Sinha, A. (2008). Analysis and estimation of variance of cross validation estimators of the generalization error: A short review, in *Proceedings of Frontiers in Applied and Computational Mathematics, 2008* (World Scientific Publishing Company), pp. 206–212.

Markatou, M., Tian, H., Biswas, S. and Hripcsak, G. (2005). Analysis of variance of cross validation estimators of the generalization error, *Journal of Machine Learning Research* **6**, pp. 1127–1168.

McLachlan, G. J. (1972). An asymptotic expansion for the variance of the errors of misclassification of the linear discriminant function, *Australian Journal of Statistics* **14**, pp. 68–72.

McLachlan, G. J. (1974). An asymptotic expansion of the expectation of the estimated error rate in discriminant analysis, *Australian Journal of Statistics* **15**, pp. 210–214.

McLachlan, G. J. (1975). Confidence intervals for the conditional probability of misallocation in discriminant analysis, *Biometrics* **32**, pp. 161–167.

McLachlan, G. J. (1976). The bias of the apparent error rate in discriminant analysis, *Biometika* **63**, pp. 239–244.

McLachlan, G. J. (1992). *Discriminant Analysis and Statistical Pattern Recognition* (John Wiley and Sons, New York).

Nadeau, C. and Bengio, Y. (2003). Inference for the generalization error, *Machine Learning* **52**, pp. 239–281.

Tian, H. (2006). *Variance estimation of the cross validation estimator of the generalization error*, Ph.D. thesis, Dept. of Biostatistics, Columbia University.

Estimation of Hazard Functions with Shape Restrictions Using Regression Splines

Mary Meyer

Department of Statistics, Colorado State University
Fort Collins, CO 80523, USA
E-mail: meyer@stat.colostate.edu

Desale Habtzghi

Department of Statistics, University of Akron
Akron, OH 44325, USA
E-mail: dh52@uakron.edu

In the estimation of distributions with time-to-event data, it is often natural to impose shape and smoothness constraints on the hazard function. Systems that fail because of wearing out might be assumed to have monotone hazard, or perhaps monotone-convex. Organ transplant failures are often assumed to have convex or bathtub-shaped hazard function. In this paper we present estimates that maximize the likelihood over a set of shape-restricted regression splines. Right censoring is a simple extension. The methods are applied to real and simulated data sets to illustrate their properties and to compare with existing nonparametric estimators.

Keywords: Constrained maximum likelihood; Convex; Convex optimization; Failure rate; Monotone; Time to event

1. Introduction and Statement of the Problem

In "time to event" data, the hazard function is a convenient measure of the risk of event at a given time, given that it has not happened yet. Specifically, if T is a positive random variable, its hazard function is

$$h(t) = \lim_{\Delta t \to 0} \frac{P[t \leq T < t + \Delta t | T \geq t]}{\Delta t}.$$

If T is continuous with density f, cumulative distribution F and $S(t) = 1 - F(t)$, then

$$h(t) = \frac{f(t)}{S(t)}.$$

While the distribution may be completely specified by any of the functions f, S, and h, the hazard function is considered to be most informative about the underlying mechanism of failure, and it is often natural to impose a shape restriction. If we assume that the hazard function is increasing, this simply means that the risk of failure, given that the unit is still functioning, increases with time. This is appropriate for modeling parts that fail because they gradually wear out. A convex or "bathtub-shaped" hazard rate may be appropriate for systems that have high early failure rates that decrease to a minimum, then begin to rise again. Often organ transplant failure is believed to have a convex hazard rate. A hazard function could be assumed to be both increasing and convex if the wear-out in the system is thought to be more dramatic as time increases.

The density is related to the hazard function by

$$f(t) = h(t) \exp\left[-\int_0^t h(u)du\right].$$

This expression may be used to obtain maximum likelihood estimates of the distribution based on assumptions concerning the hazard function. In the next section, we outline a method for maximum likelihood estimation of the hazard function under shape restrictions, where the estimates are piecewise linear with knots at the data. In section 3, a smoothed version is presented, using polynomial splines. Knots choices are discussed in section 4, and comparisons with kernel methods can be found in section 5. An extension to data with right censoring is given in section 6, with some examples and discussion.

2. Constrained Maximum Likelihood Estimation of h

Let $0 = t_0 < t_1 < t_2 < \cdots < t_n$ be realizations of failure times. The log-likelihood function can be written in terms of the values of the hazard function at the observed times as

$$\ell(h) = \sum_{i=1}^{n} \left\{\log[h(t_i)] - \int_0^{t_i} h(u)du\right\}.$$

Let $\theta_i = h(t_i)$ for $i = 1, \dots, n$ and $I_i = \int_{t_{i-1}}^{t_i} h(u)du$. Then

$$\ell(\boldsymbol{\theta}) = \sum_{i=1}^{n} \log(\theta_i) - \sum_{i=1}^{n} (n - i + 1)I_i. \tag{1}$$

We will consider estimates in the form of polynomial splines. The various forms of the splines allow us to write the integral I_i as a linear combination of the θ_i with the coefficients depending on the observed times. We can find constants c_i (depending on the observed times) so that

$$\ell(\boldsymbol{\theta}) = \sum_{i=1}^{n} \log(\theta_i) - \sum_{i=1}^{n} c_i\theta_i, \tag{2}$$

where the c_i are functions of the observations t_i. The problem is to find $\boldsymbol{\theta} \in \mathbb{R}^n$ to maximize ℓ over the appropriate set determined by the shape and smoothness assumptions.

We first consider linear splines with knots at the observed failure times. If $h(t)$ is piecewise linear with discontinuities in slope at points (t_i, θ_i), then it is easy to write

$$I_i = \frac{1}{2}(\theta_i + \theta_{i-1})(t_i - t_{i-1})$$

for $i = 1, \ldots, n$. The value $\theta_0 = h(0)$ needs to be chosen in some appropriate way, in terms of θ_1 through θ_n, and the values of c_i can be computed using the trapezoid rule. For example, suppose we can assume that h is non-decreasing. Then choosing $\theta_0 = 0$, we have the constraint set

$$\mathcal{C} = \{\boldsymbol{\theta} : 0 \le \theta_1 \le \cdots \le \theta_n\}.$$

This can be seen to be a polyhedral convex cone in $\mathbb{R}^n$, as it is the intersection of the half-spaces defined by the individual inequality constraints. Note that the choice of $\theta_0 = 0$ will maximize the likelihood and retain the constraints for any sample of times. For these choices and $d_i = t_i - t_{i-1}$, we have

$$c_i = [(n - i + 1)d_i + (n - i)d_{i+1}]/2$$

for $i = 1, \ldots, n-1$ and $c_n = d_n/2$.

In general, we can implement any shape restrictions that can be written as

$$\mathcal{C} = \{\boldsymbol{\theta} : A\boldsymbol{\theta} \ge \mathbf{0}\},$$

and for mild conditions on the $m \times n$ constraint matrix A, we again have a polyhedral convex cone. For convexity constraints, the non-zero elements of the $(n-2) \times n$ matrix A are $A_{i,i} = x_{i+2} - x_{i+1}$, $A_{i,i+1} = x_{i+2} - x_i$, and $A_{i,i+2} = x_{i+1} - x_i$, for $i = 1, \ldots, n-2$. To combine increasing and convex constraints, we add the restriction $\theta_1 \le \theta_2$ to the set of convex constraints. For convex constraints, the choice of

$$\theta_0 = \max\left\{0, \theta_1 - \frac{\theta_2 - \theta_1}{t_2 - t_1}t_1\right\}$$

will preserve the convex shape and maximize the likelihood over the assumptions.

The constrained maximum likelihood estimator for $\boldsymbol{\theta}$ maximizes (2) over the cone $\mathcal{C}$ defined by the monotone or convex constraints. This is a convex programming problem, but it may be solved using iteratively re-weighted *projections* onto convex cones. The weighted cone projection problem may be stated as follows. Given a $\boldsymbol{y} \in \mathbb{R}^n$ and a vector of weights $\boldsymbol{w} \in \mathbb{R}^n$, the weighted projection $\hat{\boldsymbol{\theta}}$ is the value that minimizes over a cone $\mathcal{C}$ the expression

$$\sum_{i=1}^{n} w_i(y_i - \theta_i)^2. \tag{3}$$

Cone projection routines are well established; here we use the algorithm of Meyer (2008a), but also see Fraser and Massam (1989). These cone projection algorithms use the "edges" of the cone, which may be defined by the columns of $A(AA')^{-1}$ if A is full row rank, as is the case for the examples in this paper. Given edges $\boldsymbol{\delta}^1, \ldots, \boldsymbol{\delta}^m$ and linearly independent vectors $\boldsymbol{v}^1, \ldots, \boldsymbol{v}^{n-m}$ orthogonal to the row space of A, the cone may alternatively be specified as

$$\mathcal{C} = \left\{ \boldsymbol{\theta} \in \mathbb{R}^n : \boldsymbol{\theta} = \sum_{j=1}^{n-m} \alpha_j \boldsymbol{v}^j + \sum_{j=1}^{m} \beta_j \boldsymbol{\delta}^j, \text{ for } \beta_j \geq 0, \; j = 1, \ldots, m \right\}.$$

Faces of the cone, indexed by subsets $J \subseteq \{1, \ldots, m\}$, are defined as

$$\mathcal{F}_J = \left\{ \boldsymbol{\theta} \in \mathbb{R}^n : \boldsymbol{\theta} = \sum_{j=1}^{n-m} \alpha_j \boldsymbol{v}^j + \sum_{j \in J} \beta_j \boldsymbol{\delta}^j, \text{ for } \beta_j > 0, \; j \in J \right\}.$$

The solution $\hat{\boldsymbol{\theta}}$ lands on a face $\mathcal{F}_J$, and can be seen to be the weighted projection of $\boldsymbol{y}$ onto the linear space spanned by the $\boldsymbol{v}^j$, $j = 1, \ldots, n-m$, and the $\boldsymbol{\delta}^j$, $j \in J$. Different realizations fall on different faces; once J is determined, the weighted cone projection reduces to an ordinary linear projection. The algorithms proceed by determining the face J through a sequence of steps.

We maximize ℓ over $\mathcal{C}$ using iteratively-reweighted cone projection. Necessary and sufficient conditions for $\hat{\boldsymbol{\theta}}$ to maximize ℓ over $\mathcal{C}$ are the Kuhn-Tucker conditions:

$$\nabla\ell(\hat{\boldsymbol{\theta}})'\hat{\boldsymbol{\theta}} = 0 \text{ and } \nabla\ell(\hat{\boldsymbol{\theta}})'\boldsymbol{\theta} \leq 0 \text{ for all } \boldsymbol{\theta} \in \mathcal{C}.$$

In other words, the gradient of ℓ at the MLE $\hat{\boldsymbol{\theta}} \in \mathcal{C}$ is orthogonal to $\hat{\boldsymbol{\theta}}$ and makes obtuse angles with all other $\boldsymbol{\theta} \in \mathcal{C}$. For our ℓ, these conditions are:

$$\sum_{i=1}^{n} \left(\frac{1}{\hat{\theta}_i} - c_i \right) \hat{\theta}_i = 0 \ \text{ and} \tag{4}$$

$$\sum_{i=1}^{n} \left(\frac{1}{\hat{\theta}_i} - c_i \right) \theta_i \leq 0 \ \text{ for all } \ \boldsymbol{\theta} \in \mathcal{C}. \tag{5}$$

We can write conditions (4) and (5) to mimic the weighted least-squares conditions:

$$\sum_{i=1}^{n} \frac{c_i}{\hat{\theta}_i} \left(\frac{1}{c_i} - \hat{\theta}_i \right) \hat{\theta}_i = 0 \ \text{ and}$$

$$\sum_{i=1}^{n} \frac{c_i}{\hat{\theta}_i} \left(\frac{1}{c_i} - \hat{\theta}_i \right) \theta_i \leq 0 \ \text{ for all } \ \boldsymbol{\theta} \in C.$$

This suggests an iteratively reweighted least-squares algorithm, where $\hat{\boldsymbol{\theta}}^{(k+1)}$ is obtained from $\hat{\boldsymbol{\theta}}^{(k)}$ using a least-squares projection of "data" $y_i = 1/c_i$ onto the constraint cone with "weights" $w_i^{(k)} = c_i/\hat{\theta}_i^{(k)}$. The algorithm may be started from any point $\boldsymbol{\theta}^{(0)}$ inside C. The proof that the algorithm is guaranteed to converge to $\hat{\boldsymbol{\theta}}$ is similar to that of Meyer and Woodroofe (2004).

Some example fits are shown in Figure 1. A data set of size $n = 100$ was simulated from a Weibull density, for which the hazard function is both increasing and convex. The piecewise linear maximum likelihood spline estimators with shape restrictions are compared to the true hazard function.

3. Smoothed Constrained Maximum Likelihood Estimation

To smooth the estimates, we consider polynomial regression splines, with $k \ll n$ knots at roughly equal t-quantiles. For monotone hazard functions, we use piecewise quadratic splines, and for estimating a convex hazard function, we use piecewise cubic splines. Monotone regression splines were introduced by Ramsay (1988) and extended to the convex case by Meyer (2008b).

A set of piecewise quadratic basis functions is defined so that each basis function is continuous with continuous first derivative and is non-decreasing over the range of the sample. The set of basis functions has the property that the set of piecewise quadratic functions with the given knots is spanned by this set of basis functions and the constant function. Further, the linear

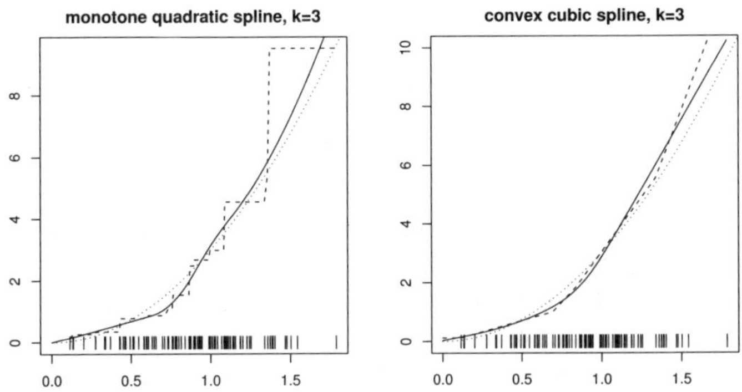

Fig. 1. Estimates of the hazard function, with data simulated from a Weibull density with shape parameter 3. The dotted curve is the true hazard function. The dashed curves are the piecewise linear spline estimates with knots at data, with monotone constraints at left and convex at right. The solid curves are the piecewise quadratic monotone spline estimate on the left and the piecewise cubic convex spline estimate on the right.

combination is non-decreasing if and only if the coefficients of the basis functions are non-negative. Similarly, for the convex case, we can define piecewise cubic basis functions so that the set of piecewise cubic functions with the given knots is spanned by this set and the constant and identity functions, and with the further property that the linear combination is convex if and only if the coefficients of the basis functions are non-negative. Note that for monotone splines, the quadratic degree is the highest for which we can find a set of basis functions with the property that the linear combination is non-decreasing if and only if the coefficients of the basis functions are non-negative. For cubic and higher, a linear combination may be increasing although some of the coefficients are positive, and hence the shape restrictions cannot be expressed by a set of linear constraints on the parameters. Similarly, for the convex case, the cubic splines are the highest degree for which the shape restrictions can be imposed through linear inequality constraints.

Examples of the monotone quadratic and convex cubic basis functions are shown in Figure 2 for three interior knots, marked with "X" along the bottom of the plot. The knots are chosen at equal time quantiles of the sample of failure times used to estimate the hazard function. A set of basis vectors $\boldsymbol{\delta}^j$, $j = 1, \ldots, m$, may be defined by evaluating the basis functions at the observed failure times; in the figure, the values are marked along the basis functions with dots. Let $\boldsymbol{\theta}$ be the vector of true hazard rate values at the observed times. To estimate $\boldsymbol{\theta}$, we maximize the likelihood over the polyhedral convex cone

$$\mathcal{C} = \left\{ \boldsymbol{\theta} \in \mathbb{R}^n : \boldsymbol{\theta} = \sum_{j=1}^{l} \alpha_j \boldsymbol{v}^j + \sum_{j=1}^{m} \beta_j \boldsymbol{\delta}^j, \text{ for } \beta_j \geq 0, \ j = 1, \ldots, m \right\}.$$

For the monotone case, $l = 1$, $\boldsymbol{v}^1$ is the constant vector, and the $\boldsymbol{\delta}$ vectors are shown in Figure 2(a). For the convex case, $l = 2$ and we can define $\boldsymbol{v}^1$ to be the constant vector and $\boldsymbol{v}^2$ to be the identity vector; the $\boldsymbol{\delta}$ vectors are shown in Figure 2(b). For the case of increasing and convex, $l = 1$, $\boldsymbol{v}^1$ is the constant vector, and $\boldsymbol{\delta}$ are those shown in Figure 2(b) plus the identity vector. The failure times shown in Figure 2 are a random sample of size $n = 100$ from a Weibull(3) density, and the knots are chosen at equal time quantiles.

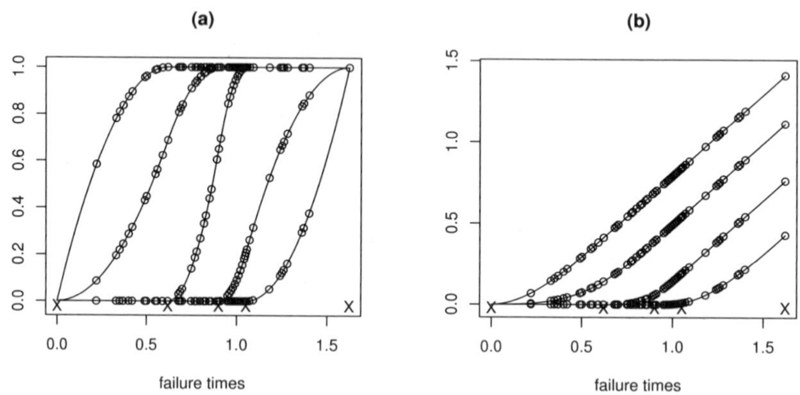

Fig. 2. (a) Monotone and (b) convex spline basis functions (curves) and spline basis vector values (circles), using three interior knots placed at equal quantiles for an observed $n = 100$ times generated from a Weibull(3) density.

The polynomial form can be used to find the c_i to describe the integral term in the likelihood function, i.e., to find the c_i such that

$$\sum_{i=1}^{n} \int_0^{t_i} h(u)du = \sum_{i=1}^{n} c_i h(t_i).$$

The iteratively re-weighted cone projection algorithm can be used to maximize the likelihood over the set of constrained regression splines. Some examples of fits are shown in Figure 3 for the same data set simulated from the Weibull(3) density. The regression splines smooth out the linear spline fits and are closer to the true function over most of the range of the data.

4. Knot Choices

The knot number and placement are to be chosen by the user. Unconstrained regression splines are known to be sensitive to these choices, but the constrained versions are more robust (Meyer, 2008b). This is demonstrated in Figure 4, where for a random sample from a Weibull (3) density, the quadratic monotone spline hazard function estimates are shown for two, three, and four interior knots. The knots are placed at equal quantiles of the observed failure times and are shown as "X" marks superimposed on the tick marks indicating the failure times.

The shape restrictions do not allow the "wiggling" associated with overfitting in nonparametric methods. The set of unconstrained spline estimates is a linear subspace of $\mathbb{R}^n$; the cone containing the constrained spline estimates is a subset of this space. It allows for all the flexibility of the regression splines within the imposed constraints.

If desired, standard methods such as AIC may be used to choose the number of knots. We use

$$AIC = -\log \text{likelihood} + d,$$

where the log likelihood is as in (2) and d is the number of the faces of the cone containing the MLE. For the fits in Figure 4, the AIC values are -71.26, -71.31, and -70.26, which are all fairly close together but indicate that three knots might be the best choice.

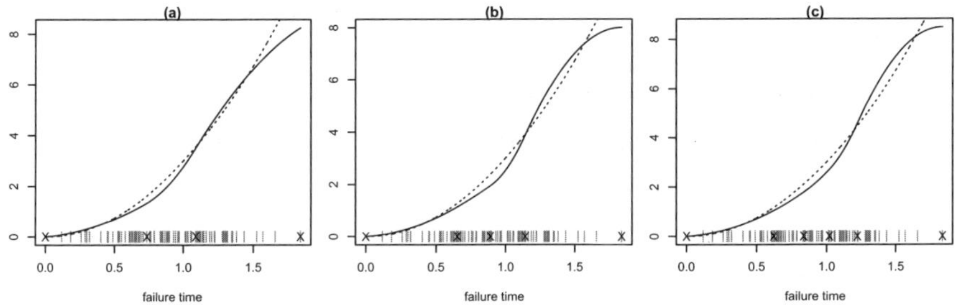

Fig. 3. Comparisons of monotone quadratic spline estimates for three choices of numbers of knots. The knots are indicated by "X" and are at equal quantiles of the observed failure times (shown as tick marks). The solid curve is the estimate and the dotted curve is the true hazard function.

Convex constraints result in even more robustness to knot choices. This is demonstrated in Figure 4, where convex cubic spline estimates for the hazard function are shown for the same data set as in Figure 3. For most of the range of the data, the estimates sit almost on top of each other and diverge only when there are but a few observations remaining. The AIC values are even closer together, and again the middle choice of three knots produces the smallest AIC.

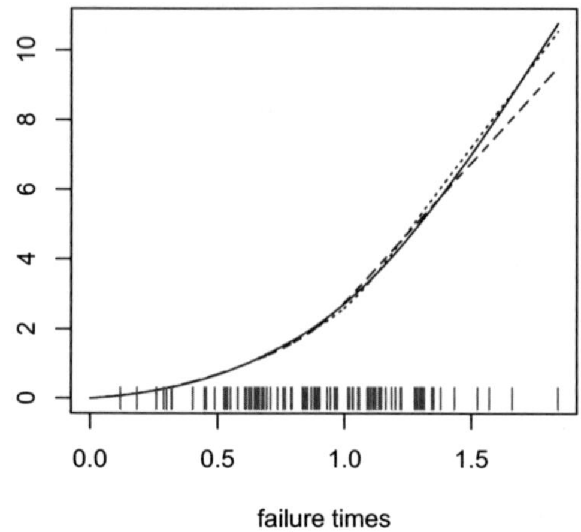

Fig. 4. Comparisons of convex cubic spline estimates for three choices of numbers of knots. For each choice, the knots are placed at equal quantiles of the observed failure times (shown as tick marks). The solid curve is the estimate for two interior knots, the dashed for three, and the dotted for four.

5. Comparing with Kernel Methods

A "naive" estimator for the hazard function is given by Silverman (1986), section 6.5:

$$\hat{h}(t) = \frac{\hat{f}(t)}{1 - \hat{F}(t)},$$

where $\hat{f}$ is the adaptive-bandwidth kernel density estimate, and $\hat{F}$ is the corresponding estimated cumulative distribution function. The hazard function estimate of Müller and Wang (1994) corrects for edge effects by using a data-adaptive variable bandwidth and is available in `R` (R Development Core Team, 2010), coded as the function `muhaz`. As an example, we look at the "time on treatment" data from Silverman. In plot (a) of Figure 5, we see the adaptive-bandwidth kernel density estimate, and in plot (b) the solid line is the corresponding naive hazard function estimate. The dashed line is the Müller-Wang `muhaz` estimate with the default options; the estimator recognizes that the tail of the hazard function is notoriously difficult to estimate, so it truncates the estimate to the time when there are ten units remaining. Finally, the dot-dash curve is the convex cubic regression spline estimator.

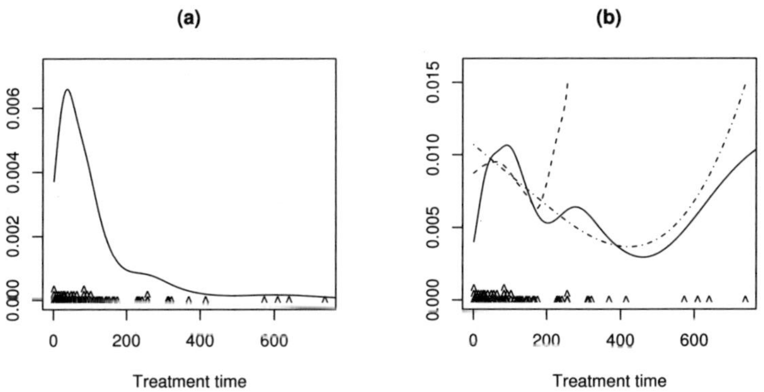

Fig. 5. (a) The variable-bandwidth kernel estimate of the density of treatment times. (b) The solid line is the naive estimator of the hazard function. The dashed line is the Müller-Wang estimator and the dot-dash line is the convex cubic regression spline estimator.

We compare the various estimates from "data sets" consisting of the quantiles of distributions to assess and compare the properties of the hazard function estimators. First, we look at quantiles from a Weibull(3) density. The quadratic monotone splines are compared with the naive and `muhaz` estimators in Figure 6. The spline estimators with two, three, and four interior knots are very close to the true hazard function until about time 1.3, which is also when approximately ten units are remaining. The naive estimator is very close to the true curve and has the best performance. The Müller-Wang estimator is a bit too large over the center range of the data.

Quantiles from a density with quadratic hazard $h(t) = (1-t)^2$ are used to compare the convex cubic regression spline estimators with the kernel versions in Figure 7. The regression spline estimators with two, three, and four interior knots are indistinguishable from the true hazard function until a time when approximately ten units are remaining. The naive estimator has poor edge performance, which is largely corrected by the `muhaz` estimator at the expense of being a bit too large over the ranges where the density is largest.

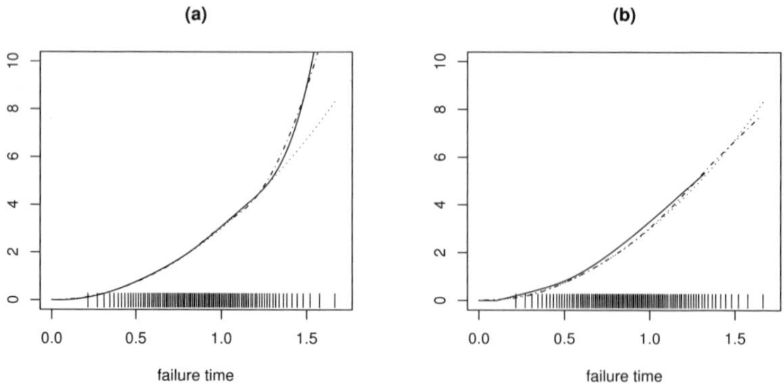

Fig. 6. Estimates of the hazard functions given a "data set" consisting of quantiles from a Weibull(3) density. The true hazard function is shown by the dotted curve. (a) Piecewise quadratic monotone estimators with two (dot-dash), three (solid), and four (dashed) interior knots. (b) The solid curve is the Müller-Wang estimate, and the dot-dash line is the naive kernel esimate.

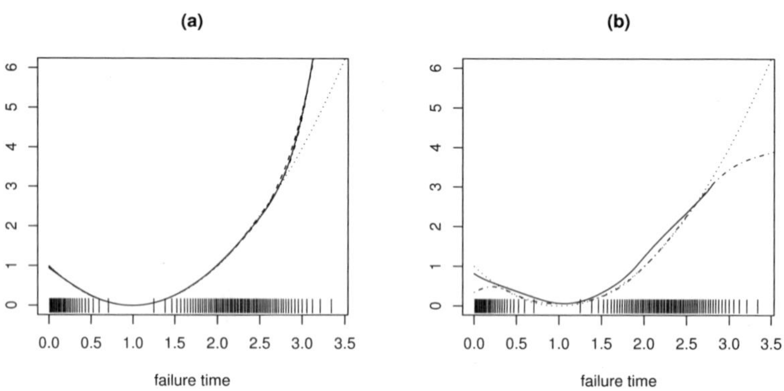

Fig. 7. Estimates of the hazard functions given a "data set" consisting of quantiles from a density with a quadratic hazard function, shown by the dotted curve. (a) Piecewise cubic convex estimators with two (dot-dash), three (solid), and four (dashed) interior knots. (b) The solid curve is the Müller-Wang estimate, and the dot-dash line is the naive kernel esimate.

6. Right Censoring and Examples

Often units are "lost" before the event occurs; this type of censoring is easily incorporated into the maximum likelihood algorithm. Let $x_1, x_2, \ldots, x_n$ denote times to failure for the n subjects under study and $z_1, z_2, \ldots, z_n$ be the corresponding censoring times. The times x_i and z_i are assumed to be independent. The observed random variables are then t_i and δ_i, where $t_i = \min(x_i, z_i)$ and $\delta_i = I(x_i \le z_i)$. The log-likelihood function is simply

$$\ell(\boldsymbol{\theta}) = \sum_{i=1}^{n} \delta_i \log(\theta_i) - \sum_{i=1}^{n} c_i\theta_i,$$

where the c_i are the same as for the uncensored case, computed using the observed t_i.

To demonstrate the methods, we simulate data and censoring from known distributions. Plot (a) of Figure 8 shows estimates of the hazard function when the failure times are from a Weibull(3) density (tick marks) and the censoring times are Weibull(6) (circles). The true hazard function is shown as the dotted curve, and the `muhaz` estimate is the dashed curve. If the hazard function is known only to be monotone, the quadratic regression spline estimate with three interior knots is shown as the dot-dash curve. The solid curve is the cubic regression spline estimator constrained to be monotone and convex.

In plot (b), the failure times are simulated from a density with quadratic hazard function (dotted curve) and the censoring times are from a decreasing triangular density on $(0, 5)$. The solid curve is the convex cubic regression spline estimate, and the dashed curve is the `muhaz` estimate.

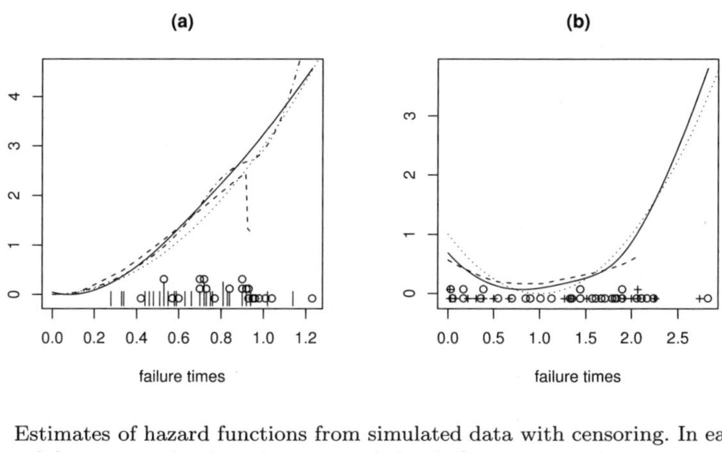

Fig. 8. Estimates of hazard functions from simulated data with censoring. In each, the true hazard function is the dotted curve, and the dashed curve is the `muhaz` estimator. The tick marks are failure times, and the circles are censoring times. (a) The x_i are from Weibull(3), the z_i from Weibull(6). The dot-dash curve is the monotone regression spline estimate, and the solid curve is constrained to be monotone and convex. (b) The x_i are from $h(t) = (1-t)^2$, the z_i from a decreasing triangular density on $(0, 5)$. The solid curve is the convex regression spline estimate.

For a "real" data example, we look at the mechanical switches data of Nair (1984). Mechanical switches wear out and so can be assumed to have an increasing failure rate. The switches are tested by using them repeatedly, and failures are observed in the units of millions of operations. There are two mechanisms by which the switch can fail, called "A" and "B"; these represent two springs in the switch. If the interest is in modeling the hazard rate for failure type A, then a failure by mechanism B may be considered a censored observation. This is a relatively small dataset, with 40 observed times of which 17 are observed failures and 23 are censoring times. The data are shown in plot (a) of Figure 9, with the type A failures shown as circles and the type B failures as "x". Three monotone regression spline estimates are also shown, for two (dot-dash), three (solid), and four (dashed) interior knots. Suppose the failure rate can also be assumed to be convex. We then obtain three cubic monotone-convex regression spline estimates, with two, three, and four interior knots.

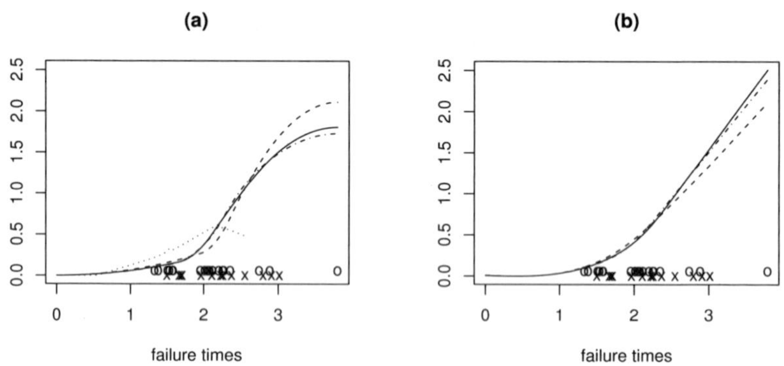

Fig. 9. Estimates of the hazard function for the switches data. (a) Monotone quadratic regression spline estimates with two (dot-dash), three (solid), and four (dash) interior knots at equal observed time quantiles. The dotted curve is the Müller-Wang adaptive kernel estimator. (b) Increasing convex cubic with two (dot-dash), three (solid), and four (dash) interior knots at equal observed time quantiles.

We use the kidney failures dataset from Klein and Moeschberger (2003) to illustrate the methods with a large data set consisting mostly of censored observations. There are 863 times to death (in years) for kidney transplant patients treated at The Ohio State University Transplant Center. The observation is censored if the patient moved from Columbus, Ohio, died from another cause, or was still alive at the end of the study. There were 140 observed deaths, most concentrated in the first year after transplant, and 723 censored observations more evenly spaced over about nine and a half years. It is reasonable to believe that the hazard function is "bathtub-shaped," or convex. The data are shown in Figure 10, with censoring times marked as gray circles, and observed failures with black plus signs. Three cubic convex regression spline estimators are shown, using three, five, and seven interior knots. The estimated functions are very close together, indicating some robustness to knot choices. The dashed curve is the Müller-Wang estimator.

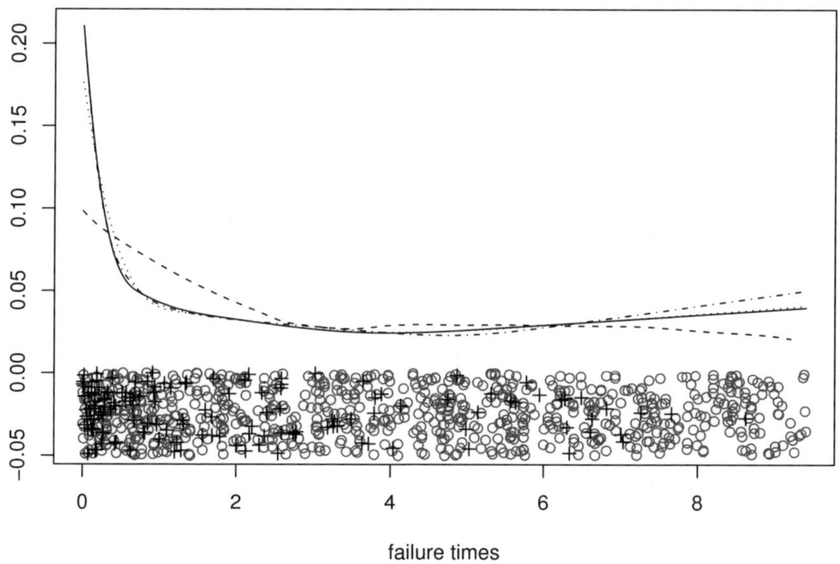

Fig. 10. Estimates of the hazard function for the kidney failure data. The dashed curve is the Müller-Wang estimator. Convex cubic estimates are shown for three (dotted), five (solid), and seven (dot-dash) knots. The times of observed failures are indicated in the jittered band by plus signs, and the censoring times are marked with circles.

We have presented an estimator for the hazard function that incorporates *a priori* information about shape and smoothness. It is a flexible and parsimonious estimator that maximizes the likelihood over a set of constrained regression splines. It behaves well in that it is very close to the true hazard function when the "data" consist of quantiles of the true hazard (except for some edge effects), compared to other available estimators.

References

Fraser, D. A. S. and Massam, H. (1989). A mixed primal-dual bases algorithm for regression under inequality constraints. application to convex regression, *Scandinavian Journal of Statistics* **16**, pp. 65–74.

Klein, J. P. and Moeschberger, M. L. (2003). *Survival Analysis; Techniques for Censored and Truncated Data, second edition* (Springer, New York).

Meyer, M. C. (2008a). An algorithm for projections onto convex cones with applications in statistical modeling, Tech. Rep. 12, Colorado State University.

Meyer, M. C. (2008b). Inference using shape-restricted regression splines, *Annals of Applied Statistics* **2**, 3, pp. 1013–1033.

Meyer, M. C. and Woodroofe, M. (2004). Maximum likelihood estimation of a unimodal density using shape restrictions, *Canadian Journal of Statistics* **32**, 1, pp. 85–100.

Müller, H. G. and Wang, J. L. (1994). Hazard rate estimation under random censoring with varying kernels and bandwidths, *Biometrics* **50**, pp. 61–76.

Nair, V. N. (1984). Confidence bands for survival functions with censored data: A comparative study, *Technometrics* **26**, pp. 265–275.

R Development Core Team (2010). *R: A Language and Environment for Statistical Computing*, R Foundation for Statistical Computing, Vienna, Austria, URL `http://www.R-project.org`, ISBN 3-900051-07-0.

Ramsay, J. O. (1988). Monotone regression splines in action, *Statistical Science* **3**, pp. 425–461.

Silverman, B. W. (1986). *Density Estimation for Statistics and Data Analysis* (Chapman & Hall, New York).

Multivariate Models and the First Four Moments

Klaus Nordhausen and Hannu Oja

Tampere School of Public Health, University of Tampere
Tampere, Finland
E-mail: klaus.nordhausen@uta.fi
E-mail: hannu.oja@uta.fi

Esa Ollila

Department of Signal Processing and Acoustics
Aalto University School of Science and Technology, Helsinki, Finland
E-mail: esollila@wooster.hut.fi

Several extensions of the multivariate normal model have been shown to be useful in practical data analysis. Therefore, tools to identify which model might be appropriate for the analysis of a real data set are needed. This paper suggests the simultaneous use of two location and two scatter functionals to obtain multivariate descriptive measures for multivariate location, scatter, skewness and kurtosis, and shows how these measures can be used to distinguish among a wide range of models that extend the multivariate normal model. The method is demonstrated with examples on simulated and real data.

Keywords: Elliptical distribution; Independent components analysis; Multivariate skewness and kurtosis; Skew-elliptical distribution

1. Introduction

Recently, several extensions of the multivariate normal model with nice tractable features have been introduced and have appeared to be useful in multivariate data analysis. The extensions we have here in mind are the family of elliptical distributions shown to be useful by Fang and Zhang (1990), the independent component model frequently used in signal processing and medical image analysis (see, e.g., Hyvärinen *et al.*, 2001), finite mixtures of multivariate elliptical distributions considered for example in McLachlan and Peel (2000), or the family of skew-elliptical distributions described in Genton (2004). Naturally, in practical data analysis, it is important to decide which model is the most appropriate one for the data and problem at hand. One can then use statistical inference tools, tests, and estimates tailored for that model. In this paper, we propose some guidelines for this task: We use two old ideas from Karl Pearson but transfer them to the multivariate case.

Karl Pearson was perhaps the first scientist to see the importance of skewness and kurtosis in model selection. In the univariate case, besides seeing these two properties as the properties measured by the standardized third and fourth moments, he also suggested measuring skewness with standardized differences between two location measures such as

$$\frac{\text{mean} - \text{mode}}{\text{standard deviation}} \quad \text{or} \quad \frac{\text{mean} - \text{median}}{\text{standard deviation}}.$$

In a similar way, kurtosis may be seen as a ratio of two univariate scale measures (see, for example, Oja, 1981).

The classical measures of skewness and kurtosis for a univariate random variable x are the standardized third and fourth moments,

$$\beta_1(x) = \frac{E\left[(x - E(x))^3\right]}{[\mathrm{Var}(x)]^{3/2}} \quad \text{and} \quad \beta_2(x) = \frac{E\left[(x - E(x))^4\right]}{[\mathrm{Var}(x)]^2}.$$

Also, β_1 can be expressed as a standardized difference of two location measures and β_2 is a squared ratio of two scale measures. To see this, first write

$$E_3(x) = E\left[\left(\frac{x - E(x)}{\sqrt{\mathrm{Var}(x)}}\right)^2 x\right] \quad \text{and}$$

$$\mathrm{Var}_4(x) = E\left[\left(\frac{x - E(x)}{\sqrt{\mathrm{Var}(x)}}\right)^2 (x - E(x))^2\right].$$

Now $E_3(x)$ is an affine equivariant location measure (based on the first three moments) and the square root of $\mathrm{Var}_4(x)$ yields an affine equivariant scale measure (based on the first four moments). Then

$$\beta_1(x) = \frac{E_3(x) - E(x)}{\sqrt{\mathrm{Var}(x)}} \quad \text{and} \quad \beta_2(x) = \frac{\mathrm{Var}_4(x)}{\mathrm{Var}(x)}.$$

Note also that, for a standardized variable $z = (x - E(x))/\sqrt{\mathrm{Var}(x)}$, $\beta_1(x) = \beta_1(z) = E_3(z)$ and $\beta_2(x) = \beta_2(z) = \mathrm{Var}_4(z)$.

The other idea of Karl Pearson we want to pick up again is Pearson's system of frequency curves (see, for example, Ord, 1986). This flexible system of distributions has been an important tool in identifying the unknown distribution of the observations or of a sample statistic. The procedure is then to use β_1 and β_2 to decide which distribution might fit best to the data. The subfamilies of probability densities in this system were originally given as solutions of a simple differential equation. The extensions to the multivariate case have proved difficult (Kotz, 1975). The approach suggested in this paper for the multivariate case is not based on any differential equation, but the idea is similar. First, multivariate measures of skewness and kurtosis are computed and, based on them, it is then decided which multivariate distribution might fit the data best.

The structure of this paper is as follows. The next section describes how two location functionals and two scatter functionals can be jointly used to obtain descriptive statistics for multivariate data. Section 3 investigates the behavior of descriptive measures for multivariate skewness and kurtosis in a wide range of models. In Section 4, model selection is demonstrated using simulated and real data. Some technical results are collected in the appendix.

2. Moments and multivariate descriptive statistics

Let now $\mathbf{x}$ be a p-variate random vector with cumulative distribution function (cdf) F. Then a vector-valued functional $\mathbf{T} = \mathbf{T}(F) = \mathbf{T}(\mathbf{x})$ is a location functional if it is affine equivariant in the sense that $\mathbf{T}(\mathbf{A}\mathbf{x}+\mathbf{b}) = \mathbf{A}\mathbf{T}(\mathbf{x}) + \mathbf{b}$ for any full rank $p \times p$ matrix $\mathbf{A}$ and any p-vector $\mathbf{b}$. A $p \times p$-matrix-valued functional $\mathbf{S} = \mathbf{S}(F) = \mathbf{S}(\mathbf{x})$ is a scatter functional if it is affine equivariant in the sense that $\mathbf{S}(\mathbf{A}\mathbf{x}+\mathbf{b}) = \mathbf{A}\mathbf{S}(\mathbf{x})\mathbf{A}^\top$, again for all $\mathbf{A}$ and $\mathbf{b}$ as defined above.

The most familiar location and scatter functionals are naturally the mean vector and covariance matrix,

$$E(\mathbf{x}) \quad \text{and} \quad \text{Cov}(\mathbf{x}) = E\left[(\mathbf{x} - E(\mathbf{x}))(\mathbf{x} - E(\mathbf{x}))^\top\right].$$

There are, however, several general families of location and scatter functionals with different properties. For a recent overview with references, see Chapter 6 of Maronna *et al.* (2006). The mean vector and the covariance matrix are functionals based on the first two moments. As in the univariate case, we can define a multivariate location functional and a multivariate scatter functional which use the first three moments and the first four moments, respectively. These functionals are

$$E_3(\mathbf{x}) = \frac{1}{p}E(r^2\mathbf{x}) \quad \text{and} \quad \text{Cov}_4(\mathbf{x}) = \frac{1}{p+2}E[r^2(\mathbf{x} - E(\mathbf{x}))(\mathbf{x} - E(\mathbf{x}))^\top],$$

where $r^2 = (\mathbf{x} - E(\mathbf{x}))^\top \text{Cov}(\mathbf{x})^{-1}(\mathbf{x} - E(\mathbf{x}))$. See Oja *et al.* (2006), for example. These two estimates can be seen as one-step M-estimates and are the natural multivariate extensions of E_3 and Var_4 given in the introduction. It is remarkable that $\mathrm{Cov}(\mathbf{x})$ and $\mathrm{Cov}_4(\mathbf{x})$ have the so called independence property: If $\mathbf{x}$ has independent components, then both $\mathrm{Cov}(\mathbf{x})$ and $\mathrm{Cov}_4(\mathbf{x})$ are diagonal matrices but of course with possibly different diagonal elements. (However, if $\mathbf{x} \sim N_p(\boldsymbol{\mu}, \boldsymbol{\Sigma})$, then $\mathrm{Cov}(\mathbf{x}) = \mathrm{Cov}_4(\mathbf{x}) = \boldsymbol{\Sigma}$. This is due to the appropriate scaling of the second matrix.) The idea in the following is to try to identify the multivariate model with two pairs of functionals, (E, Cov) and (E_3, Cov_4).

The usage of two scatter functionals, say $\mathbf{S}_1$ and $\mathbf{S}_2$, has become quite popular recently in multivariate data analysis. Cardoso (1989), in his fourth-order blind identification algorithm (FOBI), used the regular covariance matrix ($\mathbf{S}_1 = \mathrm{Cov}$) and the matrix of fourth moments ($\mathbf{S}_2 = \mathrm{Cov}_4$) for independent components analysis (ICA); Caussinus and Ruiz-Gazen (1994) used two different scatter matrices, the dispersion matrix of Gnanadesikan and Kettenring (1972) and the regular covariance matrix, for projection pursuit; and Critchley *et al.* (2008) use a one-step Tyler matrix and the regular covariance matrix in their principal axis analysis. For a general theory for the comparison of different scatter matrices, see Tyler *et al.* (2009) and the references therein. In the following, we list some of the results given in Oja *et al.* (2006) and Tyler *et al.* (2009).

Let $\mathbf{S}_1$ and $\mathbf{S}_2$ be the values of two different scatter functionals of the distribution of a p-variate random variable $\mathbf{x}$. Then let a $p \times p$ matrix $\mathbf{B}$ and a $p \times p$ diagonal matrix $\mathbf{D}$ solve the eigenvector and eigenvalue problem

$$\mathbf{S}_1^{-1}\mathbf{S}_2\mathbf{B}^\top = \mathbf{B}^\top\mathbf{D}.$$

The diagonal elements of $\mathbf{D}$ are then the eigenvalues and the rows of $\mathbf{B}$ are the eigenvectors of the matrix $\mathbf{S}_1^{-1}\mathbf{S}_2$. The directions of the eigenvectors (up to sign) are well-defined if the eigenvalues of $\mathbf{S}_1^{-1}\mathbf{S}_2$ are distinct. The eigenvectors are then unique up to a multiplication by non-zero constants. Then, for all choices of $\mathbf{B}$, the transformed observations $\mathbf{z} = \mathbf{B}\mathbf{x}$, which are called invariant coordinates, yield $\mathbf{S}_1(\mathbf{z}) = \mathbf{D}_1$ and $\mathbf{S}_2(\mathbf{z}) = \mathbf{D}_2$, where $\mathbf{D}_1$ and $\mathbf{D}_2$ are two diagonal matrices such that $\mathbf{D} = \mathbf{D}_1^{-1}\mathbf{D}_2$.

Besides two scatter matrices $\mathbf{S}_1$ and $\mathbf{S}_2$, two location functionals $\mathbf{T}_1$ and $\mathbf{T}_2$ may sometimes be used to standardize the random variable in a unique way. We then choose the matrix $\mathbf{B}$ in such a way that, if $\mathbf{z} = \mathbf{B}(\mathbf{x} - \mathbf{T}_1(\mathbf{x}))$, then

$$\mathbf{T}_1(\mathbf{z}) = \mathbf{0}, \quad \mathbf{T}_2(\mathbf{z}) \geq \mathbf{0}, \quad \mathbf{S}_1(\mathbf{z}) = \mathbf{I}_p, \quad \text{and} \quad \mathbf{S}_2(\mathbf{z}) = \mathbf{D}.$$

The standardized vector is uniquely defined if $\mathbf{T}_2(\mathbf{z}) > \mathbf{0}$ (all the components of $\mathbf{T}_2(\mathbf{z})$ are positive) and if the diagonal elements of $\mathbf{D}$ are distinct.

This then suggests multivariate descriptive measures for location, scatter, skewness and kurtosis as

- *Location:* $\mathbf{T}_1(\mathbf{x})$
- *Scatter:* $\mathbf{S}_1(\mathbf{x})$
- *Skewness:* $\mathbf{T}_2(\mathbf{z})$
- *Kurtosis:* $\mathbf{S}_2(\mathbf{z})$

These measures are moment-based if one chooses

$$\mathbf{T}_1 = E, \quad \mathbf{S}_1 = \mathrm{Cov}, \quad \mathbf{T}_2 = E_3, \quad \text{and} \quad \mathbf{S}_2 = \mathrm{Cov}_4.$$

In the univariate case, we then get the classical measures given in the introduction.

3. The moments in some multivariate statistical models

3.1. *General structures for models*

We consider multivariate location-scatter models where the observations $\mathbf{x}$ are thought to be generated by

$$\mathbf{x} = \boldsymbol{\Omega}\boldsymbol{\epsilon} + \boldsymbol{\mu}$$

with a random vector $\boldsymbol{\epsilon}$ standardized in some way. The random vector $\boldsymbol{\epsilon}$ is often used just to formulate the model, but sometimes it can be thought to be a real latent variable of its own interest. The models are then distinguished by different assumptions on the vector $\boldsymbol{\epsilon}$. The p-variate vector $\boldsymbol{\mu}$ is the location vector (parameter), and $\boldsymbol{\Sigma} = \boldsymbol{\Omega}\boldsymbol{\Omega}^\top$ is the scatter matrix (parameter). The matrix $\boldsymbol{\Omega}$ is called the transformation matrix.

In the following, we often need some matrix notation: $\mathbf{J}$ is a sign change matrix, that is, a diagonal matrix with diagonal entries ± 1; $\mathbf{P}$ is a permutation matrix obtained from an identity matrix by successively permuting its rows and/or columns; and $\mathbf{O}$ is an orthogonal matrix, that is, $\mathbf{O}\mathbf{O}^\top = \mathbf{O}^\top\mathbf{O} = \mathbf{I}$. Clearly, $\mathbf{J}$ and $\mathbf{P}$ are orthogonal.

A structure of symmetrical models is obtained with assumptions

(A1) $\boldsymbol{\epsilon} \sim N_p(\mathbf{0}, \mathbf{I})$.
(A2) $\boldsymbol{\epsilon} \sim \mathbf{O}\boldsymbol{\epsilon}$ for all orthogonal $\mathbf{O}$.
(A3) $\mathbf{P}\mathbf{J}\boldsymbol{\epsilon} \sim \boldsymbol{\epsilon}$ for all permutation matrices $\mathbf{P}$ and sign change matrices $\mathbf{J}$.
(A4) $\mathbf{J}\boldsymbol{\epsilon} \sim \boldsymbol{\epsilon}$ for all sign change matrices $\mathbf{J}$.
(A5) $-\boldsymbol{\epsilon} \sim \boldsymbol{\epsilon}$.

Skew distributions are obtained, for example, if we assume that

(B1) $\boldsymbol{\epsilon} = \sum_{i=1}^{k} p_i(\boldsymbol{\epsilon}_i + \boldsymbol{\mu}_i)$ where $k \leq p$,

$$(p_1, \ldots, p_k)^\top \sim \text{Multin}(1; (\pi_1, ..., \pi_k)),$$

and the $\boldsymbol{\epsilon}_i$ are independent and satisfy (A2).

(B2) $\boldsymbol{\epsilon} = \text{sign}(\epsilon^*_{p+1} - \alpha - \beta\epsilon^*_p)\boldsymbol{\epsilon}^*$, where $(\boldsymbol{\epsilon}^{*\top}, \epsilon^*_{p+1})^\top$ satisfies (A2).

(B3) The components of $\boldsymbol{\epsilon}$ are independent with $E(\epsilon_i) = 0$ and $\text{Var}(\epsilon_i) = 1$.

The symmetry assumptions satisfy

$$(A1) \Rightarrow (A2) \Rightarrow (A3) \Rightarrow (A4) \Rightarrow (A5).$$

It is also easy to see that the symmetric models can be seen as border cases of the asymmetric models and we have the following relationships:

$$(A1) \Leftrightarrow (A2) \; \& \; (B3) \qquad \text{and} \qquad (A2) \Rightarrow (B1) \; \& \; (B2).$$

If assumptions (A1)–(A5) or (B1)–(B3) are true, the resulting models are called models (A1)–(A5) or (B1)–(B3), respectively. In the following, we describe these models in more detail.

3.2. *Elliptical Model*

The elliptical model is obtained under assumption (A2). If assumption (A2) is true then the standardized random variable $\boldsymbol{\epsilon}$ has a spherical distribution around the origin. And if the density exists, it has the form

$$f(\boldsymbol{\epsilon}) = \exp\{-\rho(||\boldsymbol{\epsilon}||)\}$$

for some function $\rho(\cdot)$, where $||\boldsymbol{\epsilon}|| = (\epsilon_1^2 + ... + \epsilon_p^2)^{1/2}$ is Euclidean distance. The random vector $\mathbf{x}$ is then elliptically symmetric. The scatter matrix $\boldsymbol{\Sigma}$ and ρ are confounded and one of them needs to be scaled. Therefore one often assumes that $E(||\boldsymbol{\epsilon}||^2) = p$ (second moments exist) or $\text{Med}(||\mathbf{e}||^2) = \chi^2_{p,.5}$. Under this assumption, both $\boldsymbol{\mu}$ and $\boldsymbol{\Sigma}$ are uniquely defined, and $\boldsymbol{\Sigma}$ is the covariance matrix in the multivariate normal case. The transformation matrix $\boldsymbol{\Omega}$ is not well-defined.

The multivariate normal distribution $N_p(\boldsymbol{\mu}, \boldsymbol{\Sigma})$ given by assumption (A1) is naturally a member of the family of elliptical distributions with the choice $\rho(r) = \frac{p}{2}\log(2\pi) + \frac{1}{2}r^2$. The family of elliptically distributed random variables thus extends the normal model by allowing lighter as well as heavier tails while still maintaining symmetry around $\boldsymbol{\mu}$. Prominent distributions in the elliptical family are the multivariate t-distributions and the power-exponential distributions. This model is in practice the most popular extension of the multivariate model and standard multivariate gaussian procedures methods have been extended to this wider model; see, for example, Fang and Zhang (1990). Also, robust procedures often assume elliptical symmetry.

If we assume that the first four moments exist then, in the elliptic model,

$$E_3(\mathbf{z}) = \mathbf{0}, \quad \text{and} \quad \text{Cov}_4(\mathbf{z}) = c_\rho \mathbf{I}_p.$$

Here, as before, $\mathbf{z} = \mathbf{B}(\mathbf{x} - E(\mathbf{x}))$. All marginal distributions have the same kurtosis, which is a function of ρ. In the multivariate normal case, $c_\rho = 1$ and in the p-variate t_ν case, $c_\rho = (\nu - 2)/(\nu - 4)$.

3.3. *Other symmetric models*

First consider the model with assumption (A3). The model includes all elliptical distributions as well as the cases with i.i.d. symmetric components $\epsilon_1, ..., \epsilon_p$. An interesting submodel with different shapes of density contours is obtained if the density of $\boldsymbol{\epsilon}$ is of the form

$$f(\boldsymbol{\epsilon}) = \exp(-\rho(||\boldsymbol{\epsilon}||)),$$

where the norm $||\cdot||$ is any norm that satisfies the condition that $||\mathbf{z}|| = ||\mathbf{PJz}||$ for all permutations $\mathbf{P}$ and sign changes $\mathbf{J}$. This is true for any L_p norm, for example. If we again assume that the first four moments exist, then also in this model

$$E_3(\mathbf{z}) = \mathbf{0} \quad \text{and} \quad \text{Cov}_4(\mathbf{z}) = c_f \mathbf{I}_p.$$

Recall that $\mathbf{z} = \mathbf{B}(\mathbf{x} - E(\mathbf{x}))$. This means that the first four moments can not be used to distinguish model (A2) from model (A3). As in the elliptical model, both $\boldsymbol{\mu}$ and $\boldsymbol{\Sigma}$, but not $\boldsymbol{\Omega}$, are uniquely defined.

A still wider symmetric model is obtained if we assume (A4). Assuming that the first four moments exist, then

$$E_3(\mathbf{z}) = \mathbf{0} \quad \text{and} \quad \text{Cov}_4(\mathbf{z}) = \mathbf{D},$$

where $\mathbf{D}$ is a diagonal matrix. Thus model (A4) can be separated from model (A3) just by looking at the diagonal elements of $\mathbf{D}$: If the diagonal elements are not all the same, then the model (A3) is not correct any more. It is remarkable that, in model (A4), both $\boldsymbol{\mu}$ and $\boldsymbol{\Omega}$ are uniquely defined and $\boldsymbol{\Omega}^{-1}$ gives a transformation to a standardized latent variable, $\boldsymbol{\epsilon}$.

The widest symmetric model is obtained if one assumes (A5). Under the same assumptions as before,

$$E_3(\mathbf{z}) = \mathbf{0} \quad \text{and} \quad \text{Cov}_4(\mathbf{z}) = \mathbf{D},$$

where $\mathbf{D}$ is again a diagonal matrix with possibly distinct diagonal elements. Thus it is not possible to make a distinction between models (A4) and (A5). Our conjecture is that the distinction can be made if one uses three different scatter matrices, since any third scatter functional $\mathbf{S}(\mathbf{z})$ should be a diagonal matrix under (A4), whereas in (A5) it can be any symmetric positive definite matrix.

3.4. *Models with skew distributions*

Consider next the *finite mixtures of elliptical distributions* model. This is given by assumption (B1). If the first four moments exist, then in this model,

$$E_3(\mathbf{z}) = \mathbf{s} \quad \text{and} \quad \mathrm{Cov}_4(\mathbf{z}) = \mathbf{D},$$

where $\mathbf{s} \geq \mathbf{0}$ is a p-vector. If $\boldsymbol{\mu}_1 = ... = \boldsymbol{\mu}_k$, we are back in the elliptic case. Also, in the non-elliptic cases, $\mathbf{s}$ may be zero under some special conditions. The $\mathbf{D}$ matrix has at most k distinct diagonal elements. Only $\boldsymbol{\mu}$ and $\boldsymbol{\Sigma}$, not $\boldsymbol{\Omega}$, are well defined parameters. Fisher's linear subspace to discriminate between the mixture populations corresponds to the subspace spanned by the components of $\mathbf{z} = \mathbf{B}(\mathbf{x} - E(\mathbf{x}))$ that correspond to the distinct values of $\mathbf{D}$ (without any knowledge of the population membership).

Skew-elliptical distributions are given by assumption (B2). Given that the first four moments exist,

$$E_3(\mathbf{z}) = \mathbf{s} \quad \text{and} \quad \mathrm{Cov}_4(\mathbf{z}) = \mathbf{D},$$

where $\mathbf{s}$ has at most one non-zero element and the multiplicities of the two possible values of the diagonal elements of $\mathbf{D}$ are 1 and $p-1$. The corresponding $(p-1)$-subvector of $\mathbf{z} = \mathbf{B}(\mathbf{x} - E(\mathbf{x}))$ is spherically distributed, and one component of $\mathbf{z}$ absorbs all skewness. This means that a canonical form of the skew-elliptic distribution is obtained (see Proposition 4.1 of Azzalini and Capitanio, 1999).

Finally, the *independent components model* is given by (B3). If the first four moments exist, then

$$E_3(\mathbf{z}) = \mathbf{s} \quad \text{and} \quad \mathrm{Cov}_4(\mathbf{z}) = \mathbf{D},$$

where the elements of $\mathbf{s}$ and of the diagonal of $\mathbf{D}$ can be related to the classical moment-based univariate kurtosis and skewness measures if all diagonal elements of $\mathbf{D}$ are distinct: Then, $\beta_1(z_i) = ps_i$ and $\beta_2(z_i) = (p+2)D_{ii} - p + 1$. As both Cov and Cov_4 have the independence property, $\mathbf{z}$ is a latent vector of independent components. The well-known FOBI algorithm proposed by Cardoso (1989) is a solution of the ICA problem.

3.5. *Model selection*

The skewness measure $\mathbf{s}$ and the kurtosis measure $\mathbf{D}$ in the eight models of the previous section are summarized in Table 1, where $\mathcal{D}(k)$ denotes the

Table 1. Possible values of $\mathbf{s}$ and $\mathbf{D}$ in the eight models under consideration.

Model	Skewness $\mathbf{s}$	Kurtosis $\mathbf{D}$
(A1)	$\mathbf{0}$	$\mathbf{I}_p$
(A2)	$\mathbf{0}$	$\mathcal{D}(1)$
(A3)	$\mathbf{0}$	$\mathcal{D}(1)$
(A4)	$\mathbf{0}$	$\mathcal{D}(p)$
(A5)	$\mathbf{0}$	$\mathcal{D}(p)$
(B1)	$\geq \mathbf{0}$	$\mathcal{D}(k)$
(B2)	$\propto \mathbf{e}_p$ or $\propto \mathbf{e}_1$	$\mathcal{D}(2)$
(B3)	$\geq \mathbf{0}$	$\mathcal{D}(p)$

set of all positive definite $p \times p$ diagonal matrices with at most k distinct diagonal entries and $\mathbf{e}_i$ is a p-vector that has 1 in its ith position zeros elsewhere. Hence given a pair $(\mathbf{s}, \mathbf{D})$, a model selection procedure could consist of the following steps:

(1) Decide which models are possible.
(2) Look at the pairwise scatter plots.
(3) Choose the most parsimonious model.

The first step is still difficult in practice since standard errors and confidence intervals for $\mathbf{s}$ and $\mathbf{D}$ are not yet available. However, as we will show in the next section, point estimates will already give a good idea. The second step will help to distinguish models by using their pairwise shapes, which can, for example, help by distinguishing model (A2) from (A3).

4. Examples

In this section, we apply this approach to some simulated and real data. All computations are done in R 2.8.0 (R Development Core Team, 2008) using the package ICS (Nordhausen *et al.*, 2008). Naturally, all the population expected values above will be replaced by sample versions with sample means. However, general statistical inference tools (tests and estimates with confidence ellipsoids) based on $\mathbf{s}$ and $\mathbf{D}$ have not yet been developed.

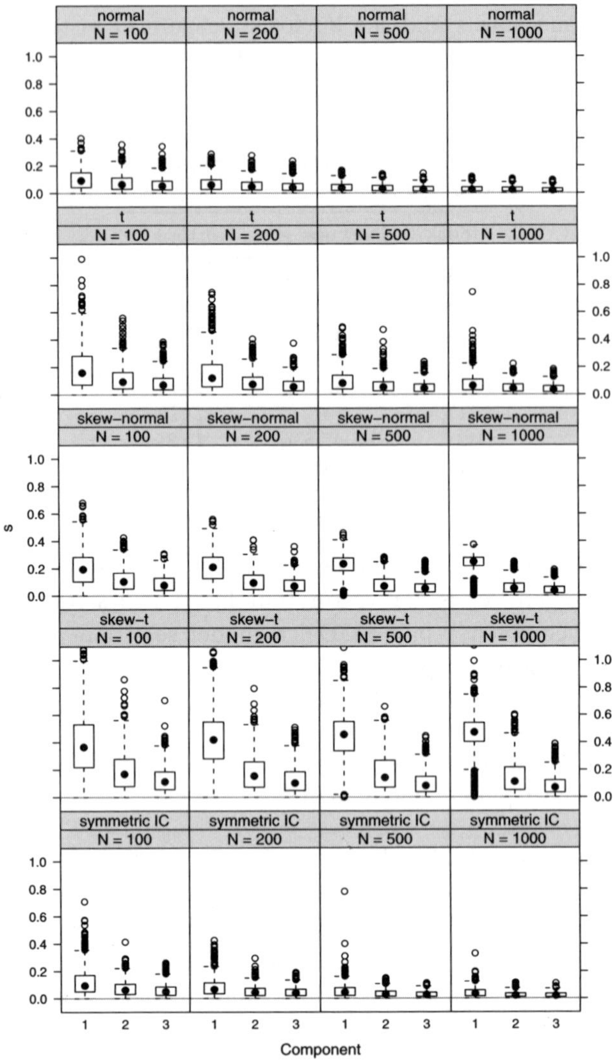

Fig. 1. Simulation results for the skewness measure s based on 1000 repetitions. Note that values of s larger than 1.1 are cut off for a better comparison between the models (this affects only the displays of the measures of the t and skew-t distributions).

4.1. *Examples with simulated data*

First, we want to evaluate the procedure in a simulation study with five simulated 3-variate settings, where the observations are obtained following our model definition (sampling $\boldsymbol{\epsilon}$ and then transforming to $\mathbf{x} = \boldsymbol{\Omega}\boldsymbol{\epsilon} + \boldsymbol{\mu}$). We choose (i) a multivariate normal model, (ii) an elliptic t_{10} model, (iii) a skew-normal model with $\alpha = 0$ and $\beta = 4$, (vi) a skew-t_{10} model with $\alpha = 0$ and $\beta = 4$, and (v) an independent component model where the three independent components have normal, t_{10}, and uniform distributions, respectively. Note that the method is affine invariant and therefore the results do not depend on the choices of $\boldsymbol{\Omega}$ and $\boldsymbol{\mu}$. We used sample sizes $N = 100$, 200, 500, and 1000 and generated 1000 samples in each case. The boxplots for the observed values of $\mathbf{s}$ are presented in Figure 1 (Table B1 contains the corresponding means and standard deviations) and the boxplots for the observed values of diag($\mathbf{D}$) are given in Figure 2 (Table B2 contains the corresponding means and standard deviations).

The estimates of $\mathbf{s}$ and $\mathbf{D}$ seem to behave as expected. The skewness measure seems to converge to $\mathbf{0}$ with increasing sample size in the symmetric models. In the skew-elliptic models, one component is always clearly sticking out and absorbing the skewness. Also, the kurtosis measures behave nicely. In the normal model, all kurtosis measures are close to one, whereas for the t_{10} distribution, they are all distinct from one but have a slower convergence and higher variation due to the heavy tails of the distribution. Note also that none of the three means is yet close to the theoretical value of 1.33. In the skew-elliptic cases, the component that absorbs all the skewness always exhibits a different kurtosis value, whereas the other kurtosis values are more or less equal, as expected. The skew-normal model can be distinguished from the skew-t_{10} distribution since the kurtosis values of symmetric components are both close to one in the normal case and both different from one in the skew-t_{10} case. In the symmetric IC model, one can clearly see that the three kurtosis values differ and that one component has heavy tails, one light tails, and one has a kurtosis of one, which suggests the normal component.

4.2. *Examples with real data sets*

Now we turn from simulated data to real data. The first data set is the famous Fisher's Iris data set, where there are four different measurements on 150 iris plants that come from three different species ($n = n_1 + n_2 + n_3$, $n_1 = n_2 = n_3 = 50$). The group membership information is ignored in the analysis. The second data set is the same subset of the Australian Institute of Sport data set (AIS) that was considered as an example of skew-normal data in Azzalini and Capitanio (1999) (the four variables are body mass index, body fat, sum of skin folds and lean body mass). In this example, we

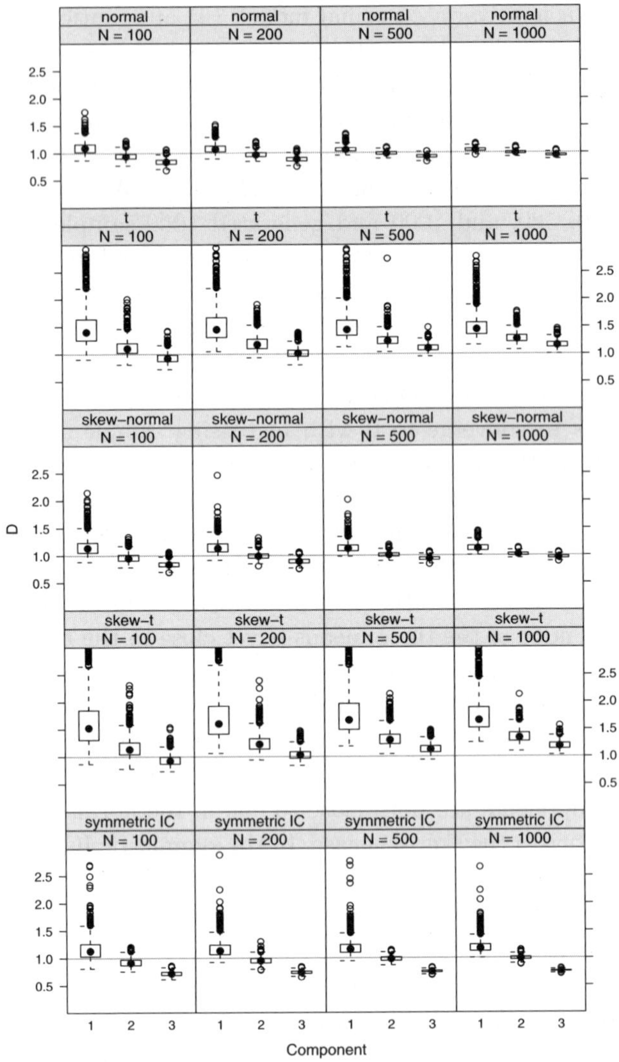

Fig. 2. Simulation results for the kurtosis measure **D** based on 1000 repetitions. Note that values of **D** larger than 3.0 are cut off for a better comparison between the models (this affects only the displays of the measures of the *t*, skew-*t* and symmetric IC distributions).

consider the data sets of all $n = 202$ athletes (men and women) and those of the $n = 100$ female athletes and $n = 102$ male athletes separately.

Table 2. Skewness and kurtosis measures for the iris and the Australian athletes data.

Data	$\mathbf{s}$	diag($\mathbf{D}$)
Iris	$(0.0613\ 0.1794\ 0.0226\ 0.1176)^\top$	(1.2074 1.0269 0.9292 0.7405)
AIS, all	$(0.4821\ 0.0928\ 0.1473\ 0.1687)^\top$	(1.7154 1.2788 0.9353 0.7659)
AIS, Female	$(0.3231\ 0.1611\ 0.1369\ 0.0654)^\top$	(1.3583 1.1320 1.0322 0.8330)
AIS, Male	$(0.5545\ 0.5274\ 0.1397\ 0.1436)^\top$	(2.1592 1.6684 1.0530 0.9456)

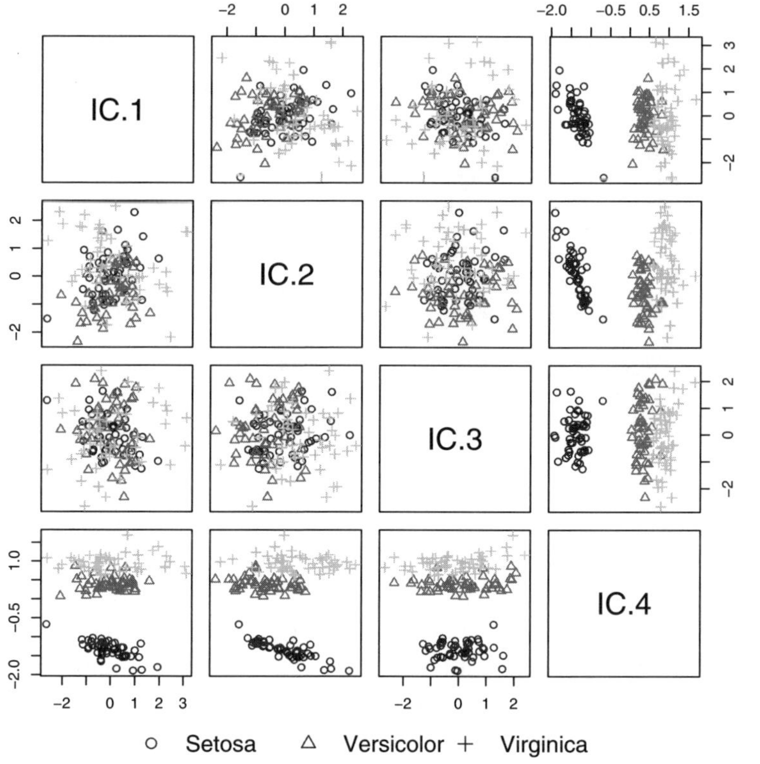

Fig. 3. Scatter plot matrix of the invariant coordinates of the iris data set.

Table 2 gives the values of $\mathbf{s}$ and the diagonal elements of $\mathbf{D}$ for these real data sets. It is obvious from these values that there are two mildly skewed components in the Iris data and that two components have a deviating kurtosis. Therefore, models (B1) and (B3) are possible candidates. In Figure 3, the scatter plot of the invariant components reveals that the

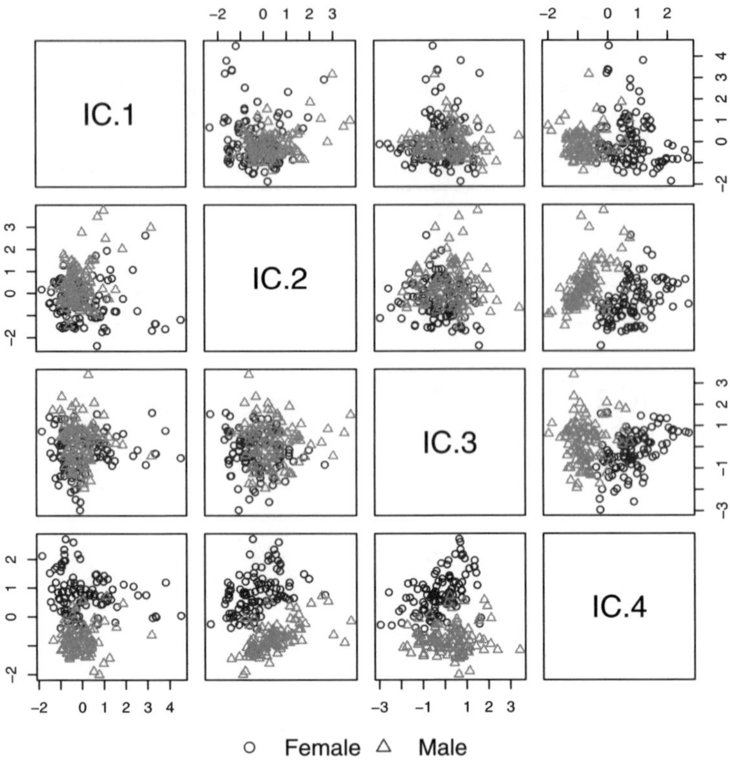

Fig. 4. Scatter plot matrix of invariant coordinates of the Australian athletes data set.

first component appears to have a different kurtosis measure to catch some slightly outlying observations, whereas the last component shows that the means of the three groups all lie on a line and that this component could be used for separation of the groups.

A skew-elliptical model for the Australian athletes data set sounds tempting since the individuals in the data set are certainly collected from a population using a kind of selective sampling. [For a discussion about the link between skewness and selective sampling, see, for example, Arellano-Valle *et al.* (2006).] However, the descriptive statistics for the full data set seem to contradict this. The reason is apparent from Figure 4, where the last component shows the two clusters of men and women. Looking in Figure 5 at only the female athletes, a skew-normal model seems much more realistic; the first component is the one that absorbs all skewness.

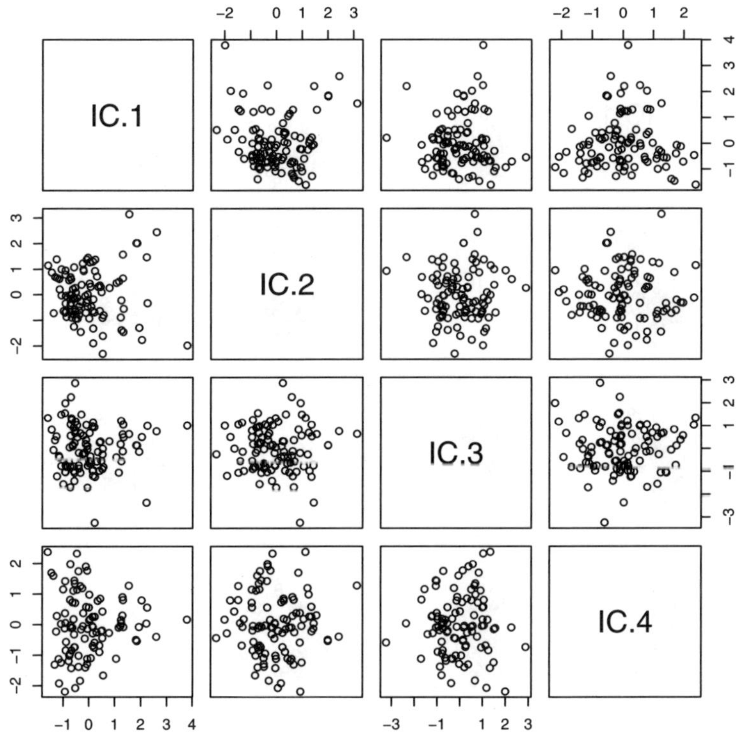

Fig. 5. Scatter plot matrix of invariant coordinates of the female Australian athletes data set.

Interpreting the results for the male Australian athletes seems more difficult. It seems the first two components are skewed. However, looking at the scatter plot matrix of the invariant coordinates as displayed in Figure 6 suggests that actually only one component (component 2) is skewed, whereas the first component might have its large values for $\mathbf{s}$ and $\mathbf{D}$ mainly because of the outliers displayed there. This data set would therefore require some further investigation, for example by replacing the nonrobust functionals here by robust functionals. This is still a topic for future research, but it might help in this case to find further evidence of whether these data are coming from model (B2).

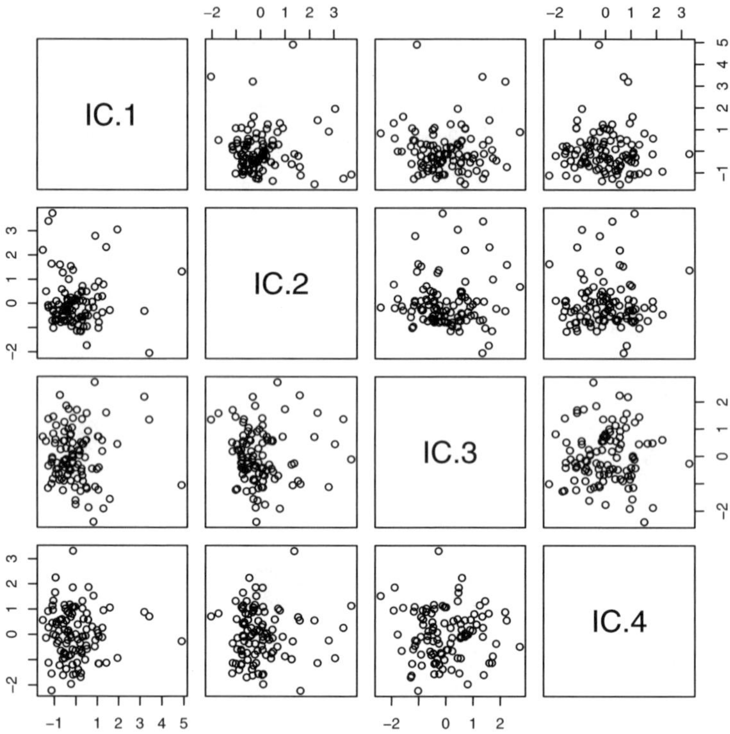

Fig. 6. Scatter plot matrix of invariant coordinates of the male Australian athletes data set.

5. Discussion

The approach proposed in the paper gives some simple tools to evaluate the model assumptions critically. Even if one is using multivariate nonparametrical tests as in Hallin and Paindaveine (2002) or Nordhausen *et al.* (2009), one can use tests and estimates tailored for a certain target distribution, e.g., the multivariate normal. Naturally, the skewness and kurtosis alone cannot distinguish among all models, but together with pairwise scatter plots as shown in Section 4.2, the choices can be narrowed down quite well. Furthermore, using a third scatter matrix may help to decide about (A4), (A5) or (B3). Also, further extensions of the model considered (e.g., skewness in several directions in the skew-elliptic model) are still possible.

Finally, note that the choice of the two location and two scatter functionals in this paper was motivated by their univariate classical counterparts $(\mu, \sigma^2, \beta_1, \beta_2)$. However, this choice assumes the existence of the first four moments and the sample statistics are highly nonrobust. It is obvious that the two location statistics and two scatter matrices used here can be replaced by other, more robust, functionals. The model selection would then follow the same rough rules. The scatter functionals should then be rescaled so that the regular covariance matrix is obtained in the multivariate normal case. In the independent components model, the scatter functionals should have the independence property.

Appendix A. Technical details

This section collects technical details that are valid when $\mathbf{S}_1$ and $\mathbf{S}_2$ are any two scatter matrices. Proofs are given only for new results.

Result A.1. If $\boldsymbol{\epsilon}$ satisfies (A4) then any scatter functional gives a diagonal matrix.

Proof of Result A.1. If $\boldsymbol{\epsilon} \sim \mathbf{Je}$, then $\mathbf{S}(\boldsymbol{\epsilon}) = \mathbf{S}(\mathbf{J}\boldsymbol{\epsilon}) = \mathbf{JS}(\boldsymbol{\epsilon})\mathbf{J}$. But $\mathbf{S}(\boldsymbol{\epsilon}) = \mathbf{A}$ where $\mathbf{A}$ is a symmetric psd matrix, and therefore $\mathbf{A}$ must equal $\mathbf{JAJ}$ for any sign change matrix $\mathbf{J}$. Let $\mathbf{J}^i$ be a sign change matrix with

$$\mathbf{J}^i_{jj} = \begin{cases} 1, & \text{if } j \neq i \\ -1, & \text{if } j = i. \end{cases}$$

Then $\mathbf{S}(\mathbf{J}^i\boldsymbol{\epsilon})$ changes the signs of the off-diagonal elements of the ith row and the ith column of $\mathbf{A}$. Therefore, those elements must be 0 in order for $\mathbf{A} = \mathbf{J}^i\mathbf{A}\mathbf{J}^i$ to hold. Since this is true for all $i = 1, \ldots, p$, $\mathbf{A}$ has to be a diagonal matrix. □

Result A.2. If $\boldsymbol{\epsilon}$ follows (A4) then the two-scatter transformation on $\mathbf{x} = \boldsymbol{\Omega}\boldsymbol{\epsilon} + \boldsymbol{\mu}$ estimates $\boldsymbol{\Omega}$ up to scale and permutation.

Proof of Result A.2. The proof follows the lines of the proof of Theorem 5 of Tyler *et al.* (2009) together with the Result A.1 and the fact that in this model,

$$\mathbf{x} = \boldsymbol{\Omega}\boldsymbol{\epsilon} + \boldsymbol{\mu} = (\boldsymbol{\Omega}\mathbf{D}^{-1}\mathbf{J})(\mathbf{J}\mathbf{D}\boldsymbol{\epsilon}) + \boldsymbol{\mu} = \boldsymbol{\Omega}^*\boldsymbol{\epsilon}^* + \boldsymbol{\mu},$$

where $\boldsymbol{\epsilon}^*$ is also part of (A4). □

Result A.3. Assume model (B1), $k = 2$, and $\boldsymbol{\mu}_1 \neq \boldsymbol{\mu}_2$. Then

$$d_{11} = \ldots = d_{p-1,p-1} > d_{pp} \quad \text{or} \quad d_{11} > d_{22} = \ldots = d_{pp} \quad \text{or} \quad d_{11} = \ldots = d_{pp}.$$

In the first two cases, the subspace corresponding to the eigenvalue with multiplicity 1 is Fisher's discriminant subspace. The result for general k is stated in Theorem 4 of Tyler *et al.* (2009).

Result A.4. If $\boldsymbol{\epsilon}$ satisfies (B2), then any scatter functional gives a diagonal matrix.

Proof of Result A.4. If $(\boldsymbol{\epsilon}^{*\top}, \epsilon^*_{p+1})^\top$ satisfies (A2) then also $\boldsymbol{\epsilon}^*$ and any of its components satisfies (A2) but with different dimension. The sign of $(\epsilon^*_{p+1} - \alpha - \beta\epsilon^*_p)$ is therefore a random sign change for the first $p-1$ components of $\boldsymbol{\epsilon}^*$. Clearly, $\mathbf{J}^i\boldsymbol{\epsilon} \sim \boldsymbol{\epsilon}$ for $i = 1, \ldots, p-1$ and

$$\mathbf{S}(\boldsymbol{\epsilon}) = \mathbf{S}(\mathbf{J}^i\boldsymbol{\epsilon}) = \mathbf{J}^i S(\boldsymbol{\epsilon})\mathbf{J}^i.$$

Applying the same reasoning as in the proof of Result A.1 shows again that $\mathbf{S}(\boldsymbol{\epsilon})$ must be a diagonal matrix. □

Result A.5. If $\boldsymbol{\epsilon}$ satisfies (B2) then

$$d_{11} = \ldots = d_{p-1,p-1} > d_{pp} \quad \text{or} \quad d_{11} > d_{22} = \ldots = d_{pp} \quad \text{or} \quad d_{11} = \ldots = d_{pp}.$$

Proof of Result A.5. Using again the fact that the first $p-1$ components of $\boldsymbol{\epsilon}$ still follow (A2), one knows that these components are exchangeable and therefore

$$\mathbf{S}(\boldsymbol{\epsilon}) = \mathbf{S}(\mathbf{P}^*\boldsymbol{\epsilon}), \quad \text{where} \quad \mathbf{P}^* = \begin{pmatrix} P & \mathbf{0} \\ \mathbf{0}^\top & 1 \end{pmatrix},$$

most hold, from which one can conclude that the first $p-1$ diagonal elements of $\mathbf{S}(\boldsymbol{\epsilon})$ must be the same. The rest follows by applying Theorem 2 of Tyler *et al.* (2009). □

Result A.6. Assume that $\boldsymbol{\epsilon}$ satisfies (B3) and $\mathbf{D}$ is based on any $\mathbf{S}_1$ and $\mathbf{S}_2$ having the independence property. If the diagonal elements of $\mathbf{D}$ are distinct, then $\boldsymbol{\epsilon}$ and $\mathbf{B}(\mathbf{x} - E(\mathbf{x}))$ differ by at most a permutation and/or change in componentwise signs and scales. For the case of non-distinct eigenvalues, see Theorem 8 of Tyler *et al.* (2009).

Appendix B. Tables

Tables B1 and B2 give true means and standard deviations for the simulation study described in Section 4.1.

Table B1. Means and standard deviations for the skewness measure **s**.

Model		N			
		100	200	500	1000
normal	s_1	0.10 (0.07)	0.07 (0.02)	0.04 (0.03)	0.03 (0.04)
	s_2	0.08 (0.05)	0.05 (0.06)	0.04 (0.02)	0.03 (0.02)
	s_3	0.06 (0.03)	0.05 (0.04)	0.03 (0.05)	0.02 (0.02)
t_{10}	s_1	0.20 (0.16)	0.15 (0.07)	0.10 (0.05)	0.08 (0.05)
	s_2	0.11 (0.13)	0.09 (0.09)	0.06 (0.04)	0.05 (0.04)
	s_3	0.08 (0.08)	0.07 (0.07)	0.05 (0.06)	0.04 (0.03)
skew-normal	s_1	0.20 (0.13)	0.21 (0.06)	0.22 (0.06)	0.24 (0.06)
	s_2	0.12 (0.11)	0.10 (0.08)	0.08 (0.05)	0.06 (0.05)
	s_3	0.09 (0.08)	0.08 (0.07)	0.06 (0.06)	0.04 (0.03)
skew-t_{10}	s_1	0.39 (0.24)	0.42 (0.14)	0.44 (0.13)	0.46 (0.10)
	s_2	0.19 (0.21)	0.18 (0.14)	0.18 (0.12)	0.15 (0.08)
	s_3	0.13 (0.19)	0.13 (0.13)	0.10 (0.09)	0.09 (0.07)
symmetric IC	s_1	0.12 (0.10)	0.08 (0.03)	0.06 (0.03)	0.04 (0.03)
	s_2	0.07 (0.07)	0.05 (0.05)	0.04 (0.02)	0.03 (0.02)
	s_3	0.06 (0.05)	0.05 (0.04)	0.03 (0.05)	0.02 (0.02)

Table B2. Means and standard deviations for the diagonal elements of the kurtosis measure **D**.

Model		N			
		100	200	500	1000
normal	d_{11}	1.10 (0.12)	1.08 (0.08)	1.06 (0.05)	1.05 (0.03)
	d_{22}	0.95 (0.07)	0.97 (0.06)	0.98 (0.03)	0.99 (0.03)
	d_{33}	0.84 (0.06)	0.88 (0.05)	0.93 (0.03)	0.95 (0.03)
t_{10}	d_{11}	1.52 (0.48)	1.56 (0.45)	1.52 (0.30)	1.51 (0.31)
	d_{22}	1.13 (0.15)	1.20 (0.14)	1.25 (0.12)	1.28 (0.09)
	d_{33}	0.94 (0.10)	1.02 (0.09)	1.11 (0.07)	1.16 (0.06)
skew-normal	d_{11}	1.17 (0.17)	1.16 (0.13)	1.14 (0.10)	1.14 (0.07)
	d_{22}	0.96 (0.08)	0.99 (0.06)	1.01 (0.04)	1.02 (0.03)
	d_{33}	0.84 (0.06)	0.89 (0.05)	0.94 (0.04)	0.96 (0.03)
skew-t_{10}	d_{11}	1.69 (0.66)	1.78 (0.61)	1.85 (0.70)	1.80 (0.51)
	d_{22}	1.17 (0.19)	1.25 (0.16)	1.33 (0.15)	1.36 (0.12)
	d_{33}	0.94 (0.11)	1.04 (0.09)	1.14 (0.09)	1.19 (0.07)
symmetric IC	d_{11}	1.18 (0.22)	1.17 (0.17)	1.20 (0.23)	1.20 (0.13)
	d_{22}	0.92 (0.08)	0.96 (0.06)	0.98 (0.04)	0.99 (0.03)
	d_{33}	0.72 (0.04)	0.74 (0.03)	0.75 (0.02)	0.75 (0.01)

References

Arellano-Valle, R. B., Branco, M. D. and Genton, M. G. (2006). A unified view on skewed distributions arising from selections. *The Canadian Journal of Statistics* **34**, pp. 581–601.

Azzalini, A. and Capitanio, A. (1999). Statistical applications of the multivariate skew-normal distribution, *Journal of the Royal Statistical Society, Series B* **61**, pp. 579–602.

Cardoso, J. F. (1989). Source separation using higher order moments, in *Proceedings of IEEE International Conference on Accoustics, Speech and Signal Processing* (Glasgow, UK), pp. 2109–2112.

Caussinus, H. and Ruiz-Gazen, A. (1994). Projection pursuit and generalized principal component analysis, in S. Morgenthaler, E. Ronchetti and W. A. Stahel (eds.), *New Directions in Statistical Data Analysis and Robustness* (Birkhäuser Verlag, Basel), pp. 35–46.

Critchley, F., Pires, A. and Amado, C. (2008). Principal axis analysis, Submitted.

Fang, K.-T. and Zhang, Y. T. (1990). *Generalized Multivariate Analysis* (Springer, New York, US).

Genton, M. G. (2004). *Skew-Elliptical Distributions and Their Applications: A Journey Beyond Normality* (Chapman & Hall / CRC, Boca Raton, US).

Gnanadesikan, R. and Kettenring, J. R. (1972). Robust estimates, residuals, and outlier detection with multiresponse data, *Biometrics* **28**, pp. 81–124.

Hallin, M. and Paindaveine, D. (2002). Optimal tests for multivariate location based on interdirections and pseudo-Mahalanobis ranks, *The Annals of Statistics* **30**, pp. 1103–1133.

Hyvärinen, A., Karhunen, J. and Oja, E. (2001). *Independent Component Analysis* (John Wiley & Sons, New York, US).

Kotz, S. (1975). Multivariate distributions at a cross road, in G. P. Patil, S. Kotz and J. Ord (eds.), *Statistical Distributions in Scientific Work*, Vol. 1 (D. Reidel Publishing, Dordrecht, Holland), pp. 247–270.

Maronna, R. A., Martin, R. D. and Yohai, V. J. (2006). *Robust Statistics - Theory and Methods* (John Wiley & Sons, Chichester, UK).

McLachlan, G. J. and Peel, D. (2000). *Finite Mixture Models* (John Wiley & Sons, New York, US).

Nordhausen, K., Oja, H. and Paindaveine, D. (2009). Signed-rank tests for location in the symmetric independent component model, *Journal of Multivariate Analysis* **100**, pp. 821–834.

Nordhausen, K., Oja, H. and Tyler, D. E. (2008). Tools for exploring multivariate data: The package ICS, *Journal of Statistical Software* **28**, pp. 1–31.

Oja, H. (1981). On location, scale, skewness and kurtosis of univariate distributions, *Scandinavian Journal of Statistics* **8**, pp. 154–168.

Oja, H., Sirkiä, S. and Eriksson, J. (2006). Scatter matrices and independent component analysis, *Austrian Journal of Statistics* **35**, pp. 175–189.

Ord, J. K. (1986). Pearson system of distributions, in *Enclyclopedia of Statistical Science*, Vol. 6 (Wiley & Sons, New York, US), pp. 655–659.

R Development Core Team (2008). *R: A Language and Environment for Statistical Computing*, R Foundation for Statistical Computing, Vienna, Austria, URL http://www.R-project.org, ISBN 3-900051-07-0.

Tyler, D. E., Critchley, F., Dümbgen, L. and Oja, H. (2009). Invariant coordinate selection, *Journal of the Royal Statistical Society, Series B* **71**, pp. 549–592.

An Empirical Study of Indirect Cross-Validation

Olga Savchuk

Department of Mathematical Sciences, Binghamton University
Binghamton, NY 13902, USA
E-mail: osavchuk@math.binghamton.edu

Jeffrey Hart† and Simon Sheather‡

Department of Statistics, Texas A&M University
College Station, TX 77843, USA
†*E-mail: hart@stat.tamu.edu*
‡*E-mail: sheather@stat.tamu.edu*

In this paper we provide insight into the empirical properties of indirect cross-validation (ICV), a new method of bandwidth selection for kernel density estimators. First, we describe the method and report on the theoretical results used to develop a practical-purpose model for certain ICV parameters. Next, we provide a detailed description of a numerical study that shows that the ICV method usually outperforms least squares cross-validation (LSCV) in finite samples. One of the major advantages of ICV is its increased stability compared to LSCV. Two real data examples show the benefit of using both ICV and a local version of ICV.

Keywords: Bandwidth selection; Cross-validation; Integrated Squared Error; Kernel density estimation; Mean Integrated Squared Error

1. Introduction

Let $X_1, \ldots, X_n$ be a random sample from an unknown density f. A kernel density estimator of f at the point x is defined as

$$\hat{f}_h(x) = \frac{1}{nh} \sum_{i=1}^{n} K\left(\frac{x - X_i}{h}\right), \tag{1}$$

where $h > 0$ is the bandwidth, and K is the kernel, which is generally chosen to be a unimodal probability density function that is symmetric about zero and has finite variance. A popular choice for K is the Gaussian kernel: $\phi(u) = (2\pi)^{-1/2} \exp(-u^2/2)$. To distinguish among estimators with different kernels, we shall refer to estimator (1) with given kernel K as a *K-kernel estimator*.

Practical implementation of the estimator (1) requires specification of the smoothing parameter h, also known as the bandwidth. Different bandwidth selection methods are reviewed and compared by Jones *et al.* (1996a,b). The two most widely-used bandwidth selectors are least squares cross-validation, proposed independently by Rudemo (1982) and Bowman (1984), and the Sheather and Jones plug-in method (Sheather and Jones, 1991). Plug-in is often preferred since it produces more stable bandwidths than does LSCV. Nevertheless, the LSCV method is still popular since it requires fewer assumptions than the plug-in method and works well when the density is difficult to estimate; see Loader (1999) and van Es (1992).

The main flaw of LSCV is high variability of the selected bandwidths. Other drawbacks include the tendency of cross-validation curves to exhibit multiple local minima, with the first local minimum being too small (see Hall and Marron, 1991), and the tendency of LSCV to select bandwidths that are much too small when the data exhibit a small amount of autocorrelation; see Hart and Vieu (1990) and Cao *et al.* (1993) for results of numerical studies. Many modifications of LSCV have been proposed in an attempt to improve its performance. These include biased cross-validation of Scott and Terrell (1987), a method of Chiu (1991), the trimmed cross-validation of Feluch and Koronacki (1992), the modified cross-validation of Stute (1992), and the method of Ahmad and Ran (2004) based on kernel contrasts.

This paper is concerned with a new modification of the LSCV method, called *indirect cross-validation* (ICV), recently proposed by the authors Savchuk *et al.* (2010). The ICV method depends on two parameters, α and σ. A main theoretical result is that at asymptotically optimal choices of α and σ, the ICV bandwidth can converge to zero at a rate $n^{-1/4}$, which is substantially better than the $n^{-1/10}$ rate of LSCV. The present paper contains the results of an empirical study of ICV. In Section 2 we provide a description of the method. Section 3 contains the details underlying the development of a practical purpose model for α and σ. Section 4 outlines the results of a numerical study that, in particular, show that ICV has greater stability in finite samples than does LSCV. In Section 5 we apply ICV and a local version of ICV to real data sets. Section 6 provides a summary of our results.

2. Description of indirect cross-validation

2.1. *Notation and definitions*

We begin with some notation and definitions that will be used subsequently. For an arbitrary function g, define

$$R(g) = \int g(u)^2\, du, \quad \mu_{jg} = \int u^j g(u)\, du,$$

where here and subsequently, integrals are assumed to be over the whole real line. The popular measures of performance of the kernel estimators (1) are integrated squared error (ISE) and mean integrated squared error (MISE). The ISE is defined as

$$ISE(h) = \int \left(\hat{f}_h(x) - f(x)\right)^2 dx, \tag{2}$$

and MISE is defined as the expectation of ISE. Assume that the underlying density f has second derivative that is continuous and square integrable. Let K be a kernel of the second order that satisfies the condition $R(K) < \infty$. A second order kernel integrates to 1, has first moment 0, and has finite, nonzero second moment. The bandwidth that asymptotically minimizes the MISE of the K-kernel estimator (1) has the following form:

$$h_n = \left\{ \frac{R(K)}{\mu_{2K}^2 R(f'')} \right\}^{1/5} n^{-1/5}. \tag{3}$$

The LSCV criterion is given by

$$\mathrm{LSCV}(h) = R(\hat{f}_h) - \frac{2}{n} \sum_{i=1}^{n} \hat{f}_{h,-i}(X_i), \tag{4}$$

where $\hat{f}_{h,-i}$ denotes the kernel estimator (1) constructed from the data without the observation X_i. A well known fact is that $\mathrm{LSCV}(h)$ is an unbiased estimator of $\mathrm{MISE}(h) - \int f^2(x)\, dx$. For this reason the LSCV method is often called *unbiased cross-validation.* Let $\hat{h}_{UCV}$ and h_0 denote the bandwidths that minimize the LSCV function (4) and the MISE of the ϕ-kernel estimator. Section 2.2 defines the ICV bandwidth, denoted as $\hat{h}_{ICV}$.

2.2. *The basic method*

The essence of the ICV method is to use different kernels at the cross-validation and density estimation stages. The same idea is exploited by the one-sided cross-validation method of Hart and Yi (1998) and Martínez-Miranda *et al.* (2009). ICV first selects the bandwidth of a second-order kernel L using least squares cross-validation. Selection kernels L used for this purpose are described in Section 2.3. The bandwidth so obtained is rescaled such that it can be used with the ϕ-kernel estimator. The rescaling constant C has the following form:

$$C = \left(\frac{\mu_{2L}^2}{2\sqrt{\pi} R(L)} \right)^{1/5}, \tag{5}$$

which is motivated by the asymptotically optimal MISE bandwidth (3). It is important that the constant C does not depend on any unknowns. Furthermore, the ICV method does not require any additional computing time compared to the LSCV method.

2.3. *Selection kernels*

We consider the family of kernels $\mathcal{L} = \{L(\,\cdot\,; \alpha, \sigma) : \alpha > 0, \sigma > 0\}$, where, for all u,

$$L(u; \alpha, \sigma) = (1 + \alpha)\phi(u) - \frac{\alpha}{\sigma}\phi\left(\frac{u}{\sigma}\right). \tag{6}$$

Note that the Gaussian kernel is a special case of (6) when $\sigma = 1$. Each member of $\mathcal{L}$ is symmetric about 0 and has the second moment $\mu_{2L} = 1 + \alpha - \alpha\sigma^2$. It follows that kernels in $\mathcal{L}$ are second-order, with the exception of those for which $\sigma = \sqrt{(1 + \alpha)/\alpha}$.

The family $\mathcal{L}$ can be partitioned into three families: $\mathcal{L}_1$, $\mathcal{L}_2$ and $\mathcal{L}_3$. The first of these is $\mathcal{L}_1 = \left\{L(\cdot; \alpha, \sigma) : \alpha > 0, \sigma < \frac{\alpha}{1+\alpha}\right\}$. Each kernel in $\mathcal{L}_1$ has a negative dip centered at $x = 0$. The kernels in $\mathcal{L}_1$ are ones that "cut-out-the-middle," some examples of which are shown in Figure 1**(a)**.

The second family is $\mathcal{L}_2 = \left\{L(\cdot; \alpha, \sigma) : \alpha > 0, \frac{\alpha}{1+\alpha} \leq \sigma \leq 1\right\}$. Kernels in $\mathcal{L}_2$ are densities that can be unimodal or bimodal. Note that the Gaussian kernel is a member of this family. The third family is $\mathcal{L}_3 = \left\{L(\cdot; \alpha, \sigma) : \alpha > 0, \sigma > 1\right\}$, each member of which has negative tails. Examples are shown in Figure 1**(b)**.

Kernels in $\mathcal{L}_1$ and $\mathcal{L}_3$ turn out to be highly efficient for cross-validation purposes but very inefficient for estimating f. This explains why we do not use L as both a selection *and* an estimation kernel.

Selection kernels in $\mathcal{L}$ are mixtures of two normal densities, which greatly simplifies computations. This fact has been utilized by Marron and Wand (1992) to derive exact MISE expressions. For kernels L it is possible to derive a closed form expression for the LSCV function. Marron and Wand (1992) also point out that, in addition to their computational advantages, normal mixtures can approximate any density arbitrarily well in various

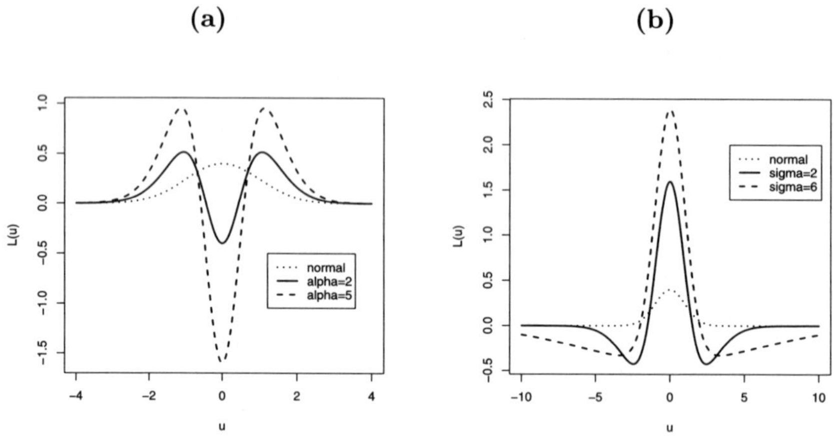

Fig. 1. **(a)** Selection kernels in $\mathcal{L}_1$ that have $\sigma = 0.5$; **(b)** Selection kernels in $\mathcal{L}_3$ with $\alpha = 6$. The dotted curve in both graphs corresponds to the Gaussian kernel.

senses. Mixtures of normals are therefore an excellent model for use in simulation studies, a fact that we take advantage of in Section 4.

3. Practical issues

In this section we address the problem of choosing the parameters, α and σ, of the selection kernel L in practice. We review some large sample theory for the ICV method and provide the theoretical results used to develop the practical-purpose model for α and σ.

3.1. *Large sample theory*

Large sample theory was developed by Savchuk *et al.* (2010) by considering the asymptotic mean squared error (MSE) of the ICV bandwidth. Their results may be summarized as follows.

(1) Under suitable regularity conditions, the ICV bandwidth is asymptotically normally distributed.
(2) The asymptotic MSE of $\hat{h}_{ICV}$ has been found for two cases: $\sigma \to 0$ (cut-out-the-middle kernels) and $\sigma \to \infty$ (negative-tailed kernels). It turns out that when the asymptotically optimal values of α and σ are used in the respective cases, the MSE converges to zero at the same rate of $n^{-9/10}$, but the limiting ratio of optimum mean squared errors is

0.752, with $\sigma \to \infty$ yielding the smaller error. In comparison, the rate at which the MSE for $\hat{h}_{UCV}$ converges to zero is $n^{-6/10}$. The subsequent theoretical results are provided for the case $\sigma \to \infty$.

(3) The relative rate of convergence of $\hat{h}_{ICV}$ to h_0 is $n^{-1/4}$, whereas the corresponding rate for $\hat{h}_{UCV}$ is $n^{-1/10}$.

(4) Values of σ which minimize the asymptotic MSE are as follows:

$$\sigma_{n,opt} = n^{3/8} A_\alpha \left[\frac{R(f)R(f'')^{13/5}}{R(f''')^2} \right]^{5/8}, \tag{7}$$

where

$$A_\alpha = 16\sqrt{\pi}\frac{2^{7/16}}{3^{5/8}}\frac{\alpha^{3/4}}{(1+\alpha)^2}\left[\frac{1}{8}(1+\alpha)^2 - \frac{8}{9\sqrt{3}}(1+\alpha) + \frac{1}{\sqrt{2}}\right]^{5/8}.$$

(5) The asymptotically optimal α is 2.4233. Remarkably, the optimal α does not depend on f.

(6) When the asymptotically optimal values of α and σ are used, the asymptotic bias and standard deviation of $\hat{h}_{ICV}$ are equal to each other and converge to zero at the same rate of $n^{-9/20}$.

The key property that allows an improvement in the ICV convergence rate is that the cross-validation kernels L are not fixed-shaped kernels but change with n through the parameter σ. A question of interest is whether it is possible to find a sequence of kernels L_n that will improve the relative $n^{-1/4}$ rate that our method attains.

3.2. *MSE-optimal α and σ*

The asymptotically optimal α is a known number, but the asymptotically optimal σ depends on f in a fairly complicated way (7). The problem of estimating the expression in square brackets in (7) is potentially as difficult or even more difficult than estimating f itself. One possibility is to use a reference estimator based on the standard normal density. However, instead of using just one reference distribution, we propose using the following five target densities, which are defined in Marron and Wand (1992):

$$\begin{aligned} \text{Gaussian density: } & N(0,1) \\ \text{Skewed unimodal density: } & \tfrac{1}{5}N(0,1) + \tfrac{1}{5}N\Big(\tfrac{1}{2}, \big(\tfrac{2}{3}\big)^2\Big) + \tfrac{3}{5}N\Big(\tfrac{13}{12}, \big(\tfrac{5}{9}\big)^2\Big) \\ \text{Bimodal density: } & \tfrac{1}{2}N\Big(-1, \big(\tfrac{2}{3}\big)^2\Big) + \tfrac{1}{2}N\Big(1, \big(\tfrac{2}{3}\big)^2\Big) \\ \text{Separated bimodal density: } & \tfrac{1}{2}N\Big(-\tfrac{3}{2}, \big(\tfrac{1}{2}\big)^2\Big) + \tfrac{1}{2}N\Big(\tfrac{3}{2}, \big(\tfrac{1}{2}\big)^2\Big) \\ \text{Skewed bimodal density: } & \tfrac{3}{4}N(0,1) + \tfrac{1}{4}N\Big(\tfrac{3}{2}, \big(\tfrac{1}{3}\big)^2\Big). \end{aligned}$$

These choices for f represent density shapes that are common in practice.

Table 1. MSE-optimal α and σ.

	Density:									
	normal		skewed unimodal		bimodal		separated bimodal		skewed bimodal	
n	α	σ	α	σ	α	σ	α	σ	α	σ
100	3.05	2.79	5.28	1.68	109.68	1.03	16.70	1.19	343.74	1.01
250	2.78	4.04	3.16	2.60	48.46	1.06	4.51	1.84	177.15	1.02
500	2.73	4.97	2.84	3.56	6.21	1.55	3.18	2.58	161.39	1.02
1000	2.69	5.97	2.75	4.49	3.73	2.12	2.84	3.54	123.78	1.03
5000	2.61	8.84	2.66	6.85	2.77	4.26	2.70	5.74	4.71	1.79
20000	2.55	12.40	2.59	9.58	2.68	6.22	2.63	8.08	2.85	3.46
100000	2.50	18.80	2.53	14.27	2.60	9.19	2.56	11.94	2.70	5.65
500000	2.47	29.54	2.49	21.88	2.54	13.65	2.50	18.07	2.62	8.39

Asymptotic results are not always reliable for practical purposes. In order to have an idea of how good choices of α and σ vary with n and f, we considered the following expression for the asymptotic MSE of the ICV bandwidth:

$$MSE(\hat{h}_{ICV}) = \left(\frac{1}{4\pi}\right)^{1/5} \frac{R(f''')^2}{R(f'')^{16/5}} n^{-3/5} \left\{ \frac{2}{25} \frac{R(f)R(f'')^{13/5}R(\rho_L)}{R(f''')^2 R(L)^{9/5}(\mu_{2L}^2)^{1/5}} + \frac{n^{-3/5}}{400} \left(\frac{R(L)^{2/5}\mu_{2L}\mu_{4L}}{(\mu_{2L}^2)^{7/5}} - \frac{3}{(4\pi)^{1/5}} \right)^2 \right\}, \tag{8}$$

where

$$\rho_L(u) = u \int L(x)L'(x+u)\, dx - 2uL'(u).$$

Expression (8) is valid for either large or small values of σ and includes a second-order bias term. Notice that the α minimizing expression (8) is not free of f.

In Table 1 we provide the MSE-optimal choices of α and σ for the target densities at eight sample sizes ranging from $n = 100$ up to $n = 500000$. It is obvious that the MSE-optimal α and σ vary greatly from one density to another, which is especially true for "small" sample sizes. However, the optimal α seems to converge to about 2.5 for each density as n increases, which fits with our observation that the optimal α is 2.4233. The optimal σ is increasing with sample size. It is remarkable that all the MSE-optimal α and σ in Table 1 correspond to kernels from $\mathcal{L}_3$, the family of negative-tailed kernels.

3.3. *Model for the ICV parameters*

We found a practical purpose model for α and σ by using polynomial regression. Our independent variable was $\log_{10}(n)$ and our dependent variables were the MSE-optimal values of $\log_{10}(\alpha)$ and $\log_{10}(\sigma)$ for the target densities. The $\log_{10}$ transformations for α and σ stabilize variability. Using a sixth-degree polynomial for α and a quadratic for σ, we arrived at the following models for α and σ:

$$\begin{aligned}\alpha_{mod} &= 10^{3.390-1.093\log 10(n)+0.025\log 10(n)^3-0.00004\log 10(n)^6},\\ \sigma_{mod} &= 10^{-0.58+0.386\log 10(n)-0.012\log 10(n)^2},\end{aligned} \tag{9}$$

which are appropriate for $100 \leq n \leq 500000$. The MSE-optimal values of $\log_{10}(\alpha)$ and σ together with the model fits are shown in Figure 2. In Table 2 we give the model choices α_{mod} and σ_{mod} for the same sample sizes as in Table 1.

Table 2. Model choices of α and σ.

n	100	250	500	1000	5000	20000	100000	500000
α_{mod}	25.20	12.77	8.24	5.71	3.23	2.66	2.66	2.62
σ_{mod}	1.39	1.89	2.37	2.95	4.83	7.21	11.22	16.98

4. Simulation study

The primary goal of our simulation study was to compare ICV with ordinary LSCV. However, we will also provide simulation results for the Sheather-Jones plug-in method.

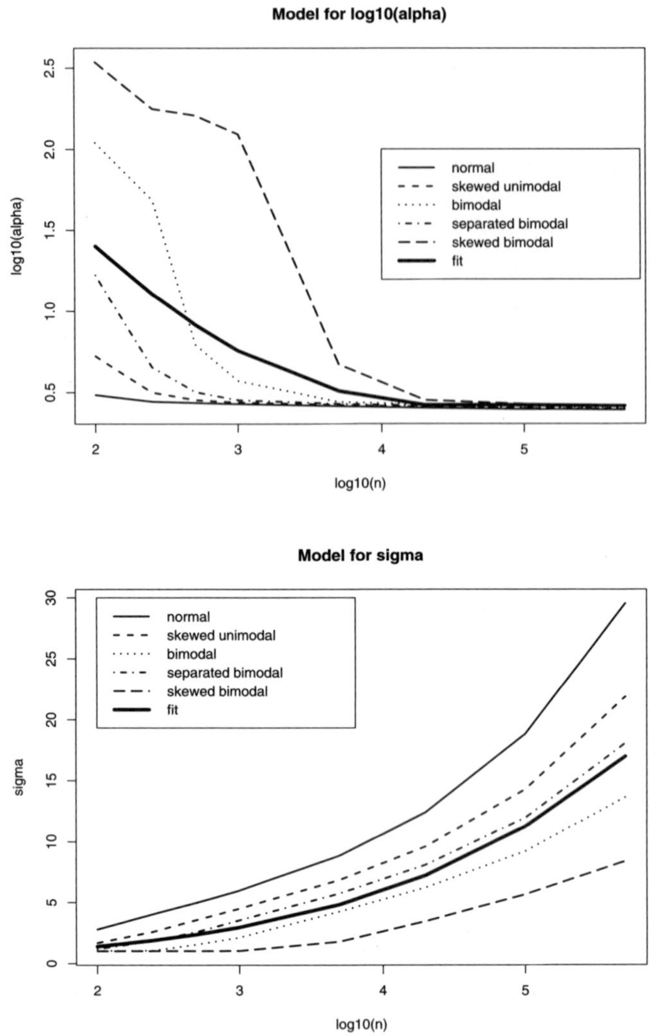

Fig. 2. MSE-optimal $\log_{10}(\alpha)$ and σ and the model fits.

We considered the four sample sizes $n = 100$, 250, 500, and 5000 and took samples from the target densities listed in Section 3.2. For each combination of density and sample size, we did 1000 replications. In all cases the parameters α and σ in the selection kernel L were chosen according to model (9).

Let $\hat{h}_0$ denote the minimizer of $ISE(h)$ for a Gaussian kernel estimator. For each sample, we computed $\hat{h}_0$, $\hat{h}^*_{ICV}$, $\hat{h}_{UCV}$, and the Sheather-Jones plug-in bandwidth $\hat{h}_{SJPI}$. The definition of $\hat{h}^*_{ICV}$ is

$$\hat{h}^*_{ICV} = \min(\hat{h}_{ICV}, \hat{h}_{OS}), \tag{10}$$

where $\hat{h}_{OS}$ is the oversmoothed bandwidth of Terrell (1990). It is arguable that *no* data-driven bandwidth should be larger than $\hat{h}_{OS}$ since this statistic estimates an upper bound for *all* MISE-optimal bandwidths (under standard smoothness conditions).

For any random variable Y defined in each replication of our simulation, we denote the mean, standard deviation, and median of Y over all replications (with n and f fixed) by $\widehat{\mathrm{E}}(Y)$, $\widehat{\mathrm{SD}}(Y)$, and $\widehat{\mathrm{Median}}(Y)$. To evaluate the bandwidth selectors, we computed $\widehat{\mathrm{E}}\big[ISE(\hat{h})/ISE(\hat{h}_0)\big]$ and $\widehat{\mathrm{Median}}\big[ISE(\hat{h})/ISE(\hat{h}_0)\big]$ for $\hat{h}$ equal to each of $\hat{h}^*_{ICV}$, $\hat{h}_{UCV}$, and $\hat{h}_{SJPI}$. We also computed the performance measure $\hat{E}\big[\hat{h}-\hat{E}(\hat{h}_0)\big]^2$, which estimates the MSE of the bandwidth $\hat{h}$.

Our main simulation results for the "normal" and "bimodal" densities, as defined in Section 3.2, are given in Tables 3 and 4 and Figures 3 and 4. Results for the other densities are available from the authors.

The reduced variability of the ICV bandwidth is evident in our study. The ratio $\widehat{\mathrm{SD}}(\hat{h}^*_{ICV})/\widehat{\mathrm{SD}}(\hat{h}_{UCV})$ ranges between 0.2103 and 0.9713 in the twenty settings considered. However, the variances of the ICV bandwidths are always higher compared to the Sheather-Jones plug-in bandwidths. It is worth noting that the ratio of sample standard deviations of the ICV and LSCV bandwidths decreases as the sample size n increases.

The mean squared distance $\widehat{\mathrm{E}}\big[\hat{h}-\widehat{\mathrm{E}}(\hat{h}_0)\big]^2$ is smaller for the ICV method than for the LSCV method in all but two cases corresponding to the Skewed bimodal density, $n = 250$ and 500. Plug-in always has a smaller value of $\widehat{\mathrm{E}}\big[\hat{h}-\widehat{\mathrm{E}}(\hat{h}_0)\big]^2$ than does ICV.

The most important observation is that the $\widehat{\mathrm{E}}\big[ISE(\hat{h})/ISE(\hat{h}_0)\big]$ values are smaller for ICV than for LSCV for all combinations of densities and sample sizes. The values of $\widehat{\mathrm{Median}}\big[ISE(\hat{h})/ISE(\hat{h}_0)\big]$ are smaller for ICV than for LSCV in all but one case, which corresponds to the Skewed bimodal density at $n = 250$, when $\widehat{\mathrm{Median}}\big[ISE(\hat{h})/ISE(\hat{h}_0)\big]$ is 1.0013 times greater for ICV than for LSCV.

Despite the fact that the LSCV bandwidth is asymptotically normally distributed (see Hall and Marron, 1987), its distribution in finite samples tends to be skewed to the left. In our simulations we have noticed that the distribution of the ICV bandwidth is less skewed than that of the LSCV

Table 3. Simulation results for the Gaussian density.

n	**LSCV**	**SJPI**	**ICV**	**ISE**
		$\widehat{\mathrm{E}}(\hat{h})$		
100	0.4452	0.3934	0.4153	0.4316
250	0.3640	0.3388	0.3494	0.3549
500	0.3109	0.2980	0.3086	0.3081
5000	0.1836	0.1899	0.1977	0.1953
		$\widehat{\mathrm{SD}}(\hat{h}) \cdot 10^2$		
100	12.3217	6.4324	6.5230	7.5201
250	8.3577	3.7174	4.4478	6.2730
500	7.1117	2.6030	3.0802	5.6350
5000	3.9008	0.6190	0.8204	3.0928
		$\widehat{\mathrm{E}}(\hat{h} - \widehat{\mathrm{E}}(\hat{h}_0))^2 \cdot 10^4$		
100	153.5291	55.9547	45.1705	
250	70.6115	16.3766	20.0568	
500	50.6085	7.7748	9.4813	
5000	16.5621	0.6679	0.7311	
		$\widehat{\mathrm{E}}\big(ISE(\hat{h})/ISE(\hat{h}_0)\big)$		
100	2.4700	1.9080	1.7218	
250	1.9159	1.5056	1.4757	
500	1.7581	1.3773	1.3610	
5000	1.4132	1.1146	1.1031	
		$\widehat{\mathrm{Median}}\big(ISE(\hat{h})/ISE(\hat{h}_0)\big)$		
100	1.3111	1.1570	1.1123	
250	1.2172	1.1041	1.0937	
500	1.2140	1.1031	1.0961	
5000	1.1091	1.0447	1.0518	

bandwidth. A typical case is illustrated in Figure 5, where kernel density estimates for the two data-driven bandwidths are plotted from the simulation with the skewed unimodal density at $n = 250$. Also plotted is a density estimate for the ISE-optimal bandwidths. Note that the ICV density is more concentrated near the middle of the ISE-optimal distribution than the density estimate for LSCV.

Figure 6 provides scatterplots of the bandwidths $\hat{h}_{UCV}$ and $\hat{h}_{ICV}$ versus $\hat{h}_0$ in the case of the Gaussian density and $n = 500$. The sample correlation coefficients are -0.52 and -0.60 for LSCV and ICV, respectively. The fact that these correlations are negative is a well-established phenomenon; see,

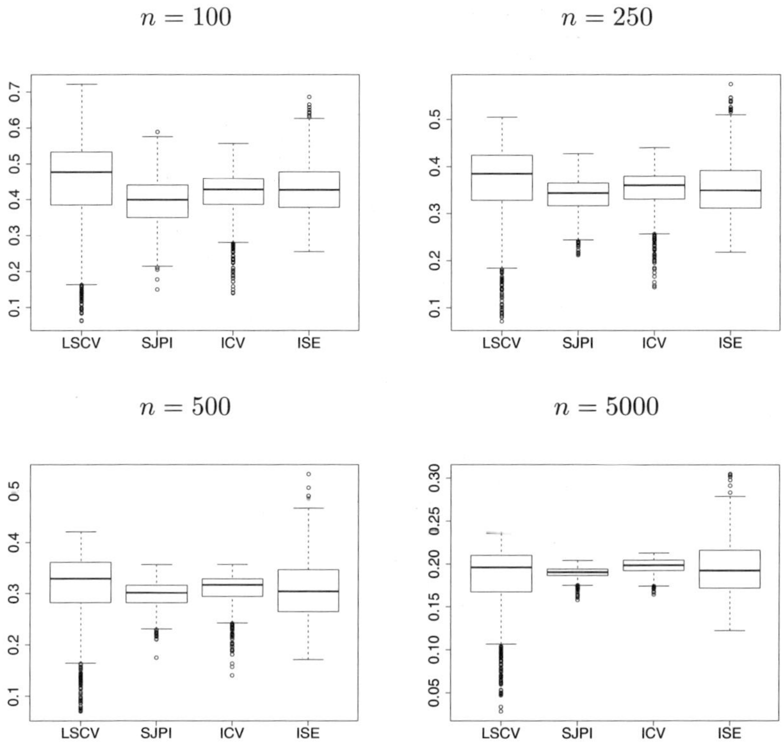

Fig. 3. Boxplots for the data-driven bandwidths in the case of the Gaussian density.

for example, Hall and Johnstone (1992). Note that the ICV bandwidths cluster more tightly about the MISE minimizer $h_0 = 0.315$.

A problem we have noticed with the ICV method is that its criterion function can have two local minima when the sample size is moderate and the density has two modes. The following example illustrates the problem. In Figure **7(a)** we have plotted three ICV curves for the case of the separated bimodal density and $n = 100$. The minimizers of the solid, dashed, and dotted lines occur at the h-values 0.2991, 2.0467, and 0.2204, respectively. For comparison, the corresponding bandwidths chosen by the Sheather-Jones plug-in method are 0.3240, 0.2508, and 0.2467. The value of $h = 2.0467$ that minimizes the dashed ICV curve is obviously too large. The local minimum at 0.1295 would yield a much more reasonable estimate. The problem of choosing too large a bandwidth from the second local minimum

Table 4. Simulation results for the bimodal density.

n	**LSCV**	**SJPI**	**ICV**	**ISE**
		$\widehat{\mathrm{E}}(\hat{h})$		
100	0.4291	0.3945	0.4196	0.3824
250	0.3136	0.3116	0.3285	0.2972
500	0.2593	0.2624	0.2745	0.2532
5000	0.1526	0.1571	0.1626	0.1548
		$\widehat{\mathrm{SD}}(\hat{h}) \cdot 10^2$		
100	13.5653	7.4443	9.5668	7.6090
250	8.4673	4.1878	6.5092	4.2943
500	5.7059	2.4444	4.2008	3.5598
5000	2.4629	0.4795	0.8146	1.9650
		$\widehat{\mathrm{E}}(\hat{h} - \widehat{\mathrm{E}}(\hat{h}_0))^2 \cdot 10^4$		
100	205.6555	56.8404	105.2554	
250	74.3324	19.6074	52.1298	
500	32.8927	6.8119	22.1647	
5000	6.1066	0.2820	1.2669	
		$\widehat{\mathrm{E}}\big(ISE(\hat{h})/ISE(\hat{h}_0)\big)$		
100	1.6995	1.3273	1.3614	
250	1.5160	1.2091	1.2874	
500	1.4167	1.1507	1.1917	
5000	2.0643	1.0684	1.0768	
		$\widehat{\mathrm{Median}}\big(ISE(\hat{h})/ISE(\hat{h}_0)\big)$		
100	1.2095	1.0874	1.1336	
250	1.1609	1.0834	1.1270	
500	1.1224	1.0607	1.0942	
5000	1.0583	1.0307	1.0365	

is mitigated by using the rule (10). Indeed, the oversmoothed bandwidths for the three samples are shown by the vertical lines in Figure 7 and are 0.7404, 0.7580, and 0.7341. Note that the problem with the ICV curve having two local minima of approximately the same value quickly goes away as the sample size increases. This is illustrated in Figure 7**(b)**, where we have plotted three criterion curves for the separated bimodal case with $n = 500$. Thus, the selection rule $\hat{h}^*_{ICV}$ given by (10), rather than just $\hat{h}_{ICV}$, appears to be useful mostly for small and moderate sample sizes.

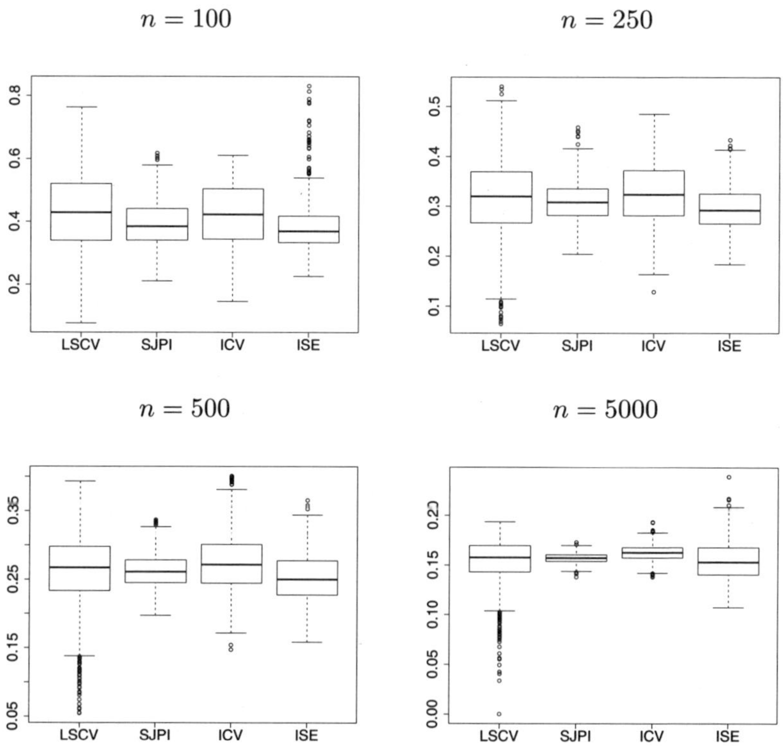

Fig. 4. Boxplots for the data-driven bandwidths in the case of the bimodal density.

5. Real data examples

In this section we show how the ICV method works on two real data sets. The purpose of the first example is to compare the performance of the ICV, LSCV, and Sheather-Jones plug-in methods. The second example illustrates the benefit of using ICV locally.

5.1. *PGA data*

In this example the data are the average numbers of putts, per round played, for the top 175 players on the 1980 and 2001 PGA golf tours. The question of interest is whether there has been any improvement from 1980 to 2001. This data set has already been analyzed by Sheather (2004) in the context of comparing the performances of LSCV and Sheather-Jones plug-in.

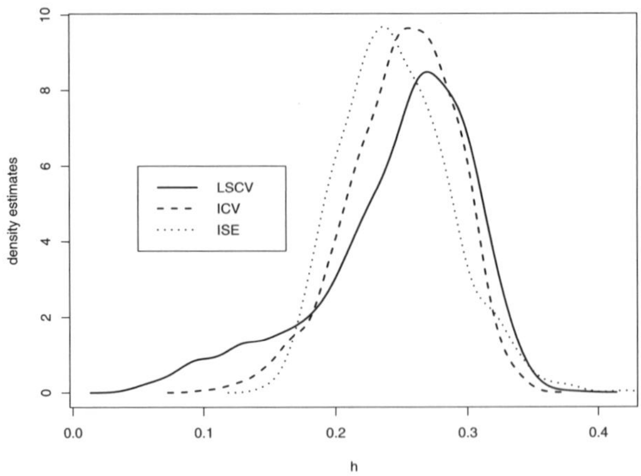

Fig. 5. Kernel density estimates for random bandwidths from the simulation with the skewed unimodal density and $n = 250$.

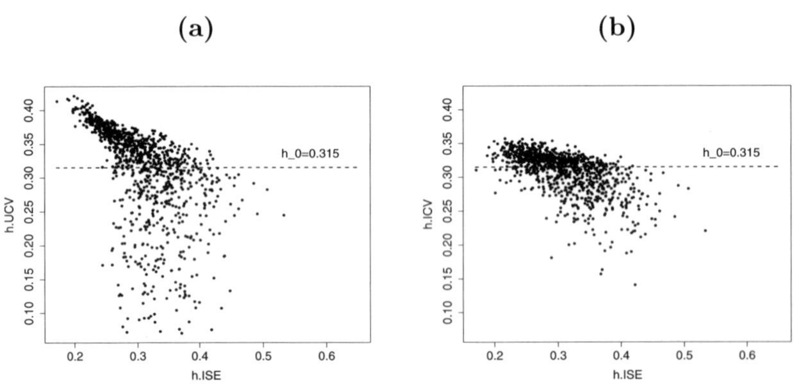

Fig. 6. Scatterplots of $\hat{h}$ vs. $\hat{h}_0$ for the case of the Gaussian density and $n = 500$, with $\hat{h}$ corresponding to the **(a)** LSCV and **(b)** ICV bandwidths.

In Figure 8 we have plotted an unsmoothed frequency histogram and the LSCV, ICV, and Sheather-Jones plug-in density estimates for a combined data set of 1980 and 2001 putting averages. The class interval size in the unsmoothed histogram was chosen to be 0.01, which corresponds to the accuracy to which the data have been reported. There is a clear indication of two modes in the histogram.

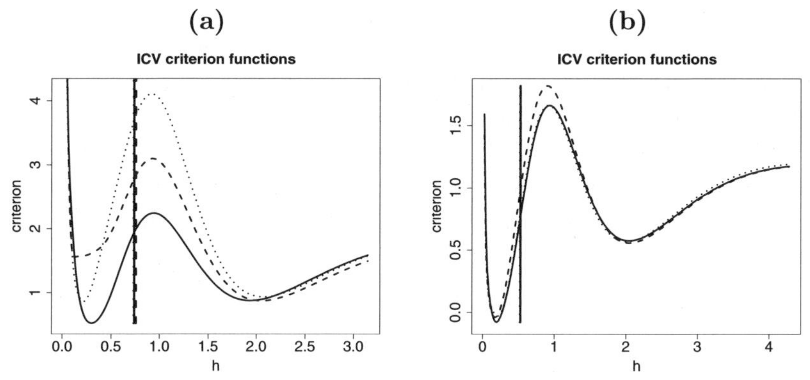

Fig. 7. Three ICV criterion functions in the case of the separated bimodal density at **(a)** $n = 100$ and **(b)** $n = 500$. The corresponding oversmoothed bandwidths are shown by the vertical lines.

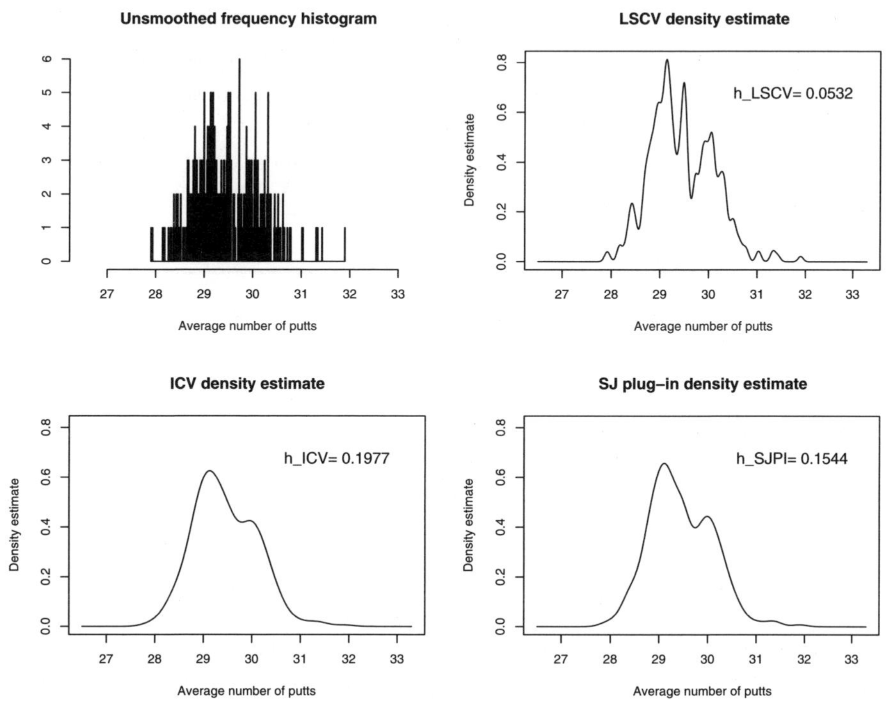

Fig. 8. Unsmoothed frequency histogram and kernel density estimates for average numbers of putts per round from 1980 and 2001 combined.

The estimate based on the LSCV bandwidth is apparently undersmoothed. The ICV and plug-in estimates look similar and have two modes, which agrees with evidence from the unsmoothed histogram and seems reasonable since the data were taken from two populations.

In Figure 9 we have plotted kernel density estimates separately for the years 1980 and 2001. ICV seems to produce a reasonable estimate in both years, whereas LSCV yields a very wiggly and apparently undersmoothed estimate in 2001.

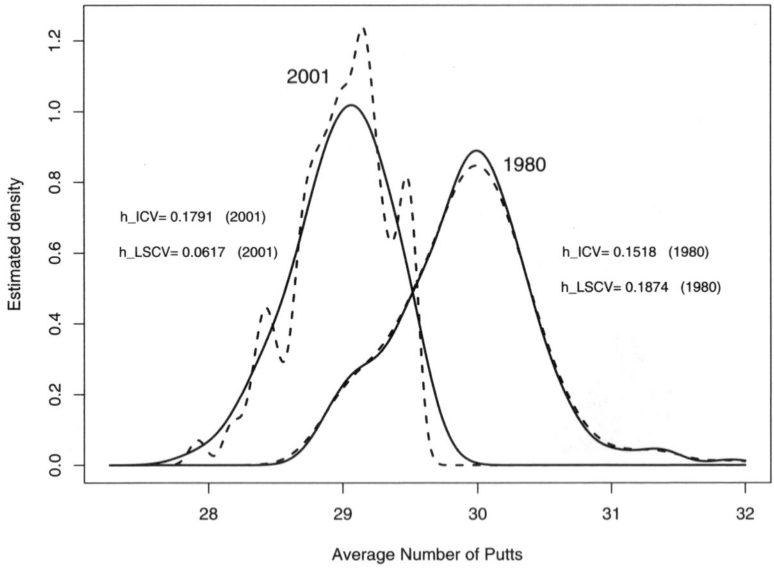

Fig. 9. Kernel density estimates based on LSCV (dashed curve) and ICV (solid curve) produced separately for the data from 1980 and 2001.

5.2. *Local ICV example*

Local cross-validation methods for density estimation, independently proposed by Hall and Schucany (1989) and Mielniczuk *et al.* (1989), consist in performing LSCV at each value of the argument x using a fraction of the data that are close to x. Allowing the bandwidth to depend on x is desirable when the smoothness of the underlying density changes sufficiently with x.

The local ICV method was introduced in Savchuk *et al.* (2010). It is different from the local LSCV method in that it uses ICV rather than LSCV for the local bandwidth selection. Another difference is that local ICV uses the first local minimizer of the local criterion function as opposed to the global minimizer of local LSCV.

The local ICV criterion function is defined as

$$ICV(x,b,w) = \frac{1}{w}\int \phi\left(\frac{x-u}{w}\right)\hat{f}_b(u)^2\,du - \frac{2}{nw}\sum_{i=1}^{n}\phi\left(\frac{x-X_i}{w}\right)\hat{f}_{b,-i}(X_i),$$

where $\hat{f}_b$ is the kernel density estimate based on a selection kernel L with a smoothing parameter b. The quantity w defines the extent to which the cross-validation is local, with a large choice of w corresponding to global ICV. Let $\hat{b}(x)$ be the first local minimum of the local ICV curve for the fixed value of x. Then the corresponding bandwidth of a ϕ-kernel estimator is defined as $\hat{h}(x) = C\hat{b}(x)$, where C is computed as in (5). Local ICV outperformed the local LSCV method in a simulated data example in Savchuk *et al.* (2010). In this paper we show how local ICV and LSCV perform in a real data example.

We analyze a data set of size $n = 517$ on the Drought Code (DC) of the Canadian Forest Fire Weather index (FWI) system. DC is one of the explanatory variables that can be used to predict the burned area of a forest in the Forest Fires data set. These data can be downloaded from the website `http://archive.ics.uci.edu/ml/datasets/Forest+Fires`. They were collected and analyzed by Cortez and Morais (2007).

We compute the LSCV, ICV and Sheather-Jones plug-in bandwidths for the DC data. The LSCV method fails by yielding $\hat{h}_{UCV} = 0$. The ICV and Sheather-Jones plug-in bandwidths are very close and produce similar density estimates. Figure 10**(a)** gives the ICV density estimate. It shows two major modes connected with a wiggly curve, which indicates that varying the bandwidth with x may yield a smoother estimate of the underlying density.

Local ICV and LSCV have been applied to the DC data. We used $w = 40$ for both methods and the selection kernel with $\alpha = 6$ and $\sigma = 6$ for local ICV. This (α, σ) choice performs quite well for unimodal densities in our simulation studies on global ICV and hence seems to be reasonable for local bandwidth selection, since locally the density should have relatively few features. Let $x_{(i)}$, $i = 1, \ldots, n$, denote the ith member of the ordered sequence of observations. The local ICV and LSCV bandwidths are found for 50 evenly spaced points in the interval $x_{(1)} - 0.2(x_{(n)} - x_{(1)}) \leq x \leq$

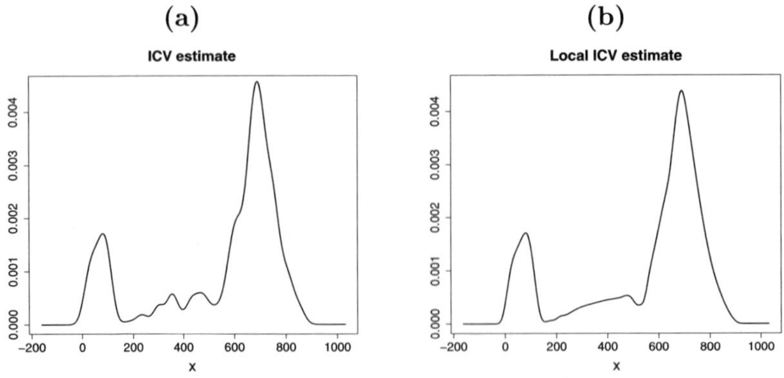

Fig. 10. Density estimates for the DC data set with **(a)** being the global ICV density estimate and **(b)** corresponding to the local ICV estimate.

$x_{(n)} + 0.2(x_{(n)} - x_{(1)})$. It turns out that, in 45 out of 50 cases, the local LSCV curve tends to $-\infty$ as $h \to 0$, which implies that the local LSCV estimate cannot be computed. All 50 local ICV bandwidths were positive. We found a smooth function $\hat{h}(x)$ by interpolating at other values of x via a spline. The corresponding local ICV estimate, given in Figure 10**(b)**, shows a smoother density estimate.

6. Summary

Indirect cross-validation is a method of bandwidth selection in the univariate kernel density estimation context. The method first selects the bandwidth of an L-kernel estimator by least squares cross-validation, then rescales this bandwidth so that it is appropriate for use in a Gaussian kernel density estimator. Selection kernels L have the form $(1+\alpha)\phi(u) - \alpha\phi(u/\sigma)/\sigma$, where $\alpha \geq 0$, $\sigma > 0$, and ϕ is the Gaussian kernel. Optimal kernels from this class yield bandwidths with relative error that converges to 0 at a rate of $n^{-1/4}$, which is a substantial improvement over the $n^{-1/10}$ rate of LSCV.

A practical-purpose model for the selection kernel parameters, α and σ, has been developed. The model is built by performing polynomial regression on the MSE-optimal values of $\log_{10}(\alpha)$ and $\log_{10}(\sigma)$ at different sample sizes for five target densities. Use of this model makes the ICV method completely automatic.

An extensive simulation study shows that in finite samples ICV is more stable than LSCV. Although both ICV and LSCV bandwidths are asymptotically normal, the distribution of the ICV bandwidths for finite n is usually more symmetric and better concentrated in the middle of the density for ISE-optimal bandwidths. Using an oversmoothed bandwidth as an upper bound for the bandwidth search interval reduces the bias of the method and prevents selecting an impractically large value of h when the criterion curves exhibit multiple local minima.

The ICV method performs well in real data examples. ICV applied locally yields density estimates that are more smooth than estimates based on a single bandwidth. Often, local ICV estimates may be found when the local LSCV estimates do not exist.

References

Ahmad, I. A. and Ran, I. S. (2004). Kernel contrasts: a data-based method of choosing smoothing parameters in nonparametric density estimation, *Journal of Nonparametric Statistics* **16**, 5, pp. 671–707.

Bowman, A. W. (1984). An alternative method of cross-validation for the smoothing of density estimates, *Biometrika* **71**, 2, pp. 353–360.

Cao, R., Quintela-del Río, A. and Vilar-Fernández, J. M. (1993). Bandwidth selection in nonparametric density estimation under dependence: A simulation study, *Computational Statistics* **8**, pp. 313–332.

Chiu, S. (1991). Bandwidth selection for kernel density estimation, *Annals of Statistics* **19**, 4, pp. 1883–1905.

Cortez, P. and Morais, A. (2007). A data mining approach to predict forest fires using meteorological data, in M. F. S. J. Neves and J. Machado (eds.), *New Trends in Artificial Intelligence, Proceedings of the 13th EPIA 2007—Portuguese Conference on Artificial Intelligence, December, Guimaraes, Portugal*, pp. 512–523.

Feluch, W. and Koronacki, J. (1992). A note on modified cross-validation in density estimation, *Computational Statistics and Data Analysis* **13**, 2, pp. 143–151.

Hall, P. and Johnstone, I. (1992). Empirical functional and efficient smoothing parameter selection, with discussion, *Journal of the Royal Statistical Society, Series B* **54**, 2, pp. 475–530.

Hall, P. and Marron, J. S. (1987). Extent to which least-squares cross-validation minimises integrated square error in nonparametric density estimation, *Probability Theory and Related Fields* **74**, 4, pp. 567–581.

Hall, P. and Marron, J. S. (1991). Local minima in cross-validation functions, *Journal of the Royal Statistical Society, Series B* **53**, 1, pp. 245–252.

Hall, P. and Schucany, W. R. (1989). A local cross-validation algorithm, *Statistics and Probability Letters* **8**, 2, pp. 109–117.

Hart, J. D. and Vieu, P. (1990). Data-driven bandwidth choice for density estimation based on dependent data, *Annals of Statistics* **18**, 2, pp. 873–890.

Hart, J. D. and Yi, S. (1998). One-sided cross-validation, *Journal of the American Statistical Association* **93**, 442, pp. 620–631.

Jones, M. C., Marron, J. S. and Sheather, S. J. (1996a). A brief survey of bandwidth selection for density estimation, *Journal of the American Statistical Association* **91**, 433, pp. 401–407.

Jones, M. C., Marron, J. S. and Sheather, S. J. (1996b). Progress in data-based bandwidth selection for kernel density estimation, *Computational Statistics* **11**, 3, pp. 337–381.

Loader, C. R. (1999). Bandwidth selection: classical or plug-in? *Annals of Statistics* **27**, 2, pp. 415–438.

Marron, J. S. and Wand, M. P. (1992). Exact mean integrated squared error, *Annals of Statistics* **20**, 2, pp. 712–736.

Martínez-Miranda, M. D., Nielsen, J. P. and Sperlich, S. (2009). One sided cross validation for density estimation, in G. N. Gregoriou (ed.), *Operational Risk Towards Basel III: Best Practices and Issues in Modeling, Management and Regulation* (John Wiley & Sons, Hoboken, New Jersey), pp. 177–196.

Mielniczuk, J., Sarda, P. and Vieu, P. (1989). Local data-driven bandwidth choice for density estimation, *Journal of Statistical Planning and Inference* **23**, 1, pp. 53–69.

Rudemo, M. (1982). Empirical choice of histograms and kernel density estimators, *Scandanavian Journal of Statistics* **9**, 2, pp. 65–78.

Savchuk, O. Y., Hart, J. D. and Sheather, S. J. (2010). Indirect cross-validation for density estimation, *Journal of the American Statistical Association* **105**, 489, pp. 415–423.

Scott, D. W. and Terrell, G. R. (1987). Biased and unbiased cross-validation in density estimation, *Journal of the American Statistical Association* **82**, 400, pp. 1131–1146.

Sheather, S. J. (2004). Density estimation, *Statistical Science* **19**, 4, pp. 588–597.

Sheather, S. J. and Jones, M. C. (1991). A reliable data-based bandwidth selection method for kernel density estimation, *Journal of the Royal Statistical Society, Series B* **53**, 3, pp. 683–690.

Stute, W. (1992). Modified cross-validation in density estimation, *Journal of Statistical Planning and Inference* **30**, 3, pp. 293–305.

Terrell, G. R. (1990). The maximal smoothing principle in density estimation, *Journal of the American Statistical Association* **85**, 410, pp. 470–477.

van Es, B. (1992). Asymptotics for least squares cross-validation bandwidths in nonsmooth cases, *Annals of Statistics* **20**, 3, pp. 1647–1657.

Extensions of Reliability Theory

Hoben Thomas

Department of Psychology, Pennsylvania State University
University Park, PA 16802, USA
E-mail: hxt@psu.edu

Arnold Lohaus[†] and Holger Domsch[‡]

Department of Psychology, University of Bielefeld
33615 Bielefeld, Germany

Individuals with matching true scores on two testing occasions are defined as stable. Individuals are unstable otherwise. Classical reliability theory assumes all individuals are stable. A proposed model assumes there is a mixed population of stable and unstable individuals. The estimated probability that individual $i = 1, 2, \ldots, n$ is stable is used to weight i's test scores, forming a weighted correlation coefficient r_w, which is reliability under the model.

Keywords: Correlation; Reliability; Test theory

1. Introduction

A fundamental problem in psychology is reliability. Reliability assesses the extent to which repeated measurements of the same individual yield the same value. In a typical reliability study, the same or a similar test is given on two occasions to the same individuals. The ordinary sample correlation coefficient r computed between the pairs of scores is the reliability coefficient.

Historically, the framework for interpreting the reliability r has been called the Spearman-Yule theory, a latent variables random-effect model (Spearman, 1910). Let Y_1 and Y_2 be the random variables on which observations are made on occasions 1 and 2. Define $Y_1 = T_1 + E_1$ and $Y_2 = T_2 + E_2$, where E_1 and E_2 are independent mean-zero common variance σ_E^2 error variables. The random variable T_1 is a "true score," independent of E_1 and E_2, with variance σ_T^2 and mean μ_T. Furthermore, $T_2 = T_1 + \alpha$, with α a fixed constant. With these assumptions, the correlation of Y_1 and Y_2, denoted as

ρ_{YY}, equals $\sigma_T^2/(\sigma_T^2 + \sigma_E^2)$. This model will be called the reliability model (RM). The same general model has appeared in other contexts, although typically independence, which is required here, has been replaced by an uncorrelated condition, especially in the historical literature (Yule, 1932). Assuming the model structure is appropriate, then $r = \hat{\rho}_{YY}$, so r (not r^2) estimates the proportion of total variance that is true score variance. For more than 100 years, this model has remained the basis for reliability interpretations in psychology.

Although sometimes referred to as an "individual differences" model because of the obvious real ("true score") differences among individuals in, for example, height or ability, this characterization is misleading because it is otherwise very restrictive. The error variance σ_E^2 is assumed to be constant for all individuals on both assessment occasions. And no allowance is made for individual perturbations in true scores between assessment occasions. This RM assumption of matching individual true scores on both assessment occasions is of primary concern here.

A mixture model will be proposed that views the population of pairs of test scores as consisting of subpopulations: Individuals with matching true scores on both testing occasions are called *stable*. Individuals without matching true scores on both testing occasions are called *unstable*. The goal is to probabilistically identify the stable individuals and use their scores as the basis for estimating ρ_{YY}. In addition, the model allows for some weakening of the constant error variance assumption. It will be seen that if the RM assumption that all individuals have matching true scores on both assessments appears inappropriate, the conventional reliability r can be a misleading estimate of ρ_{YY}. The model has the attractive feature that RM is a special case. If the RM assumption of matching true scores holds, the model returns the conventional r.

2. Classical Theory for a Fixed Individual with $T = t$

Consider an individual with a fixed $T = t$. Assume $\alpha = 0$ throughout. The model is easily altered so that $\alpha \neq 0$. Then

$$Y_{1|t} = t + E_1 \text{ for occasion 1,}$$
$$Y_{2|t} = t + E_2 \text{ for occasion 2.}$$

Because E_1 and E_2 are independent, $Y_{1|t}$ and $Y_{2|t}$ are conditionally independent given $T = t$. Expressing the model in a different form, with BV

denoting a bivariate distribution (without distributional assumptions):

$$(Y_{1|t}, Y_{2|t}) \sim BV\left(\begin{bmatrix} t \\ t \end{bmatrix}, \begin{bmatrix} \sigma_E^2 & 0 \\ 0 & \sigma_E^2 \end{bmatrix}\right).$$

Let $g_1(y_1|t)$ and $g_2(y_2|t)$ denote the conditional distributions of $Y_{1|t}$ and $Y_{2|t}$. Let $g(t)$ denote the density of T. Then the joint unconditional distribution of Y_1 and Y_2 is

$$f(y_1, y_2) = \int_{-\infty}^{\infty} g_1(y_1|t) g_2(y_2|t) g(t)\, dt.$$

Thus $f(y_1, y_2)$ is an infinite mixture distribution, with $g_1(y_1|t)g_2(y_2|t)$ representing a component with the continuous variable t, and $g(t)$ is the mixing distribution.

Furthermore, if $g_1(y_1|t)$, $g_2(y_2|t)$, and $g(t)$ are normal in distribution, then $f(y_1, y_2)$ is bivariate normal (BVN):

$$\begin{aligned}(Y_1, Y_2) &\sim BVN\left(\begin{bmatrix} \mu_T \\ \mu_T \end{bmatrix}, \begin{bmatrix} \sigma_E^2 + \sigma_T^2 & \sigma_T^2 \\ \sigma_T^2 & \sigma_E^2 + \sigma_T^2 \end{bmatrix}\right) \\ &= \int_{-\infty}^{\infty} BVN\left(\begin{bmatrix} t \\ t \end{bmatrix}, \begin{bmatrix} \sigma_E^2 & 0 \\ 0 & \sigma_E^2 \end{bmatrix}\right) g(t; \mu_T, \sigma_T^2)\, dt. \qquad (1)\end{aligned}$$

The correlation between Y_1 and Y_2 in the top term of (1) is $\sigma_T^2/(\sigma_T^2+\sigma_E^2) = \rho_{YY}$.

3. Proposed Model

Each individual i is assumed to have her own bivariate normal distribution, centered at (t_{i1}, t_{i2}), with independent marginal distributions:

$$(Y_{i1}, Y_{i2} \,|\, T_{i1} = t_{i1}, T_{i2} = t_{i2}) \sim BVN\left(\begin{bmatrix} t_{i1} \\ t_{i2} \end{bmatrix}, \begin{bmatrix} \sigma_{E_{i1}}^2 & 0 \\ 0 & \sigma_{E_{i2}}^2 \end{bmatrix}\right)$$

This expression is simply a more general version of the integrand structure in Equation (1). Viewed from the perspective of random variables,

$$Y_{io} = t_{io} + E_{io}$$

for occasions $o = 1, 2$, with $\mathrm{E}(E_{io}) = 0$ and $\mathrm{var}(E_{io}) = \sigma_{E_{io}}^2$.

3.1. *Difference scores*

Let $D_i = Y_{i1} - Y_{i2}$, $\Delta_i = T_{i1} - T_{i2}$, and $t_{i1} - t_{i2} = \delta_i$. Assume $\Delta_i \sim n(\mu_\delta, \sigma_\delta^2)$, which denotes a normal distribution with mean μ_δ and variance σ_δ^2. Furthermore, $D_i|\Delta_i = \delta_i \sim n(\delta_i, \sigma^2)$, with $\sigma^2 = \sigma_{E_{i1}}^2 + \sigma_{E_{i2}}^2$.

Recall that stable and unstable individuals are defined by

$$\mathrm{E}(Y_{i1}) = \mathrm{E}(Y_{i2}) \text{ if and only if } i \text{ is stable,}$$

and equivalently,

$$\mathrm{E}(Y_{i1}) \neq \mathrm{E}(Y_{i2}) \text{ if and only if } i \text{ is unstable.}$$

There are two models to consider, one for stable i with $\delta_i = 0$ and another for unstable i with $\delta_i \neq 0$.

For stable i, $\delta_i = 0$, so $\mathrm{E}(D_i|\Delta_i = \delta_i = 0) = 0$ and $\mathrm{var}(D_i|\Delta_i = \delta_i = 0) = \sigma^2$. Thus, $D_i|\Delta_i = \delta_i = 0 \sim n(0, \sigma^2)$ given that i is stable.

For unstable i, $\delta \neq 0$. View $\delta_i \neq 0$ as a random effect. There are two cases to consider, $\delta_i < 0$ and $\delta_i > 0$:

(1) $D_i|\Delta_i = \delta_i > 0$ implies $D_i \sim n(\mu_\delta > 0, \sigma_\delta^2)$. Then unconditional on δ_i but conditioning on i as unstable, the distribution of D_i with $\delta_i > 0$ is $n(\mu_\delta, \sigma^2 + \sigma_\delta^2)$.
(2) $D_i|\Delta_i = \delta_i < 0$ implies $D_i \sim n(-\mu_\delta < 0, \sigma_\delta^2)$. Then unconditional on δ_i but conditioning on i as unstable, the distribution of D_i with $\delta_i < 0$ is $n(-\mu_\delta, \sigma^2 + \sigma_\delta^2)$.

3.2. *Normal mixture model for the difference*

Given the above theory, the unconditional distribution of D_i is a three component normal mixture model with one stable and two unstable components with π weights, $0 < \pi_r < 1, \sum_{r=1}^{3} \pi_r = 1$:

$$D_i \sim \underbrace{\pi_1\, n(0, \sigma^2)}_{\text{stable, } \delta=0} + \underbrace{\pi_2\, n(-\mu_\delta, \sigma^2 + \sigma_\delta^2)}_{\text{unstable, } \delta<0} + \underbrace{\pi_3\, n(\mu_\delta, \sigma^2 + \sigma_\delta^2)}_{\text{unstable, } \delta>0}. \tag{2}$$

For completeness, the unconditional mean and variance of D_i are given: $\mathrm{E}(D_i) = (\pi_3 - \pi_2)\mu_\delta$ and $\mathrm{var}(D_i) = (\pi_2 + \pi_3)(\mu_\delta^2 + \sigma_\delta^2) + \sigma^2 - \mu_\delta^2(\pi_3 - \pi_2)^2$.

Equation (2) is not, of course, the only parameterization possible. But it seems to represent a reasonable compromise between what is achievable for small data sets typically encountered in reliability studies and psychological reality. A simpler but less realistic model would be to assume $\pi_2 = \pi_3$ and combine the second and third components, leading to one stable and one unstable component:

$$D_i \sim \pi n(0, \sigma^2) + (1 - \pi) n(0, \sigma^2 + \sigma_\delta^2).$$

However, scale mixtures (with common component means) tend to be more difficult to decompose, especially in small data sets, than (2), which has elements of both a scale and location mixture. Furthermore, it seems unrealistic to assume $\pi_2 = \pi_3$, as the analysis reported below will indicate.

3.3. *Posterior probability weights w_i and the weighted correlation r_w*

Let $g(d_i)$ denote the marginal (unconditional) distribution of D_i, with $g_r(d_i)$, $r = 1, 2, 3$, corresponding to the left-to-right listing of the three components in (2), so

$$g(d_i) = \sum_{r=1}^{3} \pi_r g_r(d_i).$$

The posterior probability of i belonging to the stable component is

$$w_i = p(i \text{ is stable} \mid d_i) = \frac{\pi_1 g_1(d_i)}{g(d_i)}.$$

Obtaining the w_i weights is the main focus of the analysis. These are used to weight the corresponding observed pairs (y_{i1}, y_{i2}) to return a weighted correlation, r_w, which is taken as the reliability under the model and is defined as follows. Let

$$y_{w1} = \frac{\sum_i w_i y_{i1}}{\sum_i w_i} \quad \text{and} \quad s_{w1} = \left[\frac{\sum_i w_i (y_{i1} - y_{w1})^2}{\sum_i w_i} \right]^{1/2},$$

with y_{w2} and s_{w2} defined similarly. Set

$$\text{cov} = \frac{\sum_i w_i (y_{i1} - y_{w1})(y_{i2} - y_{w2})}{\sum_i w_i}.$$

Then $r_w = \text{cov}/(s_{w1} s_{w2})$.

4. Estimation and Application Example

The difference scores $d_i = y_{i1} - y_{i2}$ are decomposed under the mixture model of (2), which yields estimates of π_1, π_2, $\pi_3 = 1 - \pi_1 - \pi_2$, μ_δ, σ^2, $\sigma^2 + \sigma_\delta^2$, and w_i. Confidence intervals, standard errors and various hypothesis tests that may be desired can be constructed with bootstrap procedures (Efron and Tibshirani, 1993). Routines to perform the main analysis are written in R (R Development Core Team, 2010), and are in the `mixtools` package (Young *et al.*, 2009) available from the `r-project.org` website.

4.1. *Example: Infant Habituation*

Habituation is a standard method of studying infant behaviors. Indeed, much of what is known about infant memory and perception rests on habituation methods. Six-month infants ($n = 51$) were habituated to a checkerboard pattern on two occasions, one week apart. On each occasion, the infant was presented with the checkerboard pattern and the length of time the infant viewed the pattern before disengaging was recorded; this denoted the end of a trial. After disengagement, another trial was presented. The procedure was implemented for eleven trials. The conventional index of habituation performance is the summed observed fixation to the checkerboard pattern over the eleven trials. Thus, an index of reliability focuses on how these fixation times, in seconds, on the two assessment occasions correlate: $r = .29$.

The proposed model is implemented by considering the d_i differences. In R, once `mixtools` is loaded using `library(mixtools)`, then with `DV` denoting the vector of 51 observed differences d_i, the R call

```
outmat<-normalmixEM(x=DV, mean.constr=c(0,"-a","a"),
                    sd.constr=c(NA,"b","b"),k=3,maxit=10000)
```

will yield output in the object `outmat` that is in Table 1. Here, `x` denotes the data vector of difference scores and `mean.constr` specifies that the first component has zero mean, followed by means of $-a$ and a. The argument `sd.constr` constrains the estimated standard deviations in the two unstable components to be equal, while `k=3` specifies three components. Although `maxit`, which specifies the number of iterations, is optional, the routine tends to be slow to converge, not an uncommon problem with EM routines, and is thus desirably set to some larger value.

The results of the habituation data analysis in Table 1 indicate that the proportion of stable infants is estimated to be 0.67. Similarly, $\hat{\mu}_\delta = 1.58$, $\hat{\sigma}^2 = 29.11$, and $\hat{\sigma}^2_\delta = 182.53$. Note that the third component associated with the mean 1.58 is essentially nonexistent. It seems plausible, therefore,

Table 1. Infant habituation data results.

	Mixture Components		
	$\delta = 0$	$\delta < 0$	$\delta > 0$
weights	.67	.33	.00
means	0	-1.58	1.58
variances	29.11	211.64	211.64

to follow up with another analysis deleting this component. Altering the arguments to `normalmixEM` in obvious ways (and ignoring `sd.constr`) yields a two-component solution with the remaining parameters the same as in Table 1.

Issuing `outmat$posterior[,1]` yields the estimated w_i weights. Combined with some elementary R code, these are used to obtain $r_w = .70$, substantially larger than $r = .29$. Under the model, r_w estimates ρ_{YY}. A bootstrap standard error for r_w is .15, while a 95% bootstrap confidence interval for ρ_{YY} ranges from .38 to .96.

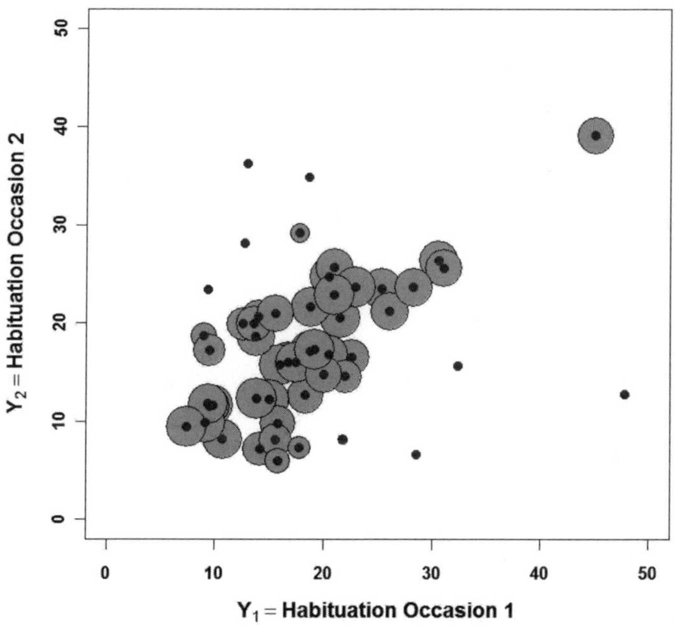

Fig. 1. Scatterplot of infant habituation data with circle size proportional to the probability of the coordinate pair being from the stable population.

Figure 1 provides a scatter plot of the habituation scores. Each point is represented by a circle. The larger the circle, the more probable the individual point is from the stable population. Not surprisingly of course, those points well away from the identity line are generally those which have a small probability of being from the stable component.

5. Discussion

The key model assumption is conditional independence: Given fixed true scores for i, Y_{i1} and Y_{i2} are independent. The assumption would not seem implausible for most settings. In fact of course, the RM model makes this assumption, as is evident from (1). Yule (1932, p. 212) assumed the error random variables shared common variance and were uncorrelated, but he made no distributional assumption. In the proposed model, E_{i1} and E_{i2} are normal in distribution and uncorrelated—which is equivalent to independence in the normal case—although the variances of each may be different for every i provided that the variance of the sum is constant over i. Although not reported here, simulations under the model suggest that it is effective with samples of size 25 or 50. How robust the model is outside of the model framework needs study.

The proposed model and procedures appear to provide a useful addition to the researcher's armamentarium of tools for assessing reliability. The model allows for weaker assumptions than RM but will return r if all $w_i = 1$. Since r_w estimates ρ_{YY} better than r in cases where the assumptions of the RM appear to be questionable, it appears to be a useful replacement for the classical reliability r.

Acknowledgments

Supported by NSF grant SES 0581772 and Deutsche Forschungsgemeinshaft grant Az Lo 337/19-2.

References

Efron, B. and Tibshirani, R. (1993). *An Introduction to the Bootstrap* (Chapman & Hall, New York).

R Development Core Team (2010). *R: A Language and Environment for Statistical Computing*, R Foundation for Statistical Computing, Vienna, Austria, URL `http://www.R-project.org`, ISBN 3-900051-07-0.

Spearman, C. (1910). Correlation calculated from faulty data, *British Journal of Psychology* **3**, pp. 271–295.

Young, D. S., Benaglia, T., Chauveau, D., Elmore, R. T., Hettmansperger, T. P., Hunter, D. R., Thomas, H. and Xuan, F. (2009). mixtools: Tools for mixture models, R package version 0.3.3.

Yule, G. U. (1932). *An Introduction to the Theory of Statistics, 10th edition* (Griffin, London).

Rank Regression under Possible Model Misspecification

Lan Wang
School of Statistics, University of Minnesota
Minneapolis, MN 55455, USA
E-mail: lan@stat.umn.edu

Rank regression offers a valuable alternative to the classical least squares approach. The use of rank regression not only provides protection against outlier contamination but also leads to substantial efficiency gain in the presence of heavier-tailed errors. This article studies the asymptotic performance of rank regression with Wilcoxon scores when the regression function is possibly misspecified. We establish that under general conditions, the Wilcoxon rank regression estimator converges in probability to a well-defined limit and has an asymptotic normal distribution. We also derive a formula for the bias of omitted variables. Besides furthering our understanding of the properties of rank regression, these theoretical results have important implications for developing rank-based model selection and model checking procedures.

Keywords: Bias of omitted variables; Model misspecification; Rank regression; Specification error; Wilcoxon procedure

1. Introduction

Wilcoxon-type rank-based procedures are fundamental for nonparametric statistical inference. The one-sample Wilcoxon rank test and two-sample Mann-Whitney-Wilcoxon test are among the most time-honored and widely applied tools for nonparametric analysis. Their great popularity comes from their validity under weak assumptions on the data, their high efficiency, and their ease of implementation. Hodges and Lehmann (1956, 1963) proved that Wilcoxon procedures sacrifice very little efficiency under normality and often lead to substantial efficiency gain with heavy-tailed errors. In the setting of regression analysis, these attractive features are inherited by Wilcoxon rank regression.

Consider a multiple linear regression model $\mathbf{Y} = \alpha \mathbf{1}_N + \mathbf{X}\boldsymbol{\beta} + \boldsymbol{\epsilon}$, where $\mathbf{Y}$ is an $n \times 1$ vector of responses, $\mathbf{1}_N$ is an $n \times 1$ vector of ones, $\mathbf{X}$ is an $n \times p$ matrix of covariates, $\boldsymbol{\beta}$ is a $p \times 1$ vector of unknown regression parameters, and ϵ is an $n \times 1$ vector of independent and identically distributed random errors. To estimate the regression parameter β, Wilcoxon rank regression minimizes

$$D(\mathbf{Y} - \mathbf{X}\boldsymbol{\beta}) = \sqrt{12}\sum_{i=1}^{n}\left[\frac{R(Y_i - X_i^\top \boldsymbol{\beta})}{n+1} - \frac{1}{2}\right](Y_i - X_i^\top \boldsymbol{\beta}), \qquad (1)$$

where $X_i^\top$ is the ith row of $\mathbf{X}$ and $R(Y_i - X_i^\top \boldsymbol{\beta})$ denotes the rank of $Y_i - X_i^\top \boldsymbol{\beta}$ among $Y_1 - X_1^\top \boldsymbol{\beta}, \ldots, Y_n - X_n^\top \boldsymbol{\beta}$. The dispersion function $D(\mathbf{Y} - \mathbf{X}\boldsymbol{\beta})$ is a nonnegative, piecewise linear, and convex function of $\boldsymbol{\beta}$. It can be minimized easily using basic linear programming techniques to obtain the Wilcoxon estimator by using the R (R Development Core Team, 2010) software developed by Terpstra and McKean (2005).

Rank regression provides a valuable alternative to classical least-squares regression. The Wilcoxon estimator is robust against possible outlier contamination, which can be explained by the fact that its influence function is bounded in the residual space. At the same time, $\widehat{\boldsymbol{\beta}}$ is 95.5% as efficient as the least squares estimator when the data are normally distributed and often outperforms the least squares estimator for nonnormal data. In a large simulation study comparing different robust regression procedures (Heiler, 1982), Wilcoxon rank regression consistently performs very well in various situations.

The distributional properties of $\widehat{\boldsymbol{\beta}}$ and related confidence region and hypothesis testing procedures were studied thoroughly by Jureăková (1971), Koul (1971), Jaeckel (1972), McKean and Schrader (1980), and Hettmansperger and McKean (1983), among others. This topic is nicely introduced in Chapter 3 of Hettmansperger and McKean (1998). For more recent developments of Wilcoxon rank regression in related areas, we refer to Chang *et al.* (1999) for high breakdown rank regression, Terpstra *et al.* (2001) and Mukherjee and Bai (2002) for R-estimation in autogression, Ozturk (2002) for rank regression for ranked set samples, Jung and Ying (2003) and Wang and Zhu (2006) for rank regression of longitudinal data, Jin *et al.* (2003) and Zhou (2005) for rank analysis of the accelerated failure time model, Chen *et al.* (2003) for rank regression in stability analysis, and Zuo and He (2006) for depth-based rank analysis of multivariate data.

Our objective is to examine the asymptotic behavior of Wilcoxon rank regression when the regression function is possibly misspecified, for example, when a linear regression function is fitted while the true regression function is nonlinear, or when some important covariates are omitted. This type of global misspecification is different from the local misspecification discussed in Section 3.9.1 of Hettmansperger and McKean (1998). There has been continuing interest in studying statistical procedures under possible global model misspecification. White (1982) discussed maximum likelihood estimation under model misspecification, Lagakos (1988) and DiRienzo and Lagakos (2001) investigated proportional hazards regression models under misspecification, Séménou (1996) studied misspecified generalized linear models, Severini (1998) derived properties of inferences in misspecified linear models, and Velilla (2001) analyzed bootstrap in misspecified regression models; see also Risch and Giuffra (1992), Gerds and Schumacher (2001), Uppal and Wang (2003), Waldorp *et al.* (2005), Boher (2006), and Hansen *et al.* (2006).

There are two major results in this paper. First, we establish that under very general conditions, the Wilcoxon rank regression estimator converges to a well-defined limit and has an asymptotic normal distribution even when the regression model is misspecified. The derived asymptotic normal distribution reduces to the well-known form in the literature when the regression function is actually correctly specified. Second, we show that the Wilcoxon rank regression estimator has an interpretation of minimizing a weighted mean-squared error loss function for specification error and we derive a formula for the bias of omitted variables, i.e., the bias that is caused when some important variables are excluded from the regression model. The first result parallels that in the classical paper of White (1981), who found that under general conditions the least squares estimator for the parameters of a misspecified nonlinear regression model converges strongly to the parameters of a weighted least squares approximation to the true model; while the second result echoes the work of Angrist *et al.* (2006) for quantile regression under misspecification.

Specification of the functional form of regression is crucial for any regression analysis. In practical data analysis, it is not unusual to have the regression function misspecified. Although the results in the paper are theoretical, they have broader implications particularly for further development of ranked-based model selection and model checking procedures. The remainder of this paper is organized as follows. Section 2 introduces notation and establishes a local quadratic approximation to the dispersion function, from which we prove the asymptotic normality of the Wilcoxon rank regression estimator under possible model misspecification. In Section 3, we give the Wilcoxon rank regression estimator a new geometric interpretation and derive an omitted variables bias formula. In Section 4, we discuss the broader impacts of the obtained theoretical results.

2. Asymptotic Normality of the Wilcoxon Rank Regression Estimator under Possible Model Misspecification

2.1. *Preliminaries and Notation*

The data are generated by

$$Y_i = m(X_i) + \epsilon_i, \tag{2}$$

where ϵ_i, $i = 1, \ldots, n$, are random errors. Similarly to Angrist *et al.* (2006), let

$$\Delta(X_i) = m(X_i) - X_i^\top \boldsymbol{\beta}, \quad i = 1, \ldots, n, \tag{3}$$

be the regression function *specification error* when the true model is given in (2) but a linear regression model is fitted.

Suppose we fit a multiple linear regression model to $\{(X_i^\top, Y_i)^\top, i = 1, \ldots, n\}$ using the Wilcoxon procedure, i.e., $\widehat{\boldsymbol{\beta}}$ minimizes the dispersion function $D(\mathbf{Y} - \mathbf{X}\boldsymbol{\beta})$ defined in (1). Let

$$S(\mathbf{Y} - \mathbf{X}\boldsymbol{\beta}) = -\bigtriangledown_{\boldsymbol{\beta}} D(\mathbf{Y} - \mathbf{X}\boldsymbol{\beta}) = \sqrt{12} \sum_{i=1}^{n} \left[\frac{R(Y_i - X_i^\top \boldsymbol{\beta})}{n+1} - \frac{1}{2} \right] X_i.$$

Then $S(\mathbf{Y} - \mathbf{X}\boldsymbol{\beta})$ is the negative of the gradient function of $D(\mathbf{Y} - \mathbf{X}\boldsymbol{\beta})$ and exists almost everywhere. For notational simplicity, we will often denote it by $S(\boldsymbol{\beta})$.

If one fits a linear regression function when the true regression function is nonlinear, does the Wilcoxon estimator $\widehat{\boldsymbol{\beta}}$ still converge asymptotically to anything meaningful? Intuition suggests that if $\widehat{\boldsymbol{\beta}}$ converges to anything at all, it should be the minimizer of the limit (or the population version) of the dispersion function. It is thus reasonable to expect that $\widehat{\boldsymbol{\beta}} \xrightarrow{p} \boldsymbol{\beta}_0 = \arg\min_{\boldsymbol{\beta}} \lim_{n\to\infty} D(\mathbf{Y} - \mathbf{X}\boldsymbol{\beta})$ under suitable conditions.

Remark 2.1. To gain more insight into the meaning of $\boldsymbol{\beta}_0$ and the validity of the consistency of $\widehat{\boldsymbol{\beta}}$, it is helpful to observe (Sievers, 1983) that minimizing $D(\mathbf{Y} - \mathbf{X}\boldsymbol{\beta})$ is equivalent to minimizing

$$\frac{1}{n(n-1)} \sum_{i=1}^{n} \sum_{j \neq i}^{n} |Z_i - Z_j|, \tag{4}$$

where $Z_i = Y_i - X_i^\top \boldsymbol{\beta} = m(X_i) - X_i^\top \boldsymbol{\beta} + \epsilon_i$. Expression (4) is often referred to as *Gini's mean difference* (Gini, 1912). It measures the variability in the data and has particular importance in economics due to its relationship with Gini's concentration index; see David (1998). Since Gini's mean difference has a U-statistic representation, it is clear that the population parameter value $\boldsymbol{\beta}_0$ minimizes $E|Z_1 - Z_2|$. If $\boldsymbol{\beta}_0$ uniquely minimizes $E|Z_1 - Z_2|$, the consistency of $\widehat{\boldsymbol{\beta}}$ for $\boldsymbol{\beta}_0$ follows from Lemma 2.2 of White (1980).

In what follows, we list some technical conditions for the asymptotic results.

Assumption 2.1. *The random errors ϵ_i, $i = 1, \ldots, n$, are independently and identically distributed with density function $f(\cdot)$ and cumulative distribution function $F(\cdot)$. The density f is absolutely continuous with finite Fisher information $0 < I(f) < \infty$, where $I(f) = \int (f'(u)/f(u))^2 f(u)\, du$ (Hettmansperger and McKean, 1998, p. 88).*

Assumption 2.2. *The covariates $X_1, \ldots, X_n$ are independently and identically distributed (i.i.d.) p-dimensional random vectors with joint cumulative distribution function $G(\cdot)$ and covariance matrix $Cov(X_i) = \boldsymbol{\Sigma}$, a positive definite matrix.*

Assumption 2.3. *$E|Z_1 - Z_2|$, where $Z_i = m(X_i) - X_i^\top \boldsymbol{\beta} + \epsilon_i$, $i = 1, 2$, has a unique minimizer at $\boldsymbol{\beta} = \boldsymbol{\beta}_0$, an interior point of the compact parameter space.*

Assumption 2.4. *Define*

$$\mathbf{B} = \sqrt{12} E \left[(X_1 - X_2) X_1^\top \int f(u + \Delta(X_1) - \Delta(X_2)) f(u)\, du \right]. \tag{5}$$

Let

$$V(X_1, X_2, X_3) =$$
$$\int [F(u + \Delta(X_1) - \Delta(X_2)) - 1/2]\, [F(u + \Delta(X_1) - \Delta(X_3)) - 1/2]\, f(u)\, du$$

and define

$$\mathbf{J} = 12 E[(X_1 - X_2)(X_1 - X_3)^\top V(X_1, X_2, X_3)]. \tag{6}$$

Assume $\mathbf{B}$ *and* $\mathbf{J}$ *are both positive definite matrices.*

Remark 2.2. Note that the dispersion function $D(\mathbf{Y} - \mathbf{X}\boldsymbol{\beta})$ is invariant to a constant change of $\mathbf{X}$; thus without loss of generality we assume the covariates are centered, i.e., $E(X_i) = 0$. Therefore $\boldsymbol{\Sigma} = E(X_i X_i^\top)$.

Remark 2.3. In the definition of the matrix $\mathbf{B}$, the expectation is taken with respect to the joint distribution of X_1 and X_2, which are two i.i.d. copies from the covariate distribution $G(\cdot)$. Note that in the case of no specification error, $\mathbf{B}$ reduces to a simple expression: $\mathbf{B} = \sqrt{12} \int f^2(u)\, du \boldsymbol{\Sigma}$.

2.2. Three steps toward asymptotic normality

The asymptotic normality of the Wilcoxon rank regression estimator under possible model misspecification is a delicate problem. Although the result we obtain is parallel to that in the classical paper of White (1981), our approach is significantly different since the Taylor expansion, which is central to White's proof, does not apply to the rank-based dispersion function.

We approximate the distribution of $\widehat{\boldsymbol{\beta}}$ by the following three steps.

- We first derive a linear approximation to the negative of the gradient function $S\left(\boldsymbol{\beta}_0 + t/\sqrt{n}\right)$.
- The linear approximation to the negative gradient function suggests naturally a quadratic approximation to the dispersion function $D(\mathbf{Y} - \mathbf{X}\boldsymbol{\beta})$. We show that this quadratic approximation is valid uniformly on a compact set.
- We then prove that the asymptotic distribution of $\widehat{\boldsymbol{\beta}}$ can be approximated by that of $\widetilde{\boldsymbol{\beta}}$, the minimizer of the quadratic approximation.

These three steps are formally stated in Lemmas 2.1–2.3 below. Our approach is similar to that of Heiler and Willers (1988) for establishing the asymptotic normality of the rank regression estimator when the model is correctly specified, which was nicely presented in Appendix A.3 of Hettmansperger and McKean (1998). The actual proof of each step is more technically involved due to the possible presence of specification error. Another minor difference is that we assume the covariates are random, rather than fixed as in Heiler and Willers (1988). Finally, the asymptotic distribution of $\widetilde{\boldsymbol{\beta}}$ is derived using the theory of U-statistics; see Theorem 2.1 below.

Lemma 2.1 below gives the linear approximation to the gradient function.

Lemma 2.1. *Under Assumptions 2.1–2.2, for any $p \times 1$ finite constant vector t,*

$$\frac{1}{\sqrt{n}} S\left(\boldsymbol{\beta}_0 + \frac{t}{\sqrt{n}}\right) = \frac{1}{\sqrt{n}} S(\boldsymbol{\beta}_0) - \mathbf{B}t + o_p(1),$$

where $\mathbf{B}$ *is defined in (5).*

Proof. Denote $Q_n = \frac{1}{\sqrt{n}}\left[S\left(\boldsymbol{\beta}_0 + t/\sqrt{n}\right) - S(\boldsymbol{\beta}_0)\right]$. Then

$$\begin{aligned}
Q_n &= \frac{\sqrt{12}}{\sqrt{n}(n+1)}\sum_{i=1}^{n}\left\{R\left[Y_i - X_i^\top\left(\boldsymbol{\beta}_0 + \frac{t}{\sqrt{n}}\right)\right] - R\left(Y_i - X_i^\top\boldsymbol{\beta}_0\right)\right\}X_i \\
&= \frac{\sqrt{12}}{\sqrt{n}(n+1)}\sum_{i=1}^{n}\sum_{j=1}^{n}\left\{I\left[Y_j - X_j^\top\left(\boldsymbol{\beta}_0 + \frac{t}{\sqrt{n}}\right) \le Y_i - X_i^\top\left(\boldsymbol{\beta}_0 + \frac{t}{\sqrt{n}}\right)\right]\right. \\
&\quad \left. - I(Y_j - X_j^\top\boldsymbol{\beta}_0 \le Y_i - X_i^\top\boldsymbol{\beta}_0)\right\}X_i \\
&= \frac{\sqrt{12}}{\sqrt{n}(n+1)}\sum_{i=1}^{n}\sum_{j=1}^{n}\left\{I\left[\epsilon_j \le \epsilon_i + (\Delta(X_i) - \Delta(X_j)) + (X_j - X_i)^\top\frac{t}{\sqrt{n}}\right]\right. \\
&\quad \left. - I\left[\epsilon_j \le \epsilon_i + (\Delta(X_i) - \Delta(X_j))\right]\right\}X_i.
\end{aligned}$$

Applying the iterated expectation formula by first conditioning on the covariates, we obtain

$$\begin{aligned}
&E(Q_n) \\
&= \frac{\sqrt{12}}{\sqrt{n}(n+1)}E\left\{\sum_{i=1}^{n}\sum_{j=1}^{n}\int\left[F\left(u + (\Delta(X_i) - \Delta(X_j)) + (X_j - X_i)^\top\frac{t}{\sqrt{n}}\right)\right.\right. \\
&\quad \left.\left. - F(u + (\Delta(X_i) - \Delta(X_j)))\right] f(u)\,du X_i\right\} \\
&= \frac{\sqrt{12}}{\sqrt{n}(n+1)}E\left\{\sum_{i=1}^{n}\sum_{j=1}^{n}\int\left[f(u + (\Delta(X_i) - \Delta(X_j)))(X_j - X_i)^\top\frac{t}{\sqrt{n}}\right.\right. \\
&\qquad\qquad \left.\left. + o\left(n^{-1/2}\right)\right] f(u)\,du X_i\right\} \\
&= -\mathbf{B}t + o(1).
\end{aligned}$$

It is sufficient to show $\mathrm{Var}(Q_n) = o(1)$, which we will do next. To simplify the notation, let

$$\begin{aligned}
W_{ij} = &\left\{I\left[\epsilon_j \le \epsilon_i + (\Delta(X_i) - \Delta(X_j)) + (X_j - X_i)^\top\frac{t}{\sqrt{n}}\right]\right. \\
&\left. - I[\epsilon_j \le \epsilon_i + (\Delta(X_i) - \Delta(X_j))]\right\}X_i.
\end{aligned}$$

Then

$$\mathrm{Var}(Q_n) = O(n^{-3})E\left[\sum_{i=1}^{n}\sum_{j\neq i}[W_{ij} - E(W_{ij})]\right]^2 = V_1 + V_2 + O(n^{-1}),$$

where

$$V_1 = O(n^{-3})\sum_{i=1}^{n}\sum_{j\neq i}\sum_{k\neq i,k\neq j} E\left\{[W_{ij} - E(W_{ij})][W_{ik} - E(W_{ik})]\right\},$$

$$V_2 = O(n^{-3})\sum_{i=1}^{n}\sum_{j\neq i}\sum_{k\neq i,k\neq j} E\left\{[W_{ij} - E(W_{ij})][W_{kj} - E(W_{kj})]\right\}.$$

Note that

$$E(W_{ij}) = \int\left[F\left(u + (\Delta(x_i) - \Delta(x_j)) + (x_j - x_i)^\top \frac{t}{\sqrt{n}}\right) - F(u + (\Delta_i - \Delta_j))\right] \times x_i f(u)\, du\, dG(x_i)\, dG(x_j),$$

and we have

$$\begin{aligned}
V_1 &= O(n^{-3})\sum_{i=1}^{n}\sum_{j\neq i}\sum_{k\neq i,k\neq j}\left\{E(W_{ij}W_{ik}) - E(W_{ij})E(W_{ik})\right\}\\
&= O(n^{-3})\sum_{i=1}^{n}\sum_{j\neq i}\sum_{k\neq i,k\neq j}\left\{\int\left[f(u + \Delta(x_i) - \Delta(x_j))(x_j - x_i)^\top\frac{t}{\sqrt{n}} + o(n^{-1/2})\right]\right.\\
&\qquad\times\left[f(u + \Delta(x_i) - \Delta(x_k))(x_k - x_i)^\top\frac{t}{\sqrt{n}} + o(n^{-1/2})\right]\\
&\qquad\times x_i^\top x_i f(u)\, du\, dG(x_i)\, dG(x_j)\, dG(x_k)\\
&\quad - \int\left[f(u + \Delta(x_i) - \Delta(x_j))(x_j - x_i)^\top\frac{t}{\sqrt{n}} + o(n^{-1/2})\right]\\
&\quad\times x_i f(u)\, du\, dG(x_i)\, dG(x_j)\\
&\quad\times\int\left[f(u + \Delta(x_i) - \Delta(x_k))(x_k - x_i)^\top\frac{t}{\sqrt{n}} + o(n^{-1/2})\right]\\
&\quad\left.\times x_i f(u)\, du\, dG(x_i)\, dG(x_k)\right\}\\
&= O(n^{-1}).
\end{aligned}$$

Similarly, we can show that $V_2 = O(n^{-1})$. □

The linear approximation in Lemma 2.1 suggests that the dispersion function $D(\mathbf{Y}-\mathbf{X}\boldsymbol{\beta})$ can be approximated by the quadratic function

$$Q(\mathbf{Y}-\mathbf{X}\boldsymbol{\beta}) = \frac{n}{2}(\boldsymbol{\beta}-\boldsymbol{\beta}_0)^\top \mathbf{B}(\boldsymbol{\beta}-\boldsymbol{\beta}_0) \\ -(\boldsymbol{\beta}-\boldsymbol{\beta}_0)^\top S(\mathbf{Y}-\mathbf{X}\boldsymbol{\beta}_0) + D(\mathbf{Y}-\mathbf{X}\boldsymbol{\beta}_0). \tag{7}$$

Lemma 2.2 below justifies this approximation.

Lemma 2.2. *Under Assumptions 2.1–2.2, for every* $\epsilon > 0$ *and for every* $c > 0$, *we have*

$$P\left[\sup_{||\boldsymbol{\beta}-\boldsymbol{\beta}_0||\le c/\sqrt{n}} |D(\mathbf{Y}-\mathbf{X}\boldsymbol{\beta}) - Q(\mathbf{Y}-\mathbf{X}\boldsymbol{\beta})| \ge \epsilon\right] \to 0$$

as $n \to \infty$, *where* $Q(\mathbf{Y}-\mathbf{X}\boldsymbol{\beta})$ *is defined in (7).*

Proof. This follows as in the proof of Theorem A.3.7 of Hettmansperger and McKean (1998). Let $D_n(t) = D\left[\mathbf{Y}-\mathbf{X}\left(\boldsymbol{\beta}_0 + t/\sqrt{n}\right)\right]$ and $Q_n(t) = Q\left[\mathbf{Y}-\mathbf{X}\left(\boldsymbol{\beta}_0 + t/\sqrt{n}\right)\right]$; then both functions are proper convex functions for $t \in R^p$. Their gradients are given by

$$\nabla D_n(t) = -\frac{1}{\sqrt{n}} S\left(\boldsymbol{\beta}_0 + \frac{t}{\sqrt{n}}\right),$$
$$\nabla Q_n(t) = -\frac{1}{\sqrt{n}} S(\boldsymbol{\beta}_0) + \mathbf{B}t.$$

Thus

$$\nabla[D_n(t) - Q_n(t)] = -\left[\frac{1}{\sqrt{n}} S\left(\boldsymbol{\beta}_0 + \frac{t}{\sqrt{n}}\right) - \frac{1}{\sqrt{n}} S(\boldsymbol{\beta}_0) + \mathbf{B}t\right].$$

Using Lemma 2.1 and the "diagonal subsequencing" argument, we have

$$\nabla[D_n(t) - Q_n(t)] \xrightarrow{a.s.} 0 \tag{8}$$

for all rational t and $n \in \widetilde{N}$, an infinite subset of positive integers defined as on page 414 of Hettmansperger and McKean (1998). Define the convex function $H_n(t) = D_n(t) - D_n(0) + t^\top S(\boldsymbol{\beta}_0)/\sqrt{n}$. Then

$$D_n(t) - Q_n(t) = H_n(t) - \frac{1}{2} t^\top \mathbf{B}t,$$
$$\nabla[D_n(t) - Q_n(t)] = \nabla H_n(t) - \mathbf{B}t.$$

By (8), $\nabla H_n(t) \xrightarrow{a.s.} \mathbf{B}t$ for all rational t and $n \in \widetilde{N}$. Also note that $H_n(\mathbf{0}) = 0 = \frac{1}{2} t^\top \mathbf{B}t|_{t=\mathbf{0}}$. Since $H_n(t)$ is convex by Theorem A.3.6. of Hettmansperger and McKean (1998), $\{H_n(t)\}_{n \in \widetilde{N}}$ converges to $\frac{1}{2} t^\top \mathbf{B}t$ almost surely uniformly on each compact subset of R^p. By the way $\widetilde{N}$ is constructed, Theorem 4 of Tucker (1967), p. 103, implies that $D_n(t) - Q_n(t) \xrightarrow{p} 0$ uniformly on each compact subset of R^p. We consider the compact subset $\{t : ||t|| \leq c\}$, where $|| \cdot ||$ denotes the Euclidean norm. This leads to the conclusion of the lemma. □

The quadratic function $Q(\mathbf{Y} - \mathbf{X}\boldsymbol{\beta})$ is minimized by

$$\widetilde{\boldsymbol{\beta}} = \boldsymbol{\beta}_0 + \frac{1}{n}\mathbf{B}^{-1}S(\mathbf{Y} - \mathbf{X}\boldsymbol{\beta}_0).$$

Lemma 2.3 below shows that the distribution of $\widehat{\boldsymbol{\beta}}$ can be well approximated by that of $\widetilde{\boldsymbol{\beta}}$.

Lemma 2.3. *Under Assumptions 2.1–2.2, we have* $\sqrt{n}(\widehat{\boldsymbol{\beta}} - \widetilde{\boldsymbol{\beta}}) \xrightarrow{p} 0$, *i.e.,* $\sqrt{n}(\widehat{\boldsymbol{\beta}} - \boldsymbol{\beta}_0) = \mathbf{B}^{-1}\frac{1}{\sqrt{n}}S(\mathbf{Y} - \mathbf{X}\boldsymbol{\beta}_0) + o_p(1)$.

Proof. We adapt the proof for the basic corollary of Hjort and Pollard (1993). Let $A_n(s) = D\left[\mathbf{Y} - \mathbf{X}\left(\boldsymbol{\beta}_0 + \mathbf{s}/\sqrt{n}\right)\right]$ and

$$B_n(s) = \frac{1}{2}s^\top \mathbf{B}s - S(\mathbf{Y} - \mathbf{X}\boldsymbol{\beta}_0)^\top \frac{s}{\sqrt{n}} + D(\mathbf{Y} - \mathbf{X}\boldsymbol{\beta}_0).$$

Then by Lemma 2.2, $A_n(s) = B_n(s) + r_n(s)$, where $r_n(s) \xrightarrow{p} 0$ uniformly on any bounded set. Note that $A_n(s)$ is minimized by $\alpha_n = \sqrt{n}(\widehat{\boldsymbol{\beta}} - \boldsymbol{\beta}_0)$ and the quadratic function $B_n(s)$ is minimized by $\alpha_n^* = \mathbf{B}^{-1}S(\mathbf{Y} - \mathbf{X}\boldsymbol{\beta}_0)/\sqrt{n}$. For any constant $c > 0$, define

$$T_n = \inf_{||s-\alpha_n^*||=c} B_n(s) - B_n(\alpha_n^*) \quad \text{and} \quad \Delta_n = \sup_{||s-\alpha_n^*||\leq c} |A_n(s) - B_n(s)|.$$

Then $\Delta_n \xrightarrow{p} 0$ as $n \to \infty$. Let s be an arbitrary point outside the ball $\{t : ||t - \alpha_n^*|| \leq c\}$. Then we can write $s = \alpha_n^* + lu$, where $l > c$ and u is a unit vector. By the convexity of $A_n(s)$,

$$\begin{aligned}
\frac{c}{l}[A_n(s) - A_n(\alpha_n^*)] &= \frac{c}{l}A_n(s) + \left(1 - \frac{c}{l}\right)A_n(\alpha_n^*) - A_n(\alpha_n^*) \\
&\geq A_n\left[\frac{c}{l}s + \left(1 - \frac{c}{l}\right)\alpha_n^*\right] - A_n(\alpha_n^*) \\
&= A_n(\alpha_n^* + cu) - A_n(\alpha_n^*) \\
&= B_n(\alpha_n^* + cu) + r_n(\alpha_n^* + cu) - B_n(\alpha_n^*) - r_n(\alpha_n^*) \\
&\geq T_n - 2\Delta_n.
\end{aligned}$$

If $\Delta_n \le \frac{1}{2}T_n$, then $A_n(s) > A_n(\alpha_n^*)$ for all s outside the ball. This implies that if $\Delta_n \le \frac{1}{2}T_n$ then the minimizer of A_n must be inside the ball. Thus

$$P(|\alpha_n - \alpha_n^*| \ge c) \le P\left(\Delta_n \ge \frac{1}{2}T_n\right) = P\left(\Delta_n \ge \frac{1}{2}\lambda c^2\right) \to 0,$$

where λ is the smallest eigenvalue of $\mathbf{B}$. Therefore, $\alpha_n = \alpha_n^* + o_p(1)$. This finishes the proof. □

2.3. *The asymptotic normality*

Lemma 2.3 suggests that the asymptotic distribution of $\widehat{\boldsymbol{\beta}}$ can be obtained through the approximation

$$\sqrt{n}(\widehat{\boldsymbol{\beta}} - \boldsymbol{\beta}_0) = \mathbf{B}^{-1}\frac{1}{\sqrt{n}}S(\mathbf{Y} - \mathbf{X}\boldsymbol{\beta}_0) + o_p(1). \tag{9}$$

We next show that $(1/\sqrt{n})S(\mathbf{Y} - \mathbf{X}\boldsymbol{\beta}_0)$ has approximately a U-statistic representation. Thus the desired asymptotic normality follows readily from the asymptotic distribution theory of U-statistics (Hoeffding, 1948); see also Lee (1990) and Chapter 6 of Lehmann (1999). Note that

$$\begin{aligned}
&\frac{1}{\sqrt{n}}S(\mathbf{Y} - \mathbf{X}\boldsymbol{\beta}_0)\\
&= \frac{\sqrt{12}}{\sqrt{n}(n+1)}\sum_{i=1}^{n}\sum_{j=1}^{n}\left\{I[\epsilon_j \le \epsilon_i + \Delta(X_i) - \Delta(X_j)] - \frac{1}{2}\right\}X_i + o_p(1)\\
&= \frac{\sqrt{3}}{\sqrt{n}(n+1)}\sum_{i=1}^{n}\sum_{j=1}^{n}\left\{I[\epsilon_j \le \epsilon_i + \Delta(X_i) - \Delta(X_j)] - \frac{1}{2}\right\}(X_i - X_j) + o_p(1)\\
&= \frac{\sqrt{3n}(n-1)}{(n+1)}U_n + o_p(1),
\end{aligned} \tag{10}$$

where

$$U_n = \frac{2}{n(n-1)}\sum_{i=1}^{n}\sum_{i<j}\left\{I[\epsilon_j \le \epsilon_i + \Delta(X_i) - \Delta(X_j)] - \frac{1}{2}\right\}(X_i - X_j)$$

is a U-statistic. Theorem 2.1 below formally states the asymptotic normal distribution of the Wilcoxon rank regression estimator under possible model misspecification.

Theorem 2.1. *Under Assumptions 2.1–2.4,*

$$\sqrt{n}(\widehat{\boldsymbol{\beta}} - \boldsymbol{\beta}_0) \xrightarrow{d} N_p(0, \mathbf{B}^{-1}\mathbf{J}\mathbf{B}^{-1}),$$

where $\mathbf{J}$ *is defined in (6).*

Proof. Let $W_i = (X_i, \epsilon_i)$. Then $W_1, \ldots, W_n$ constitute a random sample. Letting

$$\phi(W_i, W_j) = [I(\epsilon_j \le \epsilon_i + \Delta(X_i) - \Delta(X_j)) - 1/2]\,(X_i - X_j),$$

we obtain

$$U_n = \frac{1}{n(n-1)/2} \sum_{i=1}^{n} \sum_{i<j} \phi(W_i, W_j).$$

Furthermore,

$$E\left[\phi(W_i, W_j)\right] = 2E\left\{\left[P(\epsilon_j + \Delta(X_j) \le \epsilon_i + \Delta(X_i)) - 1/2\right] X_i\right\}.$$

Since β_0 minimizes $E|Z_i - Z_j|$, where $Z_i = Y_i - X_i^\top \boldsymbol{\beta} = \epsilon_i + \Delta(X_i)$, it is easy to check that $\boldsymbol{\beta}_0$ satisfies the first-order condition

$$E\left\{\left[P(\epsilon_j + \Delta(X_j) \le \epsilon_i + \Delta(X_i)) - 1/2\right] X_i\right\} = 0.$$

Thus $E[\phi(W_i, W_j)] = 0$. Let $\phi_1(w_1) = E(\phi(w_1, W_2))$, so that

$$\phi_1(w_1) = \int \left[F(\epsilon_1 + \Delta(x_1) - \Delta(x_2)) - 1/2\right](x_1 - x_2)\, dG(x_2).$$

The theory of U-statistics (through the projection method) indicates that $\sqrt{n}U_n$ has the same asymptotic distribution as $\sqrt{n}\widehat{U}_n$, where

$$\widehat{U}_n = \frac{2}{n} \sum_{i=1}^{n} \phi_1(W_1).$$

Note that $E[\phi_1(W_1)] = 0$ and

$$\begin{aligned}
\mathrm{Var}(\phi_1(W_1)) &= E(\phi_1^2(W_1)) \\
&= \int \left[F(u + \Delta(x_1) - \Delta(x_2)) - \frac{1}{2}\right]\left[F(u + \Delta(x_1) - \Delta(x_3)) - \frac{1}{2}\right] \\
&\qquad \times (x_1 - x_2)^\top (x_1 - x_3) f(u)\, du\, dG(x_1)\, dG(x_2)\, dG(x_3) \\
&= J/12.
\end{aligned}$$

We thus have $\sqrt{n}U_n \to N_p(0, J/3)$. The conclusion follows by combining this result with (9) and (10). □

Remark 2.4. When the model is correctly specified, $\mathbf{B} = \sqrt{12}\int f^2(u)\, du \boldsymbol{\Sigma}$ and

$$J = 12E\left\{\int \left[F(u) - \frac{1}{2}\right]^2 (X_1 - X_2)^\top (X_1 - X_3) f(u)\, du\right\} = \boldsymbol{\Sigma}.$$

Thus when the model is correctly specified, the asymptotic covariance matrix of $\widehat{\boldsymbol{\beta}}$ reduces to $\tau^2\boldsymbol{\Sigma}^{-1}$, where $\tau^{-1} = \sqrt{12}\int f^2(u)\,du$. This gives exactly the asymptotic distribution of the Wilcoxon rank regression estimator with correctly specified regression model; see, for example, Chapter 3 of Hettmansperger and McKean (1998).

3. Omitted Variables Bias

We next investigate the situation where some important variables are omitted in modeling the data. This issue may be compounded by the misspecification of the functional form. Consider data $\{(X_{1i}^\top, X_{2i}^\top, Y_i)^\top, i = 1,\ldots,n\}$, where $X_{1i} \in R^{p_1}$, $X_{2i} \in R^{p_2}$, and $p_1 + p_2 = p$. Assume the true model is $Y_i = m(X_{1i}, X_{2i}) + \epsilon_i$, with the ϵ_i being i.i.d. random errors, but we fit instead a Wilcoxon linear regression model with $\{(X_{1i}^\top, Y_i)^\top, i = 1,\ldots,n\}$.

To study this problem, we first derive an alternative expression for the limit of the Wilcoxon rank estimator. For this purpose, we define the specification error $\Delta(X_i,\gamma) = m(X_i) - X_{1i}^\top\gamma$, where $X_i = (X_{1i}^\top, X_{2i})^\top$. This error represents the part in the regression based on the full data that cannot be explained by a linear regression in X_{1i}.

Theorem 3.1. *Assume a Wilcoxon linear regression model is fitted to $(X_{1i}^\top, Y_i)^\top$. Then under Assumptions 2.1–2.3, the estimated coefficient of X_{1i} is a consistent estimator of γ_0, which satisfies the following expression:*

$$\gamma_0 = \arg\min_{\gamma} E\{w(X_i, X_j, \gamma_0)[\Delta(X_i,\gamma) - \Delta(X_j,\gamma)]^2\}, \tag{11}$$

where X_i and X_j are i.i.d. random variables from the covariate distribution $G(\cdot)$ and

$$w(X_i, X_j, \gamma_0) = \int_0^1 u f_e[\Delta(X_j,\gamma_0) - \Delta(X_i,\gamma_0)]\,du, \tag{12}$$

where $f_e(\cdot)$ is the pdf of $e = \epsilon_1 - \epsilon_2$, with ϵ_i, $i = 1, 2$, being i.i.d. random variables from the random error distribution $F(\cdot)$.

Proof. The proof is similar to that for Theorem 2 in Angrist *et al.* (2006). By the theory of U-statistics and an argument similar to that of White (1980) (see also Remark 2.1), the Wilcoxon rank estimator of the coefficient of X_{1i} is consistent for γ_0^*, where

$$\gamma_0^* = \arg\min_{\gamma} \left|(Y_i - X_{i1}^\top\gamma) - (Y_j - X_{j1}^\top\gamma)\right|. \tag{13}$$

We will show that γ_0^* is equivalent to γ_0 defined in (11). Note that γ_0 satisfies the FOC (first-order condition)

$$2E\{w(X_i, X_j, \gamma_0)[\Delta(X_i, \gamma) - \Delta(X_j, \gamma)](X_{j1} - X_{i1})\} = 0. \tag{14}$$

Also, γ_0^* satisfies the FOC

$$E\left\{2\left[I(Y_i - X_{1i}^\top \gamma_0^* \le Y_j - X_{1j}^\top \gamma_0^*) - \frac{1}{2}\right](X_{1i} - X_{2i})\right\} = 0. \tag{15}$$

Denote the left-hand side of (15) by I. Then by the law of iterated expectation,

$$\begin{aligned}
I &= E\left\{2\left[I(\epsilon_i + \Delta(X_i, \gamma_0^*) \le \epsilon_j + \Delta(X_j, \gamma_0^*)) - \frac{1}{2}\right](X_{1i} - X_{2i})\right\} \\
&= 2E\left\{[F_e(\Delta(X_j, \gamma_0^*) - \Delta(X_i, \gamma_0^*)) - F_e(0)](X_{1i} - X_{2i})\right\} \\
&= 2E\left\{2\int_0^1 f_e[u(\Delta(X_j, \gamma_0^*) - \Delta(X_i, \gamma_0^*))][\Delta(X_j, \gamma_0^*) - \Delta(X_i, \gamma_0^*)]\, du \right. \\
&\qquad \left. \times (X_{1i} - X_{2i})\right\} \\
&= 2E\{w(X_i, X_j, \gamma_0^*)[\Delta(X_i, \gamma) - \Delta(X_j, \gamma)](X_{i1} - X_{j1})\},
\end{aligned}$$

where in the second step $F_e(\cdot)$ denotes the distribution function of $\epsilon_i - \epsilon_j$. Thus the first-order conditions for γ_0^* and γ_0 are equivalent. Both (13) and (11) are convex minimization problems, where (13) has a unique solution γ_0^* by assumption. It follows that γ_0^* uniquely solves the first-order condition for both minimization problems. □

When Wilcoxon rank regression is applied to $(X_{1i}^\top, X_{2i}^\top, Y_i)^\top$, the estimated coefficients of $(X_1^\top, X_2^\top)^\top$ are consistent for

$$(\boldsymbol{\beta}_1^\top, \boldsymbol{\beta}_2^\top)^\top = \arg\min_{\boldsymbol{\beta}_1, \boldsymbol{\beta}_2} E\left|(Y_i - X_{1i}^\top \boldsymbol{\beta}_1 - X_{2i}^\top \boldsymbol{\beta}_2) - (Y_j - X_{1j}^\top \boldsymbol{\beta}_1 - X_{2j}^\top \boldsymbol{\beta}_2)\right|,$$

where $i \neq j$. Thus whether X_2 is included in the regression model or not may have a serious effect on the estimation and interpretation of the coefficient of X_1. We call the difference $\boldsymbol{\beta}_1 - \gamma_0$ the *asymptotic bias* of omitting the covariate X_2. Theorem 3.1 immediately leads to the formula for the asymptotic bias of omitted variables. Following the notation of Angrist *et al.* (2006), we define a remainder term: $R(X_i) = m(X_i) - X_{1i}^\top \boldsymbol{\beta}_1$, $i = 1, \ldots, n$.

Theorem 3.2. *Let* $X_i = (X_{1i}^\top, X_{2i}^\top)^\top$ *and* $X_j = (X_{1j}^\top, X_{2j}^\top)^\top$ *be independent and identically distributed copies of the vector of covariates*

$(X_1^\top, X_2^\top)^\top$. *Let* $w(X_{1i}, X_{1j}, \gamma_0)$ *be the weight function defined in (12). Then under Assumptions 2.1–2.3,*

$$\begin{aligned}\gamma_0 - \boldsymbol{\beta}_1 = & E[w(X_i, X_j, \gamma_0)(X_{1i} - X_{1j})^\top (X_{1i} - X_{1j})]^{-1} \\ & \times E\{w(X_i, X_j, \gamma_0)[R(X_i) - R(X_j)](X_{1i} - X_{1j})\}.\end{aligned}$$

Proof. Following the proof of Theorem 3.1, the parameter γ_0 satisfies the first-order condition

$$E\{w(X_i, X_j, \gamma_0)[m(X_i) - m(X_j) - (X_{1i} - X_{1j})^\top \gamma_0](X_{1i} - X_{1j})\} = 0.$$

Thus

$$\begin{aligned}\gamma &= E[w(X_i, X_j, \gamma_0)(X_{1i} - X_{1j})^\top (X_{1i} - X_{1j})]^{-1} \\ &\quad \times E\{w(X_i, X_j, \gamma_0)[m(X_i) - m(X_j)](X_{1i} - X_{1j})\} \\ &= \boldsymbol{\beta}_1 + E[w(X_i, X_j, \gamma_0)(X_{1i} - X_{1j})^\top (X_{1i} - X_{1j})]^{-1} \\ &\quad \times E\{w(X_i, X_j, \gamma_0)[R(X_i) - R(X_j)](X_{1i} - X_{1j})\}. \qquad \square\end{aligned}$$

Remark 3.1. In a frequently encountered situation, $m(X_i) = X_{1i}^\top \boldsymbol{\beta}_1 + X_{2i}^\top \boldsymbol{\beta}_2$ is a linear function in $X_i = (X_{1i}^\top, X_{2i}^\top)^\top$. In this case, $R(X_i) = X_{2i}^\top \boldsymbol{\beta}_2$.

4. Discussions

In this paper, we derive the asymptotic distribution of Wilcoxon rank regression under possible model misspecification. These theoretical results have broader impacts on rank-based model selection and model checking. In particular, to investigate the large sample properties of a model selection procedure, we need to know how the model selection criterion works when the candidate model is misspecified; to investigate the power performance of any lack-of-fit test for model checking, it is essential to understand how the test statistic behaves when the underlying model is not the null model. Recently, Wang and Li (2009) proposed a weighted Wilcoxon-type SCAD (smoothly clipped absolute deviation) procedure, i.e., a weighted Wilcoxon regression procedure with SCAD penalty. The penalty function automatically shrinks small estimated coefficients to zero. Thus the proposed method simultaneously performs variable estimation and selection. The results in this paper are applied in developing a criterion to select the tuning parameter, which when appropriately chosen can lead to a consistent variable selection procedure. There has been growing interest in studying robust model selection and model checking due to the long-standing concern that most existing procedures are excessively sensitive to data contamination;

see Sommer and Staudte (1995), Ronchetti *et al.* (1997), Müller and Welsh (2005), Heritier and Ronchetti (1994), Liu *et al.* (2005), Wang and Qu (2007), and the references therein.

The asymptotic approach in this paper may be straightforwardly extended to studying rank regression with a general score function, which minimizes the dispersion function

$$D_R(\boldsymbol{\beta}) = \sum_{i=1}^{n} a[R(Y_i - X_i^\top \boldsymbol{\beta})](Y_i - X_i^\top \boldsymbol{\beta}),$$

where $a(i) = \phi[i/(n+1)]$ is a score generated by a nondecreasing function $\phi(u)$ defined on U(0,1) such that $\int_0^1 \phi(u)\,du = 0$ and $\int_0^1 \phi^2(u)\,du = 1$. Wilcoxon rank regression corresponds to the score $\phi(u) = \sqrt{12}(u) - 1/2$. McKean and Sievers (1989) contains interesting discussions on the choice of score functions. We refer to Hettmansperger and McKean (1998) for a systematic introduction to rank regression with general scores. In rank regression, the random errors are commonly assumed to be independent and identically distributed. This assumption can be relaxed, but even if the model is correctly specified, it involves nontrivial work in asymptotic analysis; see Koul and Saleh (1993) for an extension to the time-series error case and Dixon and McKean (1996) for the heteroscedastic error case.

Acknowledgments

This research is supported by NSF grant DMS-0706842.

References

Angrist, J., Chernozhukov, V. and Fernández-Val, I. (2006). Quantile regression under misspecification, with an application to the U.S. wage structure, *Econometrica* **74**, pp. 539–563.

Boher, J. (2006). Implications of model misspecification in robust tests for recurrent events, *Lifetime Data Analysis* **12**, pp. 69–95.

Chang, W. H., W., M. J., Naranjo, J. D. and Sheather, S. J. (1999). High-breakdown rank regression, *Journal of the American Statistical Association* **94**, pp. 205–219.

Chen, Y. Q., Pong, A. and Xing, B. (2003). Rank regression in stability analysis, *Journal of Biopharmaceutical Statistics* **13**, pp. 463–479.

David, H. A. (1998). Early sample measures of variability, *Statistical Science* **13**, pp. 368–377.

DiRienzo, A. G. and Lagakos, S. W. (2001). Bias correction for score tests arising from misspecified proportional hazards regression models, *Biometrika* **88**, pp. 421–434.

Dixon, S. L. and McKean, J. W. (1996). Rank-based analysis of the heteroscedastic linear model, *Journal of the American Statistical Association* **91**, pp. 699–712.

Gerds, T. A. and Schumacher, M. (2001). On functional misspecification of covariates in the Cox regression model, *Biometrika* **88**, 2, pp. 572–580.

Gini, C. (1912). Variabilità e mutabilità, contributo allo studio delle distribuzioni e delle relazione statistiche, *Studi Economico-Giuredici dell' Università di Cagliari* **3 (part 2)**, i–iii, pp. 3–159.

Hansen, L. P., Sargent, T. J., Turmuhambetova, G. and Williams, N. (2006). Robust control and model misspecification, *Journal of Economic Theory* **128**, pp. 45–90.

Heiler, S. (1982). Robust estimates in linear regression–a simulation approach, in H. Büning and P. Naeve (eds.), *Computational Statistics. Festschrift für W. Wetzel* (De Gruyter, Berlin).

Heiler, S. and Willers, R. (1988). Asymptotic normality of r-estimates in the linear model, *Statistics* **19**, pp. 173–184.

Heritier, S. and Ronchetti, E. (1994). Robust bounded-influence tests in general parametric models, *Journal of the American Statistical Association* **89**, pp. 897–904.

Hettmansperger, T. P. and McKean, J. W. (1983). A geometric interpretation of inferences based on ranks in the linear model, *Journal of the American Statistical Association* **78**, pp. 885–893.

Hettmansperger, T. P. and McKean, J. W. (1998). *Robust nonparametric statistical methods* (Arnold, London).

Hjort, N. L. and Pollard, D. (1993). Asymptotics for minimisers of convex processes, URL `http://citeseer.ist.psu.edu/hjort93asymptotics.html`, preprint.

Hodges, J. L. and Lehmann, E. L. (1956). The efficiency of some nonparametric competitors of the *t*-test, *Annals of Mathematical Statistics* **27**, pp. 324–335.

Hodges, J. L. and Lehmann, E. L. (1963). Estimation of location based on rank tests, *Annals of Mathematical Statistics* **34**, pp. 598–611.

Hoeffding, W. (1948). A class of statistics with asymptotically normal distribution, *Annals of Mathematical Statistics* **19**, pp. 293–325.

Jaeckel, L. A. (1972). Estimating regression coefficients by minimizing the dispersion of residuals, *Annals of Mathematical Statistics* **43**, pp. 1449–1458.

Jin, Z., Lin, D. Y., Wei, L. J. and Ying, Z. (2003). Rank-based inference for the accelerated failure time model, *Biometrika* **90**, pp. 341–353.

Jung, S. H. and Ying, Z. L. (2003). Rank-based regression with repeated measurements data, *Biometrika* **90**, pp. 732–740.

Jureăková, J. (1971). Nonparametric estimate of regression coefficients, *Annals of Mathematical Statistics* **42**, pp. 1328–1338.

Koul, H. L. (1971). Asymptotic behavior of a class of confidence regions based on ranks in regression, *Annals of Mathematical Statistics* **42**, pp. 466–476.

Koul, H. L. and Saleh, A. K. M. E. (1993). R-estimation of the parameters of autoregressive models, *Annals of Statistics* **21**, pp. 534–551.

Lagakos, S. W. (1988). The loss in efficiency from misspecifying covariates in proportional hazards regression models, *Biometrika* **75**, pp. 156–160.

Lee, A. J. (1990). *U-Statistics* (Marcel Dekker, New York).

Lehmann, E. L. (1999). *Elements of Large-Sample Theory* (Springer, New York).

Liu, R. C., Markatou, M. and Tsai, C. L. (2005). Robust estimation and testing in nonlinear regression models, *International Journal of Pure and Applied Mathematics* **21**, pp. 525–552.

McKean, J. W. and Schrader, R. M. (1980). The geometry of robust procedures in linear models, *Journal of the Royal Statistical Society, Series B* **42**, pp. 366–371.

McKean, J. W. and Sievers, G. L. (1989). Rank scores suitable for the analysis of linear models under asymmetric error distributions, *Technometrics* **31**, pp. 207–218.

Mukherjee, K. and Bai, Z. D. (2002). R-estimation in autoregression with square-integrable score function, *Journal of Multivariate Analysis* **81**, pp. 167–186.

Müller, S. and Welsh, A. H. (2005). Outlier robust model selection in linear models, *Journal of the American Statistical Association* **100**, pp. 1297–1310.

Ozturk, O. (2002). Rank regression in ranked-set samples, *Journal of the American Statistical Association* **460**, pp. 1180–1191.

R Development Core Team (2010). *R: A Language and Environment for Statistical Computing*, R Foundation for Statistical Computing, Vienna, Austria, URL `http://www.R-project.org`, ISBN 3-900051-07-0.

Risch, N. and Giuffra, L. (1992). Model misspecification and multipoint linkage analysis, *Human Heredity* **42**, pp. 77–92.

Ronchetti, E., Field, C. and Blanchard, W. (1997). Robust linear model selection by cross-validation, *Journal of the American Statistical Association* **92**, pp. 1017–1023.

Séménou, M. (1996). Quantile estimation under possibly misspecified generalised linear model, *Statistics and Probability Letters* **27**, pp. 357–365.

Severini, T. A. (1998). Some properties of inferences in misspecified linear models, *Statistics and Probability Letters* **40**, pp. 149–153.

Sievers, G. L. (1983). A weighted dispersion function for estimation in linear models, *Communications in Statistics—Theory and Methods* **12**, pp. 1161–1179.

Sommer, S. and Staudte, R. G. (1995). Robust variable selection in regression in the presence of outliers and leverage points, *Australian Journal of Statistics* **37**, pp. 323–336.

Terpstra, J. and McKean, J. (2005). Rank-based analyses of linear models using R, *Journal of Statistical Software* **14**, 7, URL `http://www.jstatsoft.org/v14/i07`.

Terpstra, J. T., McKean, J. W. and Naranjo, J. D. (2001). Weighted wilcoxon estimates for autoregression, *Australian and New Zealand Journal of Statistics* **43**, pp. 399–419.

Tucker, H. G. (1967). *A Graduate Course in Probability* (Academic Press, New York).

Uppal, R. and Wang, T. (2003). Model misspecification and underdiversification,

The Journal of Finance **58**, pp. 2465–2486.

Velilla, S. (2001). On the bootstrap in misspecified regression models, *Computational Statistics and Data Analysis* **36**, pp. 227–242.

Waldorp, L. J., Huizenga, H. M. and Grasman, R. P. P. P. (2005). The Wald test and Cramér–Rao bound for misspecified models in electromagnetic source analysis, *IEEE Transactions on signal processing* **53**, pp. 3427–3435.

Wang, L. and Li, R. (2009). Weighted wilcoxon-type smoothly clipped absolute deviation method, *Biometrics* **65**, 2, pp. 564–571.

Wang, L. and Qu, A. (2007). Robust tests in regression models with omnibus alternatives and bounded influence, *Journal of the American Statistical Association* **102**, pp. 347–358.

Wang, Y.-G. and Zhu, M. (2006). Rank-based regression for analysis of repeated measures, *Biometrika* **93**, pp. 459–464.

White, H. (1980). Nonlinear regression on cross-section data, *Econometrica* **48**, pp. 721–746.

White, H. (1981). Consequences and detection of misspecified nonlinear regression models, *Journal of the American Statistical Association* **76**, pp. 419–433.

White, H. (1982). Maximum likelihood estimation of misspecified models, *Econometrica* **50**, pp. 1–26.

Zhou, M. (2005). Empirical likelihood analysis of the rank estimator for the censored aft model, *Biometrika* **92**, pp. 492–498.

Zuo, Y. and He, X. (2006). On the limiting distributions of multivariate depth-based rank sum statistics and related tests, *Annals of Statistics* **34**, 6, pp. 2879–2896.

Iterative Conditional Maximization Algorithm for Nonconcave Penalized Likelihood

Yiyun Zhang

Novartis Pharmaceuticals Corporation
Florham Park, NJ 07932, USA
E-mail: yiyun.zhang@novartis.com

Runze Li

Department of Statistics, Pennsylvania State University
University Park, PA 16802, USA
E-mail: rli@stat.psu.edu

Variable selection via penalized likelihood has received considerable attention recently. Penalized likelihood estimators with properly chosen penalty functions possess nice properties. In practice, optimizing the penalized likelihood function is often challenging because the object function may be nondifferentiable and/or nonconcave. Existing algorithms such as the local quadratic approximation (LQA) algorithm share a similar drawback of backward selection that once a variable is deleted, it is essentially left out of the final model. We propose the iterative conditional maximization (ICM) algorithm to address the aforementioned drawback. It utilizes the characteristics of the nonconcave penalized likelihood and enjoys fast convergence. Three simulation studies, in linear, logistic, and Poisson regression, together with one real data analysis, are conducted to assess the performance of the ICM algorithm.

Keywords: GLIM; ICM; LASSO; SCAD; Variable Selection

1. Introduction

Model selection has always been a fundamental issue in statistical modelling. It is well known that overfitted models lead to inefficient estimation while an underfitted model can result in serious bias. A small while adequate collection of predictors is always desirable in practice. Advanced technology in the scientific world such as genetics, astronomy, finance and computer science allows us to obtain data of very high dimension. This makes model selection more important than it has ever been. See Fan and Li (2006) for more discussion of the challenges in high-dimensional modelling.

Classical variable selection procedures can be described as best-subset-selection, i.e., comparing all possible models according to some criteria and finding the optimal subset. Such methods work relatively well with low-dimensional data. As the dimension of the data increases, it becomes numerically impossible to search all possible models exhaustively. Some modifications such as forward/backward selection or stepwise selection are practically useful sometimes, but the search is not exhaustive and hence the results may not be optimal. Furthermore, the theoretical properties of these best subset selection models are difficult to understand. Reviews of this literature can be found in, e.g., Breiman (1996), Shao (1997) and Miller (2002).

Many variable selection procedures can be understood as maximizing a penalized likelihood (Hunter and Li, 2005). Classical criteria, such as the Akaike information criterion (AIC) (Akaike, 1974) and the Bayesian information criterion (BIC) (Schwarz, 1978), correspond to applying L_0 penalties to the log-likelihood. The discontinuity of L_0 penalties results in instability and difficulties in optimization (Hunter and Li, 2005). In contrast, estimation and variable selection can be done simultaneously by properly employing a continuous but nonconcave penalty. Such methods combine shrinkage and variable selection so that their estimates possess nice properties that classical methods do not. They include bridge regression (Frank and Friedman, 1993), the least absolute shrinkage and selection operator (LASSO) (Tibshirani, 1996), and the smoothly clipped absolute deviation (SCAD) penalty (Fan and Li, 2001), among others. Because the properties of the solution are directly related to the penalty applied, we will give a detailed discussion the choice of penalty functions and the corresponding properties of the penalized estimate.

Fan and Li (2001) pointed out that in order for the estimators to be a thresholding rule (suitable for variable selection), the penalized likelihood functions have to be nondifferentiable and/or nonconcave. This imposes great challenges in optimization. Some algorithms have been developed to address this issue, including the local quadratic approximation (LQA) algorithm (Fan and Li, 2001) and its modification based on the minorize-maximize (MM) algorithm (Hunter and Li, 2005). The LQA algorithm can be viewed as an extension of the Newton-Raphson algorithm. However, to achieve numerical stability, it sets small estimates directly to zero in the iteration. Therefore the LQA algorithm shares a drawback of backward selection, that when a variable is deleted from the model it cannot enter the model again. The MM algorithm addressed this problem by applying a perturbed version of the local approximation. However, the size of perturbation is still hard to decide. We will propose an iterative conditional maximization (ICM) algorithm for finding penalized likelihood estimates that fixes the aforementioned drawbacks.

The rest of this paper is organized as follows. We first give a brief overview of the nonconcave penalized likelihood methodology in Section 2, where the existing algorithms are also reviewed. Section 3 further discusses the characterization of the nonconcave penalized likelihood estimate and introduces the ICM algorithm for general likelihood settings. Three simulation studies, including linear regression, logistic regression, and Poisson regression, are summarized in Section 4 to assess the performance of the ICM algorithm in terms of both model fitting and algorithm convergence. One real data application is also illustrated in Section 4.

2. Nonconcave Penalized Likelihood for Variable Selection

2.1. *Some Preliminaries*

Consider the data $(\mathbf{x}_1, y_1), \cdots, (\mathbf{x}_n, y_n)$ being collected identically and independently, where y_i is the response from the ith subject and $\mathbf{x}_i$ is the associated d-dimensional predictor variable. Denote the log-likelihood function by $\ell(\boldsymbol{\beta}, \boldsymbol{\phi}) = \sum_{i=1}^{n} \log f(y_i; \mathbf{x}_i^T \boldsymbol{\beta}, \boldsymbol{\phi})$, where $f(y_i; \mathbf{x}_i^T \boldsymbol{\beta}, \boldsymbol{\phi})$ is the conditional density of y_i given $\mathbf{x}_i$. Here $\boldsymbol{\beta}$ is the parameter of primary interest and $\boldsymbol{\phi}$ is the dispersion parameter. The goal of variable selection is to identify those components of $\boldsymbol{\beta}$ that are zero and to estimate the nonzero components of $\boldsymbol{\beta}$.

In some applications such as logistic regression and Poisson regression, there are no dispersion parameters or the dispersion parameter is known. In other situations such as linear regression, the dispersion parameter $\boldsymbol{\phi}$ can be estimated in advance from the full model. This allows us to write the log-likelihood function $\ell(\boldsymbol{\beta}, \boldsymbol{\phi}) \equiv \ell(\boldsymbol{\beta})$. This is a standard procedure in variable selection (Frank and Friedman, 1993; Tibshirani, 1996; Fan and Li, 2001; Hunter and Li, 2005).

Throughout this article, we assume that the regularity conditions (A)–(C) in Fan and Li (2001) are satisfied. These conditions guarantee that the log-likelihood function $\ell(\boldsymbol{\beta})$ is at least second-order differentiable and locally concave at the true unknown parameter $\boldsymbol{\beta}_0$. As a consequence, they guarantee the asymptotic normality of ordinary maximum likelihood estimates. For the sake of simplicity, we omit the details here because they are standard in the literature (see Lehmann, 1983, for example).

Adopting the approach in Fan and Li (2001), we define the nonconcave penalized likelihood function as

$$\mathcal{L}(\boldsymbol{\beta}) = \frac{1}{n}\ell(\boldsymbol{\beta}) - \sum_{j=1}^{d} p_{\lambda_j}(|\beta_j|), \tag{1}$$

where the $p_{\lambda_j}(\cdot)$ are penalty functions with regularization or tuning parameters λ_j, which control the size of the penalties. In theory, different penalties and tuning parameters can be selected for each component β_j. For example, one may not want to penalize some important parameters. For simplicity, we assume throughout this article that an equal tuning parameter λ is used, so that the penalty function in (1) can be written as $p_\lambda(\cdot)$.

With properly chosen penalties, a desirable penalized estimate that maximizes (1) should possess the following properties:

(i) *Sparsity*: The resulting estimate automatically sets small estimates to zero so that variable selection is achieved.
(ii) *Unbiasedness*: The resulting estimate is nearly unbiased when the true unknown parameter is large.
(iii) *Continuity*: The resulting estimate is continuous in data to avoid instability.

Several penalty functions have been proposed in the literature. For example, the L_q $(0 < q < 2)$ penalty, namely

$$p_\lambda(|\theta|) = q^{-1}\lambda|\theta|^q,$$

leads to bridge regression (Frank and Friedman, 1993). In particular, the L_1 penalty yields the LASSO estimator (Tibshirani, 1996). Fan and Li (2001) shed some light on the choice of penalty functions, and advocated the use of the smoothly clipped absolute deviation (SCAD) penalty, whose first derivative is given by

$$p'_\lambda(|\theta|) = \lambda\left\{ I(|\theta| \le \lambda) + \frac{(a\lambda - |\theta|)_+}{(a-1)\lambda} I(|\theta| > \lambda)\right\},$$

with $a = 3.7$ and $p_\lambda(0) = 0$. Here $I(\cdot)$ is the indicator function.

For the linear regression model, when the design matrix is orthonormal, it has been shown in Antoniadis and Fan (2001) that the solution of LASSO is sparse, continuous, but not unbiased. When the tuning parameter λ is chosen carefully, the SCAD estimate satisfies all three aforementioned requirements (Antoniadis and Fan, 2001). More insights on the choice of penalty functions will be discussed in Section 3.1.

2.2. *Optimization*

The first obstacle in applying nonconcave penalized likelihood methodology lies in maximizing the penalized likelihood function (1). In order for the penalized estimate to possess sparsity, the penalty function $p_\lambda(\cdot)$ has to be singular at origin (Fan and Li, 2001). This requirement results in great challenges in optimization because the target function is nonconcave and nondifferentiable at zero.

Fan and Li (2001) proposed a local quadratic approximation (LQA) algorithm for optimizing general nonconcave penalized likelihood. In LQA, $p_\lambda(|\beta_j|)$ is locally approximated by the quadratic function

$$\tilde{p}_\lambda(\beta_j) = p_\lambda(\beta_{j0}) + \frac{(\beta_j^2 - \beta_{j0}^2)p'_\lambda(|\beta_{j0}|+)}{2|\beta_{j0}|}$$

for β_j around the point β_{j0}. The above equation is evaluated at the $(k+1)$th iteration of the Newton-Raphson algorithm if $\hat{\beta}_j^{(k)}$ is not very close to zero. Otherwise, the resulting parameter estimator of β_j is set to 0. As a consequence, LQA suffers a drawback that when a variable is deleted at any step, it will be excluded from the final selected model. Hunter and Li (2005) modified LQA as a special minorize-maximize (MM) algorithm. It extends the idea of LQA by forming a perturbed objective function to maximize. MM algorithm fixes that drawback of LQA, but the size of perturbation remains hard to determine in practice. Recently, Zou and Li (2008) proposed a local linear approximation (LLA) algorithm instead of LQA. With the aid of LLA, the least angle regression (LARS) (Efron *et al.*, 2004) algorithm can be adopted in each optimization step, which provides fast computation of the penalized estimate. Zou and Li (2008) also suggested that it is not necessary to carry out the full iteration in LLA. The one-step estimator enjoys the same asymptotic properties as the fully iterated solution and works well enough in practice. We will propose an iterative conditional maximization (ICM) algorithm in Section 3.2 that fixes the aforementioned drawbacks of the LQA and MM algorithms. In the special case of linear regression with the L_1 penalty, the shooting algorithm of Fu (1998) is a special case of the ICM algorithm. Friedman *et al.* (2007) argued that such algorithms can outperform LARS in terms of computational speed.

2.3. *Tuning Parameter Selection*

The second burden in employing the nonconcave penalized likelihood is in choosing the tuning parameter λ, which controls the size of the final model.

When $\lambda = 0$, no penalty is applied to the likelihood function and hence the penalized likelihood estimate is the MLE. On the other hand, as λ increases, larger penalties are imposed on the model so that the resulting model becomes sparse.

The number of degrees of freedom (df) is usually used to characterize the model complexity. Fan and Li (2001) define the degrees of freedom as the trace of the approximate linear projection matrix, i.e.,

$$\mathrm{df}(\lambda) = \mathrm{tr}\left\{ \left(\nabla_{11}^2 \ell(\hat{\boldsymbol{\beta}}_\lambda) - n\Sigma(\hat{\boldsymbol{\beta}}_\lambda) \right)^{-1} \nabla_{11}^2 \ell(\hat{\boldsymbol{\beta}}_\lambda) \right\},$$

where $\nabla_{11}^2 \ell(\hat{\boldsymbol{\beta}}_\lambda)$ denotes the submatrix of $\partial^2 \ell(\hat{\boldsymbol{\beta}}_\lambda)/\partial \boldsymbol{\beta}^T \partial \boldsymbol{\beta}$ corresponding to the nonzero components of $\hat{\beta}_\lambda$ and $\Sigma(\hat{\boldsymbol{\beta}}_\lambda) = \mathrm{diag}\{p'_\lambda(\hat{\beta}_{\lambda;j})/\hat{\beta}_{\lambda;j}\}$ for j such that $\hat{\beta}_j \neq 0$.

To achieve a balance between model fitting (e.g., log-likelihood) and model complexity (degrees of freedom), Zhang *et al.* (2010) proposed choosing λ minimizing the generalized information criterion (GIC)

$$\mathrm{GIC}_{\kappa_n}(\lambda) = \frac{1}{n}\{G(\hat{\boldsymbol{\beta}}_\lambda) + \kappa_n \mathrm{df}(\lambda)\}, \tag{2}$$

where $G(\hat{\boldsymbol{\beta}}_\lambda)$ measures model fitting, $\mathrm{df}(\lambda)$ measures model complexity, and κ_n is a constant number that determines the property of the resulting estimate. In linear regression, GIC includes the BIC in Wang *et al.* (2007) as a special case with $\kappa_n = \log n$. The GCV criterion

$$\mathrm{GCV}(\lambda) = \frac{-2\ell(\hat{\boldsymbol{\beta}}_\lambda)}{n\{1 - \mathrm{df}(\lambda)/n\}^2}$$

of Fan and Li (2001), on the other hand, performs similarly to the the AIC, namely, with $\kappa_n = 2$. In our simulation studies, BIC is adopted because of its superior performance in maintaining model sparsity. That is, we use GIC with $\kappa_n = \log n$ and $G(\hat{\boldsymbol{\beta}}_\lambda) = -2\ell(\hat{\boldsymbol{\beta}}_\lambda)$.

3. An Iterative Conditional Maximization (ICM) Algorithm

3.1. *Characterizations of the Penalized Likelihood Estimates*

First we introduce some notation. Write

$$\nabla \ell(\boldsymbol{\beta}) = \frac{\partial \ell(\boldsymbol{\beta})}{\partial \boldsymbol{\beta}} \quad \text{and} \quad \nabla^2 \ell(\boldsymbol{\beta}) = \frac{\partial^2 \ell(\boldsymbol{\beta})}{\partial \boldsymbol{\beta} \partial \boldsymbol{\beta}^T}.$$

Denote the maximum likelihood estimate (MLE) by $\hat{\boldsymbol{\beta}} = \arg\max_{\boldsymbol{\beta}} \ell(\boldsymbol{\beta})$. Under the regularity conditions, the MLE satisfies the estimating equation $\nabla\ell(\hat{\boldsymbol{\beta}}) = 0$. Using Taylor expansion,

$$\ell(\boldsymbol{\beta}) \approx \ell(\hat{\boldsymbol{\beta}}) + \nabla\ell(\hat{\boldsymbol{\beta}})^T(\boldsymbol{\beta} - \hat{\boldsymbol{\beta}}) + \frac{1}{2}(\boldsymbol{\beta} - \hat{\boldsymbol{\beta}})^T\nabla^2\ell(\hat{\boldsymbol{\beta}})(\boldsymbol{\beta} - \hat{\boldsymbol{\beta}}), \tag{3}$$

where the first term is constant with respect to $\boldsymbol{\beta}$ and the second term is zero by the definition of MLE. Further, write $\hat{\mathcal{I}} = -n^{-1}\nabla^2\ell(\hat{\boldsymbol{\beta}})$, the observed information matrix. Therefore, maximizing $\mathcal{L}(\boldsymbol{\beta})$ in (1) is approximately minimizing

$$Q(\boldsymbol{\beta}) = \frac{1}{2}(\boldsymbol{\beta} - \hat{\boldsymbol{\beta}})^T\hat{\mathcal{I}}(\boldsymbol{\beta} - \hat{\boldsymbol{\beta}}) + \sum_{j=1}^{d} p_{\lambda_j}(|\beta_j|). \tag{4}$$

This is similar to the idea of least squares approximation (LSA) in Wang and Leng (2007).

Theorem 3.1. *Let $p_\lambda(\cdot)$ be a nonnegative, nondecreasing and differentiable function on $(0, +\infty)$. Furthermore, denote by $\tilde{\boldsymbol{\beta}}$ the minimizer of $Q(\boldsymbol{\beta})$ in (4). Then we have the following results.*

(i) The solution $\tilde{\boldsymbol{\beta}}$ satisfies

$$\tilde{\beta}_j = \begin{cases} 0 & \text{if } |\tilde{\beta}_j^*| \le m_j, \\ \tilde{\beta}_j^* - \dfrac{1}{\hat{\mathcal{I}}_{jj}} p'_\lambda(|\tilde{\beta}_j|)\,sgn(\tilde{\beta}_j) & \text{if } |\tilde{\beta}_j^*| > m_j, \end{cases} \tag{5}$$

where $\tilde{\beta}_j^*$ and m_j $(j = 1, \ldots, d)$ are defined by

$$\tilde{\beta}_j^* = \hat{\beta}_j - \sum_{k=1}^{d} \frac{\hat{\mathcal{I}}_{jk}}{\hat{\mathcal{I}}_{jj}}(\beta_k - \hat{\beta}_k), \tag{6}$$

$$m_j = \min_{\theta > 0}\left\{\theta + p'_\lambda(\theta)/\hat{\mathcal{I}}_{jj}\right\}, \tag{7}$$

respectively.

(ii) When $p'_\lambda(\beta_j)$ is continuous on $(0, +\infty)$, the $\tilde{\beta}_j$ are continuous in the $\hat{\beta}_j$ if and only if the minimum of $|\theta| + p'_\lambda(|\theta|)/\hat{\mathcal{I}}_{jj}$ is attained at zero for all $j = 1, \ldots, d$.

(iii) When $p'_\lambda(\beta_j)$ is continuous on $(0, +\infty)$, and if $p'_\lambda(\beta_j) \to 0$ as $|\beta_j| \to +\infty$, then

$$\tilde{\beta}_j = \tilde{\beta}_j^* - p'_\lambda\left(\left|\tilde{\beta}_j^*\right|\right) sgn\left(\tilde{\beta}_j^*\right)/\hat{\mathcal{I}}_{jj} + o\left\{p'_\lambda\left(\left|\tilde{\beta}_j^*\right|\right)\right\} \tag{8}$$

for all j such that $\tilde{\beta}_j \ne 0$.

The proof is outlined in the appendix. Theorem 3.1 gives some insights on how to choose penalty functions. Part (i) suggests that the minima m_j, $j = 1, ..., d$, should be positive in order for some $\tilde{\beta}_j$ to be set to zero when they are small. This implies that $p'_\lambda(\cdot)$ should be positive around zero, and hence the penalty function is singular at the origin. Part (ii) gives the sufficient and necessary condition for the solution to be continuous. Finally, note that when $\tilde{\beta}_j^*$ is unbiased for β_j, part (iii) concludes that $\tilde{\beta}_j$ is approximately unbiased for β_j if $p'_\lambda(\theta) \to 0$ as $|\theta| \to +\infty$, which means that the penalty function $p_\lambda(\cdot)$ should be bounded. These conditions are satisfied by many properly chosen penalties, including SCAD. Other commonly used penalties such as the hard thresholding penalty, the L_q penalties $(0 < p < 2)$, and LASSO meet some but not all conditions.

3.2. *The ICM Algorithm*

Theorem 3.1 suggests the ICM algorithm for maximizing the nonconcave penalized likelihood. Notice that the algorithm essentially maximizes the penalized likelihood function with respect to its major terms up to first two orders. This is a common recipe, as in the Newton-Raphson algorithm, and is good enough in practice.

Step 1: Choose the maximum likelihood estimate $\tilde{\boldsymbol{\beta}}^{(0)} = \hat{\boldsymbol{\beta}}$ as the starting point. Compute $\mathbf{m} = (m_1, ..., m_d)^T$ according to (7).

Step 2: Start with $m = 1$. Update:

$$\tilde{\beta}_j^{*(m)} = \tilde{\beta}_j^{(m-1)} - \sum_{k=1}^{d} \frac{\hat{\mathcal{I}}_{jk}}{\hat{\mathcal{I}}_{jj}} \left(\tilde{\beta}_k^{(m-1)} - \tilde{\beta}_k^{(0)} \right) \tag{9}$$

and

$$\tilde{\beta}_j^{(m)} = \begin{cases} 0 & \text{if } \left|\tilde{\beta}_j^{*(m)}\right| \le m_j, \\ \tilde{\beta}_j^{*(m)} - \dfrac{1}{\hat{\mathcal{I}}_{jj}} p'_\lambda\left(\left|\tilde{\beta}_j^{(m-1)}\right|\right) \operatorname{sgn}\left(\tilde{\beta}_j^{(m-1)}\right) & \text{if } \left|\tilde{\beta}_j^{*(m)}\right| > m_j \end{cases} \tag{10}$$

for $j = 1, 2, ..., d$.

Step 3: Repeat step 2 until convergence.

Note that instead of updating $\tilde{\boldsymbol{\beta}}^{(m)}$ using $\tilde{\boldsymbol{\beta}}^{(m-1)}$ in step 2, we can actually update the $\tilde{\beta}_j^{(m)}$ one by one, using the updated $\tilde{\beta}_1^{(m)}, ..., \tilde{\beta}_j^{(m)}$ and $\tilde{\beta}_{j+1}^{(m-1)}, ..., \tilde{\beta}_d^{(m-1)}$ to update $\tilde{\beta}_{j+1}^{(m)}$. Simulation studies also show this can ensure faster convergence.

Remark 3.1. From Theorem 3.1(iii), the minimum of $|\theta| + p'_\lambda(|\theta|)/\hat{\mathcal{I}}_{jj}$ needs to be attained at 0 to ensure the continuity of the penalized estimate. Therefore, in addition to the shape of $p'_\lambda(\cdot)$, the properties of the solution are also decided by the quantity $\hat{\mathcal{I}}_{jj}^{-1}$. In linear regression, $\hat{\mathcal{I}} = (\frac{1}{n}\mathbf{X}^T\mathbf{X})^{-1}$, so we can standardize the covariates so that $\hat{\mathcal{I}}_{jj} = 1$ for all $j = 1, ..., d$. This means that the continuity of the penalized estimate is guaranteed by choosing a penalty $q_\lambda(\cdot)$ such that the minimum of $|\theta| + q'_\lambda(|\theta|)$ is attained at 0. The commonly used penalties LASSO and SCAD with $a = 3.7$ both satisfy this condition. In other applications such as logistic regression and Poisson regression, $\hat{\mathcal{I}}$ is related to the true unknown parameter $\boldsymbol{\beta}_0$. Hence the penalty can be adjusted accordingly. For example, after calculating $\hat{\mathcal{I}}_{jj}$, we can use different penalties $p_j(\cdot)$ for each component and hold $p'_j(|\theta|)/\hat{\mathcal{I}}_{jj} = q'_\lambda(|\theta|)$. Another possible modification is to use $p_\lambda(\theta) = \min_j\{\hat{\mathcal{I}}_{jj}\}q_\lambda(\theta)$. Both adjustments ensure that the minimum of $|\theta| + p'_\lambda(|\theta|)/\hat{\mathcal{I}}_{jj}$ is attained at 0, so that the penalized estimate is continuous. See Zhang (2006) for more detailed discussion on this issue.

4. Numerical Studies

In this section, simulation studies are conducted to assess the performance of the ICM algorithm. Adapting the model setting from Tibshirani (1996) and Fan and Li (2001), 1000 Monte Carlo samples are drawn from the following three models:

(1) *Linear regression:* $y|\mathbf{x}, \boldsymbol{\beta} \sim \mathcal{N}(\mathbf{x}^T\boldsymbol{\beta}, 1)$.
(2) *Logistic regression:* $y|\mathbf{x}, \boldsymbol{\beta} \sim \text{Bernoulli}\{\mu(\mathbf{x}^T\boldsymbol{\beta})\}$, where

$$\mu(\mathbf{x}^T\boldsymbol{\beta}) = \frac{\exp(\mathbf{x}^T\boldsymbol{\beta})}{1 + \exp(\mathbf{x}^T\boldsymbol{\beta})}.$$

(3) *Poisson regression:* $y|\mathbf{x}, \boldsymbol{\beta} \sim \text{Poisson}\{\mu(\mathbf{x})\}$, where

$$\mu(\mathbf{x}^T\boldsymbol{\beta}) = \exp(\mathbf{x}^T\boldsymbol{\beta}).$$

In all three models, $\mathbf{x}$ is an 8-dimensional random vector. The first 6 components of $\mathbf{x}$ are multivariate normal with mean 0 and covariance matrix $\Sigma = (\sigma_{ij})$, in which $\sigma_{ij} = 0.5^{|i-j|}$, and the last 2 components of $\mathbf{x}$ are generated from independent Bernoulli distributions with success probability 0.5. For the linear and logistic regression models, $\boldsymbol{\beta} = (3, 1.5, 0, 0, 2, 0, 0, 0)^T$; for the Poisson regression model, $\boldsymbol{\beta} = (1.2, 0.6, 0, 0, 0.8, 0, 0, 0)^T$. Sample sizes $n = 50$ and 100 are considered for linear regression. For the other two models, slightly larger sample sizes are adopted, with $n = 200$ and 400 for logistic regression and $n = 100$ and 200 for Poisson regression.

4.1. *Model Fitting and Model Sparsity*

Although the ICM algorithm and the commonly used LQA algorithm are both used to solve for the penalized estimate, the ICM algorithm actually solves the approximation (4) rather than the original penalized log-likelihood function (1). In practice, it is of interest to examine the differences between the solutions of different algorithms. To this end, we find the SCAD solution via both the ICM algorithm and the LQA algorithm and compare the model fitting and sparsity of both solutions in finite samples. The tuning parameter λ is selected using BIC. In addition, classical variable selection procedures, including AIC and BIC, are performed for comparison. Moreover, the oracle results, i.e., model fittings with only the truly nonzero predictors, are reported as benchmarks.

The model error (ME) can be defined by

$$\mathrm{ME}(\hat{\boldsymbol{\beta}}) = E_{\mathbf{x}}\left[\mu(\mathbf{x}^T\boldsymbol{\beta}) - \mu(\mathbf{x}^T\hat{\boldsymbol{\beta}})\right]^2,$$

where the expectation is taken over a new observation of the covariate vector $\mathbf{x}$, and $\mu(\mathbf{x}) = E(y|\mathbf{x})$, which depends on $\mathbf{x}$ through $\mathbf{x}^T\boldsymbol{\beta}$. Using Taylor expansion around the true parameter value $\boldsymbol{\beta}_0$, the model error can be approximated by

$$\mathrm{ME}(\hat{\boldsymbol{\beta}}) \approx (\hat{\boldsymbol{\beta}} - \boldsymbol{\beta}_0)^T E_{\mathbf{x}}\left[\{\mu'(\mathbf{x}^T\boldsymbol{\beta}_0)\}^2\mathbf{x}\mathbf{x}^T\right](\hat{\boldsymbol{\beta}} - \boldsymbol{\beta}_0),$$

where the middle term can be obtained in advance via simple calculation or Monte Carlo simulation. To compare the model error, we report the median of the relative model error (MRME). Here relative model error is defined to be the ratio of the model error of a penalized estimator to that of the unpenalized MLE under the full model.

The selected model can be underfitted, correctly fitted, or overfitted. It is underfitted if at least one parameter that should not be zero is selected as zero, it is overfitted if at least one parameter that should be zero is not selected as zero, and it is correctly fitted if it is neither under- nor overfitted. To investigate whether the penalized methods overfit and how often they select the smallest correct model, we report the average number of correct (C) and incorrect (I) zero components in the final estimates. The percentages of underfitted, correctly fitted, and overfitted cases in 1000 replications are also reported. Note that a selected model can be both under- and overfitted at the same time. Therefore the proportions of these three situations do not necessarily add up to 1.

The MRME and proportion of underfitted, correctly fitted, and overfitted results are summarized in Tables 1 to 3. In all three models, the

Table 1. Simulation results for the linear regression model: $n = 50$ and $n = 100$.

Method	MRME (%)	Zeros C	Zeros I	Under (%)	Correct (%)	Overfitted (%) 1	2	3	4	5
n=50:										
LQA	37.33	4.81	0.00	0.0	88.2	7.2	2.7	1.5	0.4	0.0
ICM	37.09	4.82	0.00	0.0	88.5	7.0	2.6	1.4	0.4	0.1
AIC	73.30	4.03	0.00	0.0	38.0	35.8	18.7	6.3	1.2	0.0
BIC	47.47	4.67	0.00	0.0	72.4	23.5	3.3	0.8	0.0	0.0
Oracle	32.29	5.00	0.00	0.0	100.0	0.0	0.0	0.0	0.0	0.0
n=100:										
LQA	35.90	4.92	0.00	0.0	94.2	4.4	1.1	0.3	0.0	0.0
ICM	36.17	4.92	0.00	0.0	93.9	4.6	1.2	0.3	0.0	0.0
AIC	73.73	4.11	0.00	0.0	40.4	37.6	16.3	4.3	1.4	0.0
BIC	42.79	4.83	0.00	0.0	84.8	13.8	1.2	0.2	0.0	0.0
Oracle	33.26	5.00	0.00	0.0	100.0	0.0	0.0	0.0	0.0	0.0

Table 2. Simulation results for the logistic regression model: $n = 200$ and $n = 400$.

Method	MRME (%)	Zeros C	Zeros I	Under (%)	Correct (%)	Overfitted (%) 1	2	3	4	5
n=200:										
LQA	30.31	4.88	0.01	1.1	88.9	9.2	0.9	0.2	0.0	0.0
ICM	27.95	4.85	0.01	0.8	86.9	10.9	1.4	0.2	0.1	0.0
AIC	66.41	4.13	0.00	0.2	39.4	39.1	17.4	3.1	0.9	0.1
BIC	31.12	4.88	0.01	1.2	87.7	11.3	0.4	0.1	0.0	0.0
Oracle	24.75	5.00	0.00	0.0	100.0	0.0	0.0	0.0	0.0	0.0
n=400:										
LQA	30.73	4.94	0.00	0.0	93.9	5.8	0.3	0.0	0.0	0.0
ICM	29.22	4.93	0.00	0.0	93.2	6.3	0.5	0.0	0.0	0.0
AIC	67.38	4.20	0.00	0.0	42.3	38.8	15.3	3.3	0.3	0.0
BIC	30.98	4.93	0.00	0.0	93.5	6.2	0.3	0.0	0.0	0.0
Oracle	28.05	5.00	0.00	0.0	100.0	0.0	0.0	0.0	0.0	0.0

performances of SCAD penalized estimates calculated via LQA and ICM are similar to each other. Because the BIC tuning parameter selector is adopted, SCAD performs similarly to the classic BIC in both controlling model errors and identifying the correct subsets. The model errors of SCAD are close to those of the oracle procedures, indicating that the model fitting is good.

Table 3. Simulation results for the Poisson regression model: $n = 100$ and $n = 200$.

Method	MRME (%)	Zeros C	Zeros I	Under (%)	Correct (%)	Overfitted (%) 1	2	3	4	5
n=100:										
LQA	40.53	4.32	0.00	0.0	49.1	35.3	13.9	1.7	0.0	0.0
ICM	47.43	4.33	0.00	0.0	52.2	33.3	10.3	3.5	0.7	0.0
AIC	68.07	4.16	0.00	0.0	42.6	35.9	17.6	3.2	0.5	0.2
BIC	39.58	4.84	0.00	0.0	85.5	13.0	1.4	0.1	0.0	0.0
Oracle	30.96	5.00	0.00	0.0	100.0	0.0	0.0	0.0	0.0	0.0
n=200:										
LQA	42.14	4.55	0.00	0.0	63.8	28.0	7.6	0.6	0.0	0.0
ICM	49.41	4.61	0.00	0.0	69.6	22.8	6.8	0.8	0.0	0.0
AIC	71.38	4.22	0.00	0.0	44.3	37.6	14.5	3.5	0.1	0.0
BIC	42.76	4.91	0.00	0.0	91.4	8.5	0.1	0.0	0.0	0.0
Oracle	37.55	5.00	0.00	0.0	100.0	0.0	0.0	0.0	0.0	0.0

4.2. *Tests of Convergence and Computing Time*

Next, we assess the convergence of the LQA and ICM algorithms for each model. Both algorithms carry out iterative procedures to solve for the penalized estimate. In our implementation, the algorithm is considered converged if $\sum_{j=1}^{d} |\hat{\beta}_j^k - \hat{\beta}_j^{k+1}| < 10^{-9}$. Otherwise, a maximum of $s_{\max}$ iterations are conducted. In our simulation, $s_{\max} = 200$ is used for linear and logistic regression. For Poisson regression, convergence is usually slower, and hence $s_{\max} = 500$ iterations are allowed. In our simulations, we define the rate of convergence as the proportion of simulations where convergence criterion is met within $s_{\max}$ iterations. The rates of convergence are reported for both the LQA and ICM as "rate (%)" in Table 4 for linear, logistic, and Poisson regressions, respectively. When the algorithm converges within $s_{\max}$ iterations, the average number of iterations performed is also reported in Table 4.

In order to provide a more direct comparison of the computational burden, we try to compare the average computing time for the LQA and ICM. More specifically, we record the time used to solve the SCAD solution in each iteration, $t_{i;\text{LQA}}$ and $t_{i;\text{ICM}}$ for the LQA and ICM, respectively. The ratios of the computing time, $\sum t_{i;\text{LQA}}/\sum t_{i;\text{ICM}}$, are also reported in Table 4 for different simulation scenarios.

From the results in Table 4, we see that in general, the ICM algorithm requires a decent number of iterations to reach convergence. In the case of logistic regression, the LQA algorithm converges slightly faster than the ICM algorithm, but both algorithms converge in a small number of itera-

Table 4. Convergence and computing time: Linear, logistic, and Poisson regression models.

Algorithm	Rate (%)	Steps	Time Ratio	Rate (%)	Steps	Time Ratio
			Linear regression			
		$n = 50$			$n = 100$	
LQA	99.0	108.18	4.07	99.2	108.35	4.67
ICM	99.8	18.79	–	100.0	18.61	–
			Logistic regression			
		$n = 200$			$n = 400$	
LQA	100.0	10.44	0.86	100.0	10.35	0.65
ICM	100.0	31.34	–	100.0	27.68	–
			Poisson regression			
		$n = 100$			$n = 200$	
LQA	82.1	206.67	15.65	80.1	214.11	21.61
ICM	99.3	70.98	–	100.0	57.66	–

tions. However, ICM converges much faster than LQA in linear regression and Poisson regression. As a consequence, the ICM algorithm provides a better overall computational performance than the LQA algorithm.

4.3. *A Real Data Application*

To illustrate the proposed procedure, we apply the nonconcave penalized methodology to a real dataset that was collected by the General Hospital Burn Center at the University of Southern California. The dataset consists of 981 observations. The response Y is whether patients survived their burns or not. Four variables are considered: X_1 = age; X_2 = sex; X_3 = log(burn area + 1); and X_4 = oxygen (0–normal, 1–abnormal). Quadratic terms of X_1 and X_3 and all two-way interactions are also included to fit a logistic regression model. All the variables are standardized so that they have mean 0 and standard deviation 1. We apply the SCAD procedure to choose important variables. With the tuning parameter selected by the BIC selector, $\lambda = 0.6276$, the SCAD penalized estimate suggests that the intercept, X_1, X_3^2, and X_1X_3 should be involved. Because the quadratic term X_3^2 and the interaction between X_1 and X_3 are selected, we also include the lower order term X_3, and conclude the model

$$\log\left\{\frac{P(Y=1|\mathbf{x})}{1-P(Y=1|\mathbf{x})}\right\} = 5.81 - 11.42Z_1 - 4.78Z_2 - 1.67Z_3 + 9.07Z_4,$$

where Z_1–Z_4 are the standardized versions of X_1, X_3, X_3^2, and X_1X_3, respectively. This model suggests that people who are older or who have larger burn area are less likely to survive, and the effects of age and burn area are not additive.

5. Conclusion

In this paper, we study the ICM algorithm for calculating the nonconcave penalized estimate for variable selection. The ICM algorithm maximizes the penalized likelihood function up to the first two orders. It fixes the drawbacks of the commonly used LQA algorithm and enjoys very fast convergence. Our simulation studies demonstrate that the finite sample performance of the penalized estimates calculated via the ICM algorithm is very satisfactory in terms of both model accuracy and sparsity. Although we only discuss the ICM algorithm for classical likelihood settings, its application is not limited. For example, it can be naturally extended to the cases of the Cox proportional hazards model, partial linear models, and other parametric and semiparametric regression models.

Appendix A. Proof of Theorem 3.1

Proof. To prove (i), write

$$\begin{aligned}\frac{\partial Q(\boldsymbol{\beta})}{\partial \beta_j} &= \sum_{k=1}^{d} \hat{\mathcal{I}}_{jk}(\beta_k - \hat{\beta}_k) + p'_\lambda(|\beta_j|)\mathrm{sgn}(\beta_j) \\ &= \hat{\mathcal{I}}_{jj}\left[\beta_j - \hat{\beta}_j + \sum_{k\neq j}\frac{\hat{\mathcal{I}}_{jk}}{\hat{\mathcal{I}}_{jj}}(\beta_k - \hat{\beta}_k) + \frac{1}{\hat{\mathcal{I}}_{jj}}p'_\lambda(|\beta_j|)\mathrm{sgn}(\beta_j)\right].\end{aligned}$$

Using the notation $\tilde{\beta}_j^*$ as in (6), the above equation can be simplified as

$$\frac{\partial Q(\boldsymbol{\beta})}{\partial \beta_j} = \hat{\mathcal{I}}_{jj}\left\{\left[\beta_j + \frac{1}{\hat{\mathcal{I}}_{jj}}p'_\lambda(|\beta_j|)\mathrm{sgn}(\beta_j)\right] - \beta_j^*\right\}.$$

Denote the minimizer of $Q(\boldsymbol{\beta})$ by $\tilde{\boldsymbol{\beta}}$. In the function $\partial Q(\boldsymbol{\beta})/\partial\beta_j$, set $\beta_k = \tilde{\beta}_k$ for all $k \neq j$ and regard it as a function of β_j, written as $Q'_j(\beta_j)$. Remember that $m_j = \min_{\theta>0}\{\theta + p'_\lambda(\theta)/\hat{\mathcal{I}}_{jj}\}$. Note that $\hat{\mathcal{I}}_{jj}$ is positive under regularity conditions. It is obvious that if $0 \leq |\beta_j^*| < m_j$, the sign of $Q'_j(\beta_j)$ should be the same as that of β_j and hence $\tilde{\beta}_j = 0$. On the other hand, if $|\beta_j^*| \geq m_j$, $\tilde{\beta}_j$ should be chosen such that $Q'_j(\beta_j) = 0$. Therefore (5) holds.

From (i), it suffices to establish the continuity of $\tilde{\beta}_j$ at the point where $|\tilde{\beta}_j^*| = m_j$. Note that at this point, $\tilde{\beta}_j$ is the minimizer of $\theta + p'_\lambda(\theta)/\hat{\mathcal{I}}_{jj}$ for

$\theta > 0$. Therefore when $\left|\tilde{\beta}_j^*\right| = m_j$, $\tilde{\beta}_j = 0$ if and only if this minimizer is attained at zero. Hence (ii) holds.

Part (iii) follows immediately from (i) and the continuity of $p'_\lambda(\cdot)$. □

Acknowledgments

Zhang's research was supported by National Science Foundation grant DMS 0348869 as a research assistant. Li's research was supported by NIDA and NIH grants R21 DA024260 and P50 DA10075.

References

Akaike, H. (1974). A new look at the statistical model identification, *IEEE Transactions on Automatic Control* **19**, pp. 716–723.

Antoniadis, A. and Fan, J. (2001). Regularization of wavelet approximations (with discussion), *Journal of the American Statistical Association* **96**, pp. 939–967.

Breiman, L. (1996). Heuristics of instability and stabilization in model selection, *Annals of Statistics* **24**, pp. 2350–2383.

Efron, B., Hastie, T., Johnstone, I. and Tibshirani, R. (2004). Least angle regression, *Annals of Statistics* **32**, pp. 407–499.

Fan, J. and Li, R. (2001). Variable selection via nonconcave penalized likelihood and its oracle properties, *Journal of the American Statistical Association* **96**, pp. 1348–1360.

Fan, J. and Li, R. (2006). Statistical challenges with high dimensionality: Feature selection in knowledge discovery, *Proceedings of the International Congress of Mathematicians (M. Sanz-Sole, J. Soria, J.L. Varona, J. Verdera, eds.) Vol. III* , pp. 595–622.

Frank, I. and Friedman, J. (1993). A statistical view of some chemometrics regression tools, *Technometrics* **35**, pp. 109–148.

Friedman, J., Hastie, T., Höfling, H. and Tibshirani, R. (2007). Pathwise coordinate optimization, *Annals of Applied Statistics* **1**, pp. 302–332.

Fu, W. (1998). Penalized regressions: The bridge versus the LASSO, *Journal of Computational and Graphical Statistics* **7**, 3, pp. 397–416.

Hunter, D. and Li, R. (2005). Variable selection using MM algorithms, *Annals of Statistics* **33**, pp. 1617–1642.

Lehmann, E. L. (1983). *Theory of Point Estimation* (Wadsworth & Brooks/Cole, Pacific Grove, California).

Miller, A. (2002). *Subset Selection in Regression*, 2nd edn. (Chapman & HALL/CRC, New York).

Schwarz, G. (1978). Estimating the dimension of a model, *Annals of Statistics* **19**, 2, pp. 461–464.

Shao, J. (1997). An asymptotic theory for linear model selection, *Statistica Sinica* **7**, pp. 221–264.

Tibshirani, R. (1996). Regression shrinkage and selection via LASSO, *Journal of the Royal Statistical Society, Series B* **58**, pp. 267–288.

Wang, H. and Leng, C. (2007). Unified LASSO estimation via least squares approximation, *Journal of the American Statistical Association* **102**, pp. 1039–1048.

Wang, H., Li, R. and Tsai, C.-L. (2007). Tuning parameter selectors for the smoothly clipped absolute deviation method, *Biometrika* **94**, pp. 553–568.

Zhang, Y. (2006). *Variable Selection via Penalized Likelihood and Iterative Conditional Minimization Algorithm*, Master's thesis, Department of Statistics, The Pennsylvania State University.

Zhang, Y., Li, R. and Tsai, C.-L. (2010). Regularization parameter selections via generalized information criterion, *Journal of the American Statistical Association* **105**, pp. 312–323.

Zou, H. and Li, R. (2008). One-Step Sparse Estimates in Nonconcave Penalized Likelihood Models, *Annals of Statistics* **36**, pp. 1509–1533.

AUTHOR INDEX